AF722886

Design of Enterprise Systems

Theory, Architecture, and Methods

Design of Enterprise Systems

Theory, Architecture, and Methods

Ronald E. Giachetti

CRC Press
Taylor & Francis Group
Boca Raton London New York

CRC Press is an imprint of the
Taylor & Francis Group, an **informa** business

CRC Press
Taylor & Francis Group
6000 Broken Sound Parkway NW, Suite 300
Boca Raton, FL 33487-2742

© 2010 by Taylor and Francis Group, LLC
CRC Press is an imprint of Taylor & Francis Group, an Informa business

No claim to original U.S. Government works

Printed in the United States of America on acid-free paper
10 9 8 7 6 5 4 3

International Standard Book Number: 978-1-4398-1823-7 (Hardback)

This book contains information obtained from authentic and highly regarded sources. Reasonable efforts have been made to publish reliable data and information, but the author and publisher cannot assume responsibility for the validity of all materials or the consequences of their use. The authors and publishers have attempted to trace the copyright holders of all material reproduced in this publication and apologize to copyright holders if permission to publish in this form has not been obtained. If any copyright material has not been acknowledged please write and let us know so we may rectify in any future reprint.

Except as permitted under U.S. Copyright Law, no part of this book may be reprinted, reproduced, transmitted, or utilized in any form by any electronic, mechanical, or other means, now known or hereafter invented, including photocopying, microfilming, and recording, or in any information storage or retrieval system, without written permission from the publishers.

For permission to photocopy or use material electronically from this work, please access www.copyright.com (http://www.copyright.com/) or contact the Copyright Clearance Center, Inc. (CCC), 222 Rosewood Drive, Danvers, MA 01923, 978-750-8400. CCC is a not-for-profit organization that provides licenses and registration for a variety of users. For organizations that have been granted a photocopy license by the CCC, a separate system of payment has been arranged.

Trademark Notice: Product or corporate names may be trademarks or registered trademarks, and are used only for identification and explanation without intent to infringe.

Library of Congress Cataloging-in-Publication Data

Giachetti, Ronald E.
Design of enterprise systems : theory, architecture, and methods / Ronald E. Giachetti.
p. cm.
Includes bibliographical references and index.
ISBN 978-1-4398-1823-7 (hardcover : alk. paper)
1. Systems engineering. 2. Software engineering. 3. Management information systems. I. Title.

TA168.G53 2010
658.4'038011--dc22 2009044118

Visit the Taylor & Francis Web site at
http://www.taylorandfrancis.com

and the CRC Press Web site at
http://www.crcpress.com

Contents

Preface

Design is the defining characteristic of engineering. Each engineering discipline is strongly associated with the artifact it designs. In mechanical, electrical, and civil engineering the artifact is a physical, technological system. This is not so for engineers who design enterprise systems. What distinguishes the design of enterprise systems from other designed systems is twofold: first the design artifact – the enterprise system – is not tangible, and second, enterprise systems include humans as part of the artifact, not as users of the artifact.

The term "enterprise system" has taken on a narrow meaning of only the information system an organization uses. Research and project experience has taught us that to design a good enterprise system, we need to adopt a much broader understanding of enterprise systems. The greater view of enterprise systems is inclusive of the processes the system supports, the people who work in the system, and the information content of the system. Hereafter, I shall use the term *enterprise systems* to refer to not just the information system, but to the enterprise itself. This use of the term encompasses companies (both manufacturing and service), not-for-profits, and government organizations.[1] Adopting this view encourages us to analyze the integrated system and focus on the relationships between the components that can lead to better system designs, and consequently, project success.

In many ways, the abstract characteristics of enterprises make their design much more challenging than the design of physical objects. First, the fact that enterprises have both physical and abstract aspects makes the representation of the system difficult. There are countless ways to model a process, which is only a single part of an enterprise. In fact, there are countless ways to even define what a process is, highlighting the abstract nature of dealing with enterprise systems. Second, enterprises include a human element, and we have as yet found equations that model humans as accurately as we can model stress/strain, mechanical loads, or the flow of electrons. While civil, mechanical, and electrical engineers can base design decisions on analysis that draws from physics, chemistry, and other basic sciences, the enterprise engineer must also include sociology and psychology to understand the human element. Moreover, enterprise design does not occur at a single point in time like the design of most systems. Instead, enterprises evolve over time and are constantly changing, or are constantly *being designed.*

In practice, many different people with backgrounds in many different disciplines contribute to the design of an enterprise. It is not only engineers but people from management, organizational theory, accounting, finance, business process design, psychology, and sociology that all study, analyze, or design parts of the enterprise system. Anybody that makes decisions to change the current enterprise to achieve some preferred structure or performance is a designer. What is problematic is that the knowledge of enterprise design is fragmented. Each specialty is important for the design of enterprises, but what is lacking is the holistic and systems-wide perspective to integrate the specialized knowledge of separate aspects of the enterprise to achieve a globally optimized enterprise. What is really needed is a new term, *enterprise engineering* to describe the discipline that focuses on the design of enterprise systems. Enterprise engineers need to have knowledge in a broad area of enterprise systems. Enterprise engineers will not have the same depth of knowledge in each

[1] The word "organization" also fits this criteria, but we reserve this word to later specify the organizational structure of an enterprise.

sub-specialty as their counterparts in each contributing field. What enterprise engineers specialize in is integration: the process of making subsystems work together harmoniously in a way that optimizes the performance of the entire enterprise. The integration of many disparate systems, processes, people, and resources is specialist knowledge – the knowledge of the enterprise engineer.

The purpose of this book is to describe an enterprise engineering methodology. Because enterprise systems are exceedingly complex, encompassing many independent domains of study, students must first be taught how to think about enterprise systems. This book takes a system-theoretical perspective of the enterprise, and describes a systematic approach, called an enterprise design method to design the enterprise. The design method demonstrates the principles, models, methods, and tools needed to design enterprise systems. The book details the enterprise engineering process from initial conceptualization of an enterprise to its final design.

Intended Audience

The book is aimed at three groups: engineering students, business students, and working professionals. My intention is that this book fills a need for greater design content in engineering curricula by describing how to design enterprise systems. Inclusion of design is also critical for business students, since they should realize the importance their decisions may have on the long-term design of the enterprises they work with. Forrester[2] uses the analogy of an airplane to explain the difference between enterprise design and enterprise operation. There are two groups of people: those who design the airplane and those, the pilots, who fly the airplane. The pilot's success depends on the airplane designer to create a good airplane. Most all textbooks in related areas such as operations management, manufacturing systems, etc. are aimed toward the operator or manager of the enterprise, not the designer. Somebody must design an enterprise system; so there is a need to collect the theory, models, tools, and methods to design enterprise systems in a single book.

Organization

To present all the topics needed to analyze and design an enterprise is a tremendous undertaking; what I have decided to focus on is the life cycle of an enterprise from initial conceptualization to final design. This includes the design models, tools, and methodology to design an enterprise. The book uses an enterprise reference architecture that contains three views: process, organization, and information. The enterprise architecture is used to provide a model to understand how the parts of the enterprise fit together. The enterprise architecture is also used to organize the last four parts of the book.

The book is organized into six parts as follows:

- Part I establishes the foundation for enterprise engineering and the remainder of the book. Chapter 1 describes the history of thought leading up to enterprise engineering. It describes the knowledge required to do enterprise engineering, a classification of enterprises, and the types of enterprise engineering projects. Chapter 2 describes systems

[2]Forrester, J.W., Designing the Future, Universidad de Sevilla, Sevilla, Spain, December 15, 1998.

theory and how it can be used to understand, analyze, and design enterprises. Chapter 3 describes modeling concepts because building models is required for the analysis and design of enterprises. Chapter 4 reviews engineering design theory and the enterprise design methodologies. Chapter 5 reviews enterprise architectures. Chapter 6 presents the enterprise design methodology that is used in this book.

- Part II contains three chapters describing the preliminaries and initiation of an enterprise project. Chapter 7 describes strategy as a guide for defining how the enterprise should be designed. Chapter 8 discusses problem formulation and requirements engineering. Chapter 9 describes how to generate and evaluate alternatives.
- Part III focuses on the process view of the enterprise. Chapter 10 describes how to model processes. Chapter 11 reviews queueing theory, which is used to analyze processes. Chapter 12 describes approaches to analyze and design processes.
- Part IV focuses on the information view. Chapter 13 describes how to model the information structure of the enterprise. Chapter 14 describes a design methodology, SQL, and normalization.
- Part V focuses on the organization view. Chapter 15 discusses organizational theory and describes an approach to organizational design.
- Part VI describes the integration of the three views of process, organization, and information. Chapter 16 defines five types of enterprise integration, the technical architecture of systems, and integration technologies. Chapter 17 describes techniques for integrating the three enterprise views. It summarizes the design methodology by showing how all the independent subsystems come together.

Website

The author maintains a Website at http://web.eng.fiu.edu/Ronald/ that contains:

- Powerpoint slides for each chapter.
- Templates to support enterprise engineering activities, including: Project Charter, Business Case, Cost-Benefit Analysis, Risk Management Assessment, Problem Analysis, Requirements Analysis, Alternative Matrix, Evaluation Matrix, Process Design, Information Design, Organization Design, Interface List, and more.
- Project case studies that can be assigned to students as semester-long projects to accompany the text.
- Quiz questions for each chapter.
- Business Process Analyzer software for download.

Book Features

The book has the following special features:

- Focus on doing enterprise engineering – The book encourages the reader to apply the enterprise system design concepts and techniques. For each major technique, the book provides examples and explains best practices. The review questions and exercises, and accompanying projects all emphasize the practice of enterprise engineering.
- Business Process Analyzer – A spreadsheet and VBA implementation of multi-class queueing networks using the parameter-decomposition method for GI/G/n queues that allows students to model a business process and estimate the performance measures of cycle time, waiting time, and resource utilization.
- Coverage of information modeling – The book using entity-relationship models and shows how to create the models, normalize them, and write SQL.
- A project-based approach – The book uses the enterprise system design method that consists of seven phases. An accompanying project lets the reader understand the inputs, activities, and outputs of each life-cycle phase.
- Coverage of architecture – The enterprise architecture provides a high-level design of the enterprise and guides all other system projects.
- Chapter on integration – A chapter describes how to integrate the three architectural views with each other and with enterprise technologies.
- Instructor resources – Available to instructors on the Website are an accompanying project book, PowerPoint files for each chapter, quizzes, and exam questions.
- Student resources – Available to students on the Website are accompanying templates, checklists, forms, and models to support the enterprise engineering process.

Acknowledgments

I started collecting information and writing this book in 2005. Since that time I have had useful discussions, suggestions, and outright help from many different people. I would like to acknowledge my colleagues at FIU and the many students who took my courses (at FIU in Miami; at Tecnolgico de Monterrey (ITESM), Chihuahua, Mexico; at Universidad del Norte in Barranquilla, Colombia; at the University of Technology, Kingston, Jamaica; and the Peru Catholic University, Lima, Peru), and did projects with me, or did their dissertation work with me. I would like to specifically acknowledge the help of Lixiang Jiang who worked with me to develop the queueing network approach to analyzing business processes; Duane Truex a colleague and friend from the business school who introduced me to a new way of viewing methods and collaborated on the ERP short course for the U.S. Air Force; Ching-Sheng Chen who I frequently discussed enterprise systems theory with; Mario Kim for helping to design the cover artwork and drawing the abstraction of the horses for me; and to the Ph.D. students I supervised who helped advance my research: Ramakrishnan Sundaram, Oscar Sáenz, Jose Rojas, Rene Amaya, Heriberto Garcia, Alba Nuñez, Bertha Arteta, Maria Paula Hernandez, Chris Ellis, Giacomo Boria, and Sergio Hernández.

Author Biography

Ronald E. Giachetti, Ph.D., is an Associate Professor of Engineering Management at Florida International University (FIU) in Miami, Florida. Prior to joining FIU in 1998, he worked at the National Institute of Standards and Technology in Gaithersburg, Maryland. He conducts research in enterprise systems, operations research, and information systems. He has completed projects for government agencies including the National Science Foundation (NSF), U.S. Air Force, and National Aeronautics and Space Administration (NASA). For the U.S. Air Force at Wright Patterson he and a colleague developed a short course on Enterprise Resource Planning (ERP) Project Management. At NASA he worked with the Ames Research Laboratory and developed an enterprise integration methodology to help them address their information integration challenges. He has also completed projects with industry including Carnival Cruise Lines, Royal Caribbean Cruise Lines, Americatel, Baptist Healthcare and KoolSmiles Dentistry. These industry projects focused on business process improvement. He has published over 50 journal articles, book chapters, and conference papers on this work. At FIU he teaches courses in enterprise systems and operations research to both undergraduate and graduate students. He has received the IIE Teacher of the Year award three times at FIU. He teaches in the graduate Engineering Management program on campus and through FIU's Global Programs Office. He has taught graduate students in Mexico, Jamaica, Peru, and Colombia. He has a Ph.D. in Industrial Engineering from North Carolina State University, an MS in Manufacturing Engineering from Polytechnic University, and a BS in Mechanical Engineering from Rensselaer Polytechnic Institute.

Part I

Enterprise Engineering

1

Enterprise Engineering

> *"The world hates change, yet it is the only thing that has brought progress."* – Charles F. Kettering (1876-1958), inventor and head of General Motors Research, 1920-1947.

This chapter introduces enterprise engineering as the discipline concerned with the design of enterprises. It provides an overview of what enterprise engineering entails and how enterprise engineering projects are initiated and managed. This chapter reviews the historical background, intellectual development, and industry trends that have led to a need for enterprise engineering. After completing this chapter, you should be able to:

- Define enterprise engineering and explain why it can be considered a separate discipline.
- Categorize enterprises according to their type.
- Describe how enterprise engineering projects are conducted.
- Discuss the intellectual developments that have contributed to modern thought on enterprise systems.
- Explain how enterprise engineering is related to systems engineering.
- Describe the requisite skills and knowledge for enterprise engineering.
- Assess an enterprise environment and how it affects the enterprise.

1.1 Definition of Enterprise Engineering

Enterprise Engineering is defined as the body of knowledge, principles, and practices to design an enterprise. The key of the definition is *to design*, which is considered the characteristic, defining activity of engineering. Moreover, an enterprise is not designed just once, but an enterprise is, to varying degrees, redesigned many times until its eventual retirement. Enterprise engineering is a subdiscipline of systems engineering in that it addresses the entire life-cycle of an enterprise.

Enterprises have of course existed for millennia. The word might be new, certainly the technology employed by enterprises is new, but the organization of man and machines in pursuit of some common goal is not new. So, if enterprises have always existed, how have they been designed? To a large extent, in the past, the enterprise was not viewed as a whole system that could be rationally designed. More likely, parts or subsystems of the enterprise were designed in isolation without a holistic perspective of how the parts would work in the entire system. Enterprises would come into being and change not as a result of a conscious, purposeful design effort, but due to ad hoc, sometimes short-term decisions made individually by many different people. Many businesses that had grown quickly paid little

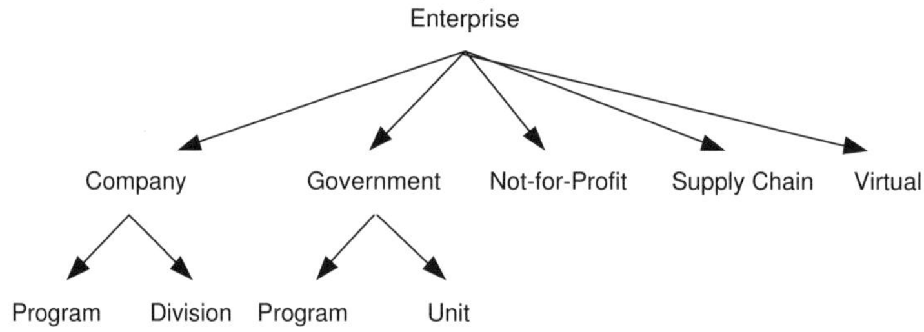

FIGURE 1.1
Enterprise systems.

attention to the design of their processes and systems. Porter [25] cites a study by Ernst and Young that found in many companies the key business processes were designed long before there was information technology. Even though these companies had new IT systems, they continued to follow the paper-based business processes but with IT. What this study and other observations indicate is that for many enterprises their current structure does not reflect the results of a rationale design, but the result of a multitude of small changes made over time without consideration of the overall enterprise.

1.1.1 Enterprise Systems

An *enterprise* is a complex, socio-technical system that comprises interdependent resources of people, information, and technology that must interact with each other and their environment in support of a common mission.[1] As a system, it is the interactions that are important to the enterprise behavior. Interactions include such activities as coordination of functions, sharing of information, and allocation of resources. The enterprise is a socio-technical system in that it involves people and technology, it is an open system in that it interacts with its environment, and it is purposeful in that it has goals that it works towards accomplishing.

We use the term enterprise because it encompasses all types of organizations: companies, government, not-for-profit, supply chains, virtual enterprises, as well as parts of a company such as division or program. The term organization is not used because later we use it to describe one of the views of the enterprise. Figure 1.1 shows the classification of different types of enterprises. The enterprise types are:

- Companies defined as commercial enterprises that pursue a market position to return long-term profits to the business owners. Frequently when we talk about enterprises we are referring to companies. A large company will have many semi-independent business units that are also enterprises in their own right.
    - A division is a semi-independent business unit that usually focuses on a single market.
    - A program is a temporary, but long-duration collection of activities to produce a particular product, system, or service. The program enterprise usually will have a single business process for developing and delivering the project, and it will usually encompass the entire life-cycle of the product it produces. Program enterprises

[1] In the software and business literature, the term enterprise system frequently has a more narrow definition of a large, enterprise-wide information system such as ERP.

are common in the defense industry (e.g., the B1-B program) or the automotive industry (e.g., Chrysler's LH platform).

- Government organizations at all levels (national, state or region, city or town) are enterprises. Government enterprises pursue goals to provide a service or product to citizens. Government organizations can be subdivided into programs or units. A program is defined the same as in businesses. Governments are also subdivided into units, and each unit is an enterprise in its own right.
- Not-for-profit organizations are charities, funds, and volunteer organizations. Non-profit enterprises pursue goals to improve society.
- Supply chain, which are the collaboration of multiple, separate companies involved in the development and delivery of a product or service.
- Virtual enterprises defined as enterprises that are quickly created to exploit a market opportunity and then once the opportunity passes the enterprise dissolves. Hollywood makes use of temporary enterprises to create films [13]. It could also be a temporary, not-for-profit enterprise such as Hurricane Relief for Haiti, or a temporary government enterprise such as the U.S. Federal Government's $700 billion bailout plan for banks, which will require a small organization to execute and then disband once the money is distributed.

These enterprises have the following in common:

- All enterprises are man-made systems, where a system is an integrated collection of components (people and technology).
- All enterprises use resources of people, material, machines, information, and knowledge.
- All enterprises produce a product, provide a service, or do both.
- All enterprises have customers who derive value from the product or service.
- All enterprises have a goal: in for-profit enterprises the goal is profitability and growth. Other enterprise types may have different goals.
- All enterprises interact with their environment. They obtain raw materials, labor, and other resources from their environment, they compete with other enterprises, they collaborate with other enterprises (e.g., in supply chains), and they are subject to changes in the political, social, economic, and technical aspects of their environment.

1.1.2 Enterprise Engineer

The definition of enterprise engineering casts the enterprise as a product. If an enterprise is a product, then it must have a designer. It is unusual for a single person to design an enterprise, instead it usually involves many people from different disciplines during all phases of the design project. We shall call this group the enterprise engineering team. In addition to the enterprise design that results from a project, enterprise design is an ongoing, incremental activity. All managers are constantly making enterprise design decisions. Sometimes managers consciously consider their decision's impact on the long-term design of the enterprise; they consider whether the current decision will set a precedent that will become established policy in the enterprise. Other times, decisions are made without forethought of how they impact enterprise design. Consequently, managers are also enterprise designers because the decisions they make change the enterprise. A central argument of this book is that enterprise design should be a conscious, purposeful endeavor, and managers should regularly review their systems from multiple perspectives to ascertain whether they are meeting enterprise needs.

Now that we established that enterprise design is done primarily on projects, we need to describe the job titles of some of the people who would be part of the enterprise design team. These are:

- *Business Systems Analyst* is a person who identifies and analyzes business problems, and generates system requirements. The Business Systems Analyst is a person who needs to be conversant in both the business domain language (e.g., accounting) as well as understand the technology that might be used in that domain. There are similar job titles of Business Analyst, System Analyst, and Process Analyst for people with similar job descriptions.

- *Enterprise Architect* is a person who develops a holistic view of the enterprise's strategy, processes, information, and organizational structure – usually delivered as the enterprise architecture.

- *System Architect* is a person who creates the high-level design of a technical system. Here, system means a subsystem of the enterprise such as an ERP system, or the accounting system. The difference in scope is what differentiates a system architect from an enterprise architect. If the system in question is software, then the title might be software architect.

- *Project Manager* is the person responsible for accomplishing all project objectives. The project manager would play a role in identifying all team members, planning the project, supervise team members, monitor project progress, and is responsible for all project deliverables.

- *System Designer* is the person who designs one or more parts of the system. The system designer is a technical person who generates the specifications for how the system will work.

- *Change Manager* is the person responsible for the change management plan. All new enterprise systems involve change, and the change manager plans the change, supervises the change management sub-team, monitors the progress of the change, and is responsible for the successful execution of the change management plan.

- *System Engineer* is a technical person who is involved with the overall process of defining, developing, operating, maintaining, and ultimately replacing quality systems. System engineers concentrate on the integration of all the system components and the entire system life-cycle.[2] A system engineer also has broad knowledge of all the other disciplines involved in the project. The depth of this knowledge is described as being two questions deep [23]: First, what technology is relevant to the problem at hand? and second, What is the telephone number of an engineer who knows this technology in depth?

- *Application Developer* is a technical person who builds new software systems. There are many variations depending on the technology including: web application developer, Java developer, ERP developer, etc.

The list of job titles and descriptions conveys the breadth of the skills required for an enterprise engineering project.

---

[2]INCOSE http://www.incose.org/educationcareers/careersinsystemeng.aspx

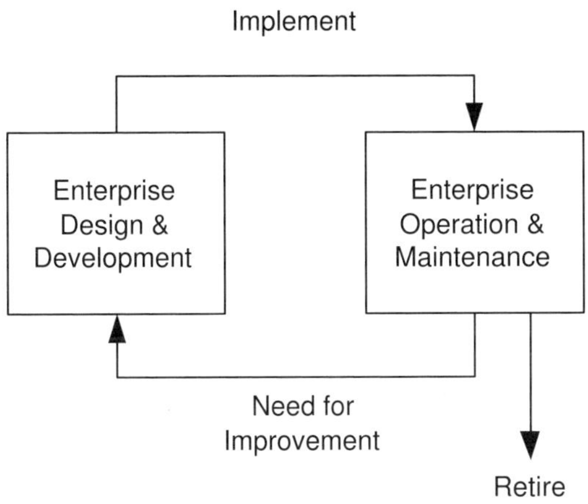

FIGURE 1.2
Relationship between enterprise development and operation.

1.1.3 Enterprise Life-Cycle

The concept of a life-cycle is central to systems engineering. Whether the system is a satellite, power plant, mass transit, or any other large and complex system it has a life-cycle. The enterprise *life-cycle* describes the history of the enterprise from the initial concept of a business in the mind of an entrepreneur, through a series of phases as the enterprise grows, until the business venture ends. The enterprise life-cycle consists of three general, distinct stages: development, deployment, and operation. Development covers the engineering phases to create an enterprise system, deployment is the change management process to implement the enterprise system, and operation is the management of the enterprise system and its continuous improvement. These three stages are described by more precise phases, where a *phase* is a step in the life-cycle that delineates the project's progress. How a life-cycle is broken up into phases differs with each methodology, but we can identify the typical enterprise life-cycle phases as:

1. System identification – The system boundaries, purpose, and project scope are defined.
2. Analysis – The system problems are analyzed; requirements are generated.
3. Design – The system design is generated.
4. Construction – The system is built.
5. Implementation – The system is implemented and deployed into its environment.
6. Operation and Maintenance – The system is operated and maintained.
7. Decommission– The system is retired.

Phases 1 through 4 are part of the enterprise development methodology. These are the phases done during an enterprise engineering project to develop a new enterprise or enterprise subsystem. Phase 5 is the deployment phase when the enterprise is implemented and deployed into its environment. Phase 6 is the operation of the enterprise until it is decommissioned in phase 7. Figure 1.2 shows that an enterprise might cycle through many development projects before being decommissioned.

1.1.4 Enterprise Design Method

To design an enterprise is a complex undertaking involving a large group of people over a long period of time. In order to be successful, the project team members need to work together in an effective and efficient manner. To accomplish this, we view enterprise design as a problem. The problem can be posed as:

> Given a project goal, determine the problem scope and problems; analyze them and generate requirements; generate alternatives, evaluate them, and choose the best alternative; design the enterprise system; implement the design; and maintain the design with periodic updates to improve the enterprise.

To solve such complex problems engineers use methodologies. A methodology is a staged problem-solving approach. A general problem-solving methodology is:

1. Scope the problem.
2. Design the solution.
3. Evaluate the solution.
4. Satisfactory? – If not, then repeat.

The goal of an enterprise design methodology is to specify and develop a "best" system solution that will demonstrably satisfy the requirements of all enterprise stakeholders. A *stakeholder* is any person who can affect or is affected by the achievements of the enterprise's goal [12]. As social systems, enterprises tend to have many stakeholders. When there are many stakeholders, just to define what the enterprise should be doing and what its goals are is a difficult task. One goal of an enterprise methodology is to provide a systematic approach to understanding all the stakeholder perspectives, understanding the problem situations facing the enterprise, and determining a strategy for meeting the stakeholder needs. This is done in phases 1 and 2 of the problem-solving methodology presented above.

In addition to understanding the design problem, a methodology needs to guide the team on how to generate solutions (or designs) to satisfy the requirements. This is done in phase 3 of the problem-solving methodology.

Once a design is proposed, there needs to be a way to evaluate how good the design is. Again, a methodology must include activities, techniques, and tools to help the team evaluate the feasibility, economics, and merits of proposed design solutions. This is done during phase 4 to test the solutions.

The enterprise design methodology is iterative, and it takes several iterations of understanding, analyzing, designing, and evaluating before arriving at a final enterprise design. Once a design is selected, the methodology needs to help the team construct the system, which involves a combination of building new systems, acquiring systems, training people, generating documentation, and other activities needed to create the enterprise system. Implementation is a matter of going live with the system; change management is the critical activity here, and frequently the determinant of overall project success.

The enterprise design methodology supports projects of varying scopes. Enterprises must interact with other enterprises. For example, businesses are organized into supply chains, and it is recognized that supply chain interactions and performance are important to each individual supply chain member. An enterprise design method supports both the design of an enterprise subsystem, an entire enterprise, as well as the design of large, complex supply chains. Later in Chapter 4 we discuss at length various enterprise design methodologies.

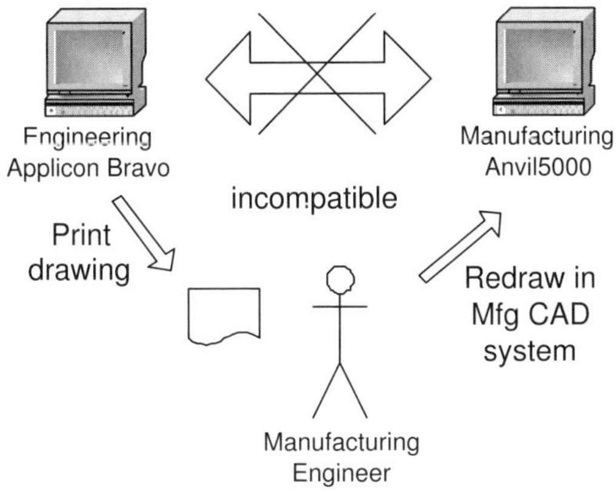

FIGURE 1.3
Result of poor acquisition decisions when there is no enterprise architecture.

1.1.5 Enterprise Architecture

Seldom is the entire enterprise designed in a single project. More likely, there are many smaller projects that design one or more parts of the overall enterprise. For this reason, we say the design of an enterprise never really ends. Enterprises need to evolve and change over their life-cycle. The problem becomes how do you ensure all the separate design projects will lead to some desired vision of the enterprise? It has occurred more than once where one part of an enterprise is designed only to find out it does not work well with other parts of the enterprise. For example, the author's first industry job was in an aerospace company. At the time the manufacturing division was moving aggressively into computer-integrated manufacturing. They acquired a computer-aided design (CAD) system called Anvil5000 based on a wire-frame modeling technique to support their operations. The CAD system let them define work instructions and generate numerical control programs for the machines. Soon after, the engineering division acquired a new CAD system from Schlumberger called Bravo. This CAD system had good solid modeling capabilities that supported various types of engineering analysis. The two CAD systems were incompatible and it soon became apparent the types of problems that would occur. Designs developed in Bravo by engineering would have to be completely redrawn in Anvil5000, resulting in a wasted duplication of efforts (see Figure 1.3). The result was problems of maintaining consistency between design documentation and the manufacturing documentation. What had happened is that each division developed their systems without consideration of how it would work in the overall enterprise. Eventually, it was decided that manufacturing would switch over to Bravo, essentially abandoning its earlier investments in equipment, training, and process changes associated with the purchase of Anvil5000.

To avoid these types of problems, the enterprise needs to provide an overall enterprise design to guide all other projects. An enterprise architecture fulfills this role. An *enterprise architecture* is a high-level design of the enterprise. The enterprise architecture specifies an enterprise-wide view of the processes, information, and organization of the enterprise and how the three views are integrated. Whenever, a small enterprise project is embarked on, the project deliverables should conform to the enterprise architecture. In this way, the enterprise can ensure that all these projects contribute towards enterprise-wide improvement and attainment of the enterprise goals. Returning to the aforementioned aerospace company, if they had an enterprise architecture, it would have described the needed information flow, contained principles for interoperability, and provided a holistic plan that would have highlighted the impending integration problems of the CAD selection decisions.

In today's rapidly changing world, having an enterprise architecture may be insufficient to guide enterprise design. Consequently, an important design goal is to design the enterprise so that it can easily change to meet future requirements. This leads to the many "-ilities" to describe various means of change including flexibility, agility, and scalability. The idea is that an enterprise is designed so that it can not only react swiftly to unexpected changes in its environment, but also seize the opportunities offered by change for its competitive advantage [13, 3]. Enterprises that can change themselves (redesign) to adapt to unforeseen changes are more likely to succeed.

1.1.6 Enterprise Engineering Projects

It is obvious that enterprise engineering can be applied to new enterprises. A small start-up business requires a complete design of its systems. Less clear is how to apply enterprise engineering to existing enterprises. Existing enterprises are continuously being changed via either projects or continuous improvement programs. So, while for many enterprise engineers it is unlikely they will encounter in their entire career a situation in which they can design a complete enterprise from a clean slate, the redesign of existing enterprise subsystems is a very common project. Typical enterprise design projects are:

1. Strategy-initiated project. This project type stems from the strategic plans of the organization. A project performed by a charity organization in Fort Lauderdale, Florida provides an example of an enterprise design project that was done to implement a strategic vision. The charity provides food for the poor, elderly citizens of the city. A question was asked of them if they provide fresh produce. The answer was that they currently did not have the capability, but provision of fresh produce was part of their strategic vision. To add this capacity, they were developing a project plan to develop the processes for acquiring, for cold storage, and for distribution of fresh produce. This is enterprise design. The charity needs to develop the business processes, the organization, and all the subsystems needed to have this capability.
2. Subsystem design. As enterprises grow, contract, or change there is a need to design or redesign enterprise subsystems. These subsystems include:
   (a) Subsystems defined by view: Analyze and design a subsystem that involves one view of the enterprise, such as the information, process, or organizational view.
   (b) Subsystems defined by process: Analyze and design an end-to-end business process. For example, the order fulfillment process.
   (c) Subsystems defined by organizational unit: Analyze and design a single organizational unit of the enterprise. This can be a division, department, or other unit. For example, a company that is entering the Asian market for the first time might need to design the entire Asian division including the organization, information systems, and processes.
3. Reengineering or other large-scale transformation projects. Business Process Reengineering (BPR), or sometimes just reengineering, describes a project to essentially replace an existing enterprise system with a new enterprise system. What distinguishes BPR from other projects is the project team purposefully looks for completely different ways to design the system in order to achieve dramatic improvements in system performance. This is different than many projects

that seek incremental improvement by making small changes to the existing system. During the 1990s, BPR was popularized by Hammer and Champy [16] – it is still practiced today, but not with the zeal exhibited in the 1990s.

4. Enterprise information system. An enterprise information system is a large, enterprise-wide information system to partially automate one or more business functions. This category includes Enterprise Resource Planning (ERP) systems, Supply Chain Management (SCM) systems, Data Warehouses, Customer Relationship Management (CRM) systems, and e-commerce systems. These enterprise information systems represent not only technology, but also a change in business processes, information structure and flow, and organizational structure. For this reason, they are not just a technology project but an enterprise engineering project.
5. Continuous improvement. Most companies institute some type of continuous improvement programs (six sigma, total quality management (TQM), or kaizen). The small decisions and changes made in continuous improvement are design changes to the organization and consequently should be guided by enterprise architectures that describe what the company wants to look like.
6. Supply chain project. A supply chain describes the relationships between companies that trade with each other. To set up a supply chain relationship requires a project that impacts the companies' information systems, processes, and organization. It is therefore an enterprise engineering project.

1.2 Need for Enterprise Engineering

Enterprises have long been studied by researchers in the management sciences, engineering, social sciences, and information sciences. Typically, each discipline would study a single subsystem of the enterprise or study the enterprise from only a single perspective. For example, industrial engineers traditionally consider only the production subsystem. Moreover, they predominantly emphasize efficient operations of the production subsystems. Organizational scientists mostly investigate the structure of the organization. Behavioral scientists study the decision-making in the enterprise or how the interaction between workers, management policies, and the work environment affect productivity. What many researchers and industry leaders now see is there is a lack of an overall, all-encompassing view of the enterprise. Our knowledge is compartmentalized in separate disciplines. What is needed is an enterprise-wide view to understand the problems facing enterprises as a whole and not separately.

Unfortunately, the nature of our university system encourages researchers and students to stay narrowly focused in their specialty. The reason is threefold. First, it is easier for researchers to make significant contributions in narrowly defined domains than in a larger, broader context. Second, the amount of knowledge has increased so much that it is becoming difficult to master more than a few topics. Third, most universities are structured such that each discipline is a different department, physically located together, with a curriculum mostly taught within the department. This structure limits cross-disciplinary learning and reinforces the developed specialization of the students.

The problem is that enterprises have become more complex, the environment they operate in changes quickly, and the competition they face has increased. For this reason, many industry leaders, researchers, and policymakers have raised the call to develop enterprise engineering as a discipline (see Sáenz et al. [30]). The term used is not necessarily enterprise engineering. For example, Towill [34] says there is a need for what he calls a "business systems engineer" who uses a systematic approach to design new business processes and to redesign existing business processes to maximize customer value and the business's performance. Likewise, Leung et al. [18] at IBM argue for the need of business process engineers. Rouse [28] makes similar arguments that industrial engineering needs to expand their focus to the entire business enterprise including external entities such as suppliers, vendors, and distributors. Liles [19] argues for enterprise engineers who know "how to design and improve all elements associated with the total enterprise through the use of engineering and analysis methods and tools to more effectively achieve its [the enterprise's] goals and objectives."

There are several calls from outside of the engineering discipline for the need of enterprise engineering. Davenport and Short [11] describe the need of engineers who analyze and design business processes and know how to apply IT to improve them. Also, the same author later argues how commercially available information systems, often called enterprise systems, are just that, large information systems [10]. These systems often lack an enterprise model or are not well implemented because of the lack of enterprise engineering principles. Consequently, there is a need for a discipline that studies the large picture of the entire enterprise.

Martin [20] specifically states the need for enterprise engineering and defines enterprise engineering as consisting of a series of change management methods. In this view, enterprise engineering is needed to constantly change the enterprise so that it can meet new challenges and prosper. Alter [1] argues instead of enterprise information systems we need to conceptualize them as work systems. He remarks most system implementation failures are not technological in nature but due to not designing the information system for the enterprise. Again, he is arguing against the technical focus that dominates the implementation of large information systems, and instead the need to have multiple perspectives that see these systems are part of the larger enterprise.

These authors recognize a need to transition from a situation in which enterprises are evolved in an ad hoc fashion to a systematic, or an engineering approach, to the design of enterprises. To accomplish this, an enterprise engineering discipline needs three foundations:

1. Enterprise integration knowledge.
2. Enterprise architecture.
3. Enterprise methodology.

The first foundation, enterprise integration knowledge, describes the ways that the parts of the enterprise can be coordinated and integrated so that they work together as a harmonious whole. This knowledge is not part of the sub-disciplines of enterprise engineering. The second foundation, enterprise architecture, is needed to provide a unifying view of the enterprise design. The third foundation, enterprise methodology, describes the formal engineering approach to design an enterprise. Included in the methodology are the methods, techniques, and tools for each phase of the enterprise design process.

What will an enterprise engineering discipline deliver? For one thing, it will provide the holistic view required to design an enterprise. While the many subsets of what is defined as enterprise engineering is not new, the integrated knowledge and focus on enterprise design is novel. Both the enterprise architecture and the methodology will lead to better enterprise designs and more successful enterprise engineering projects. Nowadays, when enterprises are purposefully designed they are usually done based on experience and judgment. Clearly some designs are better than others; however, except for specific cases there is a lack of rationalization justifying a particular enterprise design. The enterprise engineering knowledge will provide the rationale to answer why and how enterprises are designed.

1.2.1 Enterprise Engineering Compared to Systems Engineering

Enterprise engineering is closely related to systems engineering. The International Council on Systems Engineering (INCOSE) defines systems engineering as, "an interdisciplinary approach and means to enable the realization of successful systems." Systems engineering entails the entire life-cycle of a product from determining customer needs, documenting the requirements, designing the system, testing, and then deployment. Systems engineering integrates all the disciplines and specialty groups into a team effort forming a structured development process that proceeds from concept to production to operation. Systems engineering considers both the business and the technical needs of all customers with the goal of providing a quality product that meets the user needs.

Traditional systems engineering focuses on technical systems like airplanes, satellites, or it deals with construction projects like dams and nuclear power plants (see for example, pp. 38-39 in [6] as well as [31]). The systems engineering approach is to decompose the complicated design problem into smaller sub-problems that when solved can be integrated together seamlessly to arrive at a solution to the original system design problem. Systems engineering addresses the problems of how to identify the system requirements, how to manage large-scale engineering projects with many stakeholders, and how to design complex systems.

Enterprise engineering is systems engineering but limited to the design of the enterprise, a socio-technical system. The basic ideas of systems engineering carry over to this domain: the life-cycle phases, development methodology, and tools. But we must adapt some systems engineering concepts, discard some, and add new ones. Most of the changes are due to the following two main differences between enterprises and technical systems:

- The human in the enterprise is a significant, distinguishing characteristic of enterprise systems [35].
- Enterprise design is a continuous process, not limited to single projects.

A consequence of the human component of the enterprise is that enterprise engineers need to consider how the people work in the organization. Enterprise engineers require background knowledge in the organizational sciences, sociology, and psychology. The fact that enterprises are constantly evolved makes it different than other systems that are designed and built as part of a single project. As a result, enterprise architecture plays an important role in aligning all the small developments in the enterprise so that they all contribute to the overall enterprise design goals.

In addition to systems engineering, we believe the root disciplines include industrial engineering, systems engineering, organization sciences, psychology, sociology, information sciences, and the management sciences. The study of enterprise engineering involves understanding this vast domain knowledge and how it can be applied toward the design of enterprises [19].

1.2.2 Skills and Knowledge of Enterprise Engineers

We say the design of an enterprise is done by an enterprise engineer. Engineering is a discipline with a body of knowledge that can be taught. The knowledge and skills required to do enterprise engineering are as follows:

1. Systems thinking defined as the ability to visualize enterprises as systems and use systems theory in the understanding, analysis, and design of enterprises.
2. Domain knowledge of the business, whether manufacturing, healthcare, transportation, energy, or any other domain.

TABLE 1.1
Twenty Greatest Engineering Achievements

1. Electrification	11. Highways
2. Automobile	12. Spacecraft
3. Airplane	13. Internet
4. Water supply and distribution	14. Imaging
5. Electronics	15. Household appliances
6. Radio and television	16. Health technologies
7. Agricultural mechanization	17. Petroleum and petrochemical technologies
8. Computers	18. Laser and fiber optics
9. Telephone	19. Nuclear technologies
10. Air conditioning and refrigeration	20. High-performance materials

3. Mathematics is a foundation of all engineering fields.
4. Modeling skills are especially important for enterprise engineering because the artifact of study (i.e., the enterprise) can only be understood, analyzed, and designed through models.
5. Analysis defined as the ability to approach a problem in a structured way with an inquiring mind.

The list is non-exhaustive but touches on what we consider the more important skills an enterprise engineer should possess.

1.3 The Enterprise Environment

In section 1.1 we described an enterprise as a system. In chapter 2, we will explore in greater depth precisely what it means to be a system; one characteristic that will be discussed is that an enterprise is an open system. As an open system, an enterprise interacts with its environment and *cannot* be understood in isolation of its environment. For this reason, it is essential to understand the enterprise's context (environment) in order to understand the enterprise as a system [29]. The enterprise environment is one characterized by dynamism – continuous, rapid, and often unpredictable change. While it is often difficult to fully appreciate how our world is changing while the changes are happening, the following main themes are ongoing: technological development, globalization, and population growth. In this section, we explore in some depth these global trends and how they affect the operation and therefore the design of enterprises.

Technology Development

The development of technology includes the improvement of previous technologies as well as the invention of new technologies. The National Academy of Engineering surveyed professional engineering societies to identify the 20 greatest engineering achievements (see Table 1.1). These new inventions greatly changed society, and in our context, enterprises.

Several of the technologies listed in Table 1.1 have greatly changed transportation. Only a few generations ago traveling across just the U.S. was difficult, never mind overseas travel. The first person to drive cross-country in a car was Dr. Horatio Nelson Jackson and his mechanic Sewall Crocker in 1903. They left San Francisco and arrived in New York 63 days later, spending some $8,000 in the process.[3] They succeeded despite the fact that there were no gas stations in 1903 and less than 150 miles of paved roads in the U.S. Today, the same cross-country trip takes about five days if you exclusively drive during daylight hours. Flying was not an option in 1903 since in that same year the airplane was only first being tested by Orville and Wilbur Wright in North Carolina.

To travel across the Atlantic or any body of water was done by boat, taking one week to cross from the U.S. to Europe. Not until 1927 did Charles Lindbergh make a transatlantic flight from New York to Paris that took him 33 hours. Commercial airlines did not really start until after World War II. The first jet airliner was the Boeing 707, which was introduced in 1959.[4] Nowadays, it is both relatively inexpensive and quick to travel from one part of the world to another. A businessman can fly directly from New York to Tokyo, have a business meeting, and be back in New York the next day.

In addition to the easy movement of people, the advances in communication and information technology make it possible to send large amounts of data around the world in seconds. Compare this to the first telephone call made by Alexander Graham Bell in 1876 in which he told his assistant, "Mr. Watson, come here, I want you." In 1915, Bell was able to repeat that same request, but over the first transcontinental telephone lines connecting the U.S. east coast to the west coast. Today, we do not think twice about calling nationally or even internationally.

Much of today's information flow is in electronic format over computer networks. Computer networking is a relatively new technology – the ARPANET, the predecessor to the Internet, only had 213 nodes in 1981 [15]. The first publicly available access to the Internet was provided by Delphi in 1992. However, growth in usage has been exponential such that nowadays almost half the population in the industrialized world have access to the Internet. Cisco forecasts monthly Internet traffic to grow from 4234 PB (pera-bytes) in 2007 to 43,551 PB in 2012 [7]. The Internet has made possible enterprises, such as Amazon.com, that exclusively sell merchandise online. Many airports, universities, and even the downtown sections of some cities have wireless wide-area networks so that anybody with a computer (or other IT device) can access the Internet.

A technology trend is the distinction between telecommunication technology and computer technology is becoming blurred. What was once exclusively a telecommunication device such as cellphones has now also become a computer device as well. People can now view webpages on their mobile phones. Likewise, computers now serve as telecommunication devices. For example, the software technology Skype enables inexpensive telephone calls using a personal computer and the Internet.

Another, related technology change is the advent of mobile technology, which has had profound impact on world cultures, especially in countries where telecommunications was under-developed. In countries such as India, cellular telephones have enabled people to "leap-frog" the conventional land-based telephone lines to acquire telecommunications service. Mobile technologies has made information technology ubiquitous because it is no longer found in defined places: no longer must telephone calls be made from the home or the office. For example, email is no longer limited to office hours because workers can send and receive emails even while vacationing on a Caribbean beach (whether they want to is another matter).

The change in technology has changed our lifestyles, our social interactions, our culture, and of course how enterprises interact with their environment. For example, IBM has teams that work together but rarely meet face-to-face. A manager might be located in Miami, the

[3]Peter Firmrite, Long before the interstates, there was a Winton, *San Francisco Chronicle*, June 16, 2003.

[4]Heppenheimer, T.A. *Turbulent Skies: The History of Commercial Aviation.* New York: John Wiley, 1995.

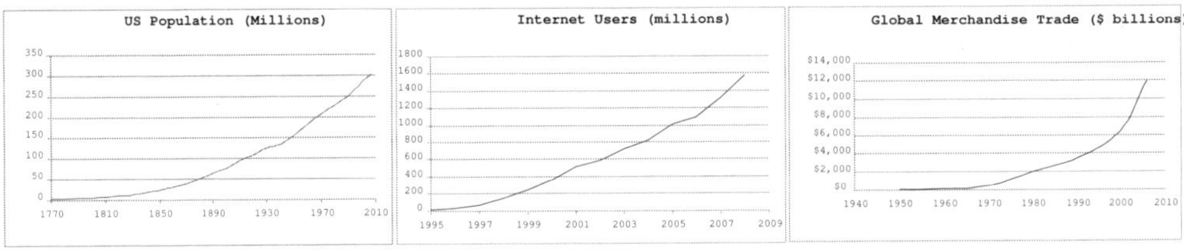

FIGURE 1.4
Examples of exponential growth.

systems analysts might be in Ohio, and the client might be in Barbados. They are able to work together due to the low cost, speed, and ease of communication and travel. In India's Kerala state, the economist Robert Jensen found that the fishermen adopted mobile phones, which attests to the ubiquity of new technology. Using mobile telephones the fishermen increased market efficiency by improving the flow and availability of information leading to a reduction in wasted fish catch, increased profits, and the lowering of consumer prices.[5] These are but two examples of how technology has affected enterprises.

Globalization

Globalization is a term describing how barrier after barrier to the world-wide flow of materials, information, people, and knowledge are disappearing or no longer becoming relevant. Especially pertinent to enterprises is the growth in global trade, emergence of a uniform global market, and ability to source labor and materials globally.

Figure 1.4 shows that global merchandise trade has grown from $62 billion in 1950 to $9 trillion in 2004. In many nations, the percentage of the gross domestic product attributed to imports and exports has grown (see Figure 1.4). The tremendous growth of global trade is due to technological advances, political enablers, economic enablers, and social factors. Technological advances have simultaneously decreased the cost and increased the speed of the physical transportation of goods, transmission of information, and capital. Trade agreements made through organizations such as the World Trade Organization (WTO) and regional agreements such as the North American Free Trade Agreement (NAFTA) have reduced tariffs. Also, there has been a convergence in standards so that products can more easily be sold in different countries.

The globalization of trade has an immediate, clear impact on enterprises. Today's enterprise competes in a global marketplace, which for most industries means greater competition on price, quality, performance, and speed. To illustrate, consider Ford, which competes in its home market of North America against foreign brands from Japan, Korea, Germany, and elsewhere. In 2008, Ford's North American market share was 15.2% compared to Toyota which enjoyed a 16.4% market share. Ford also competes outside of North America; in 2008 more than half of Ford's sales are from outside of North America.[6]

The example of Ford raises questions about the nationality of a corporation. Ford is considered a U.S. corporation because it is headquartered in the U.S. and files U.S. financial report. Yet it operate facilities around the globe selling to customers around the globe. Ford is an example of a multi-national corporation. Some multi-national corporations retain an ethnocentric perspective of their home country, in which they think and operate as if they were in their home country. Others adopt a global perspective in which they act

[5]Cellphones bridge the digital divide, *The Times of India*, Jan 29, 2006.
[6]2008 Ford Annual Report.

independently of geography, working according to local languages, customs, and traditions. ABB, headquartered in Switzerland but operating in more than 100 countries is an example of the later perspective.

We can say that globalization has a disrupting influence in that it is changing the world. Placing a value statement on the change is more difficult and depends on the context and perspective adopted. In Thomas L. Friedman's best-selling book, *The World is Flat,* he suggests that globalization is beneficial for all people. Such a generality greatly simplifies a highly complex phenomenon. Globalization has led to increased worldwide trade and also increased wealth in many nations, but that wealth is unevenly distributed [33]. The increase of global trade has contributed to the increased inequality within a country between the richest and poorest [8] [9]). The Gini coefficient, a measure of income inequality in which the higher the coefficient the less equal the society, shows the U.S. going from 0.395 in 1974 to 0.463 in 2007. In the U.S. the median full-time male workers' income has been essentially unchanged between 1978 and 2007 [2]. However, in any complex system, there are many factors that influence income disparity – in the U.S. this includes immigration, de-unionization, changing family structures, relative size of agricultural work sector, and other changes. Technology also contributes to the disparity in that more educated, higher-income individuals reap greater benefits from technology.

Other effects of globalization include the erosion of local distinctions and culture (see [27]) with the paradox of people simultaneously feeling greater affinity towards their own group. This may be a benefit to many enterprises because it reduces the need to customize products for different markets. If the same product or service can be sold worldwide without change, then economies of scale can be achieved to reduce overall operating costs.

Population Growth

The world population has grown tremendously in recent human history. It has grown from 2.5 billion people in 1950 to 6 billion in 2000. Further growth is projected such that the world population will be 7 billion in 2012.[7] Even with a decreasing growth rate, total population will continue to grow for the next generation. The repercussions of this growth are many. Concerning enterprise, a growing population means more people demanding products and services. Moreover, concomitant with population growth has been an increase in the quality of life for much of the world. More and more citizens of what were previously poor countries have greater incomes and are using their income to consume more and more resources. However, the second effect of the growing population is that the products we consume are built from raw materials and resources that are ultimately limited.

The world has seen rapidly rising costs for food, raw materials, and fuel.[8] This is a natural result of having limited resources sought after by a rapidly growing population. Some authors have suggested that the world will reach these limits on resources as predicted by Thomas Malthus in his *Essay on the Principle of Population* in 1798.

What is needed is to control world population and to better utilize resources. These challenges will affect enterprises in many ways. For example, the high cost of fuel has made some U.S. manufacturers look at sourcing parts locally to avoid the high cost of transporting goods from locations in Asia.[9] Changes in commodity prices are causing enterprise to consider substitutes for various materials, rethink how they utilize resources, and pay greater attention to sustainability.

[7] Source: U.S. Census Bureau, International Database, June 2008 Update, www.census.gov.

[8] Jeffrey D. Sachs, How to end the global food shortage, *Time Magazine*, April 24, 2008; Vivienne Walt, The world's growing food-price crisis, *Time Magazine*, February 27, 2008.

[9] Stung by Soaring Transport Costs, Factories Bring Jobs Home Again, by Timothy Aeppel, *Wall Street Journal* June 13, 2008; Page A1; China's Outsourcing Appeal Dimming, by Ariana Eunjung Cha, *Washington Post*, 8 September 2008.

Global Environment's Effect on Enterprises

The trends of technology development, globalization, and population growth all interact and affect the design of enterprises. Some of the ways in which these trends affect enterprise are:

- Global competition means that local markets are essentially international markets. Enterprises must compete for market share against both local and international enterprises.
- Information and knowledge flow freely across borders. In some cases this causes problems because of the limited protection in some nations for intellectual property. It also means that competitive advantage is often temporary because competitors can learn and copy best practices from around the world. On the beneficial side, it means that enterprises can operate globally far more efficiently and better integrate their supply chains.
- Markets change rapidly so enterprises need a quick payback on investments before the market opportunity, technology, or environment changes. Enterprises must continuously grow and improve in order to compete in rapidly changing environments.
- Population growth and limited resources have made sustainability the keyword for many industries. *Sustainability* means the adoption of practices so that natural resources are not depleted in a way that reduces the ability to continue those practices.

In addition to these global trends that affect all enterprises, each enterprise will be affected by more local, industry-specific or geographic-specific forces.

It is difficult for us to envision how the world will change in our lifetime. Most of the trends outlined above are exponential. Unfortunately, according to the science-fiction writer Robert Heinlein, most people can only extrapolate that the current level of technology will continue or slightly change, few people will even consider progress at the current rate, and even fewer are willing to make predictions for a continuation of exponential progress, which is precisely what the world has experienced over the past 50 years.[10] Gordon Moore, a co-founder of Intel, predicted that computing power would double approximately every two years – this is exponential growth, which has proven accurate to date!

As this section opened, it was declared that the enterprise environment is dynamic. In the past century, technology changed, political thought changed, society changed, and of course the enterprises in society changed. We cannot foresee what future changes will occur. What we can foresee is that there will be change and that the change is happening far faster than it ever has in the history of human existence. What this calls for are enterprises that are agile so that they can constantly change and improve with their environment. An agile enterprise scans its environment and understands how it is changing, how those changes may affect its operations, and makes design choices in order to work within that environment. In Chapter 4 on enterprise design methods, we discuss agility in greater depth.

1.4 History of Enterprise Engineering

Early work addressing the design of enterprise is to be found in many different disciplines. In this section, we outline some of the main intellectual influences on modern-day enterprise engineering.

[10]Heinlein, R.A., *Expanded Universe*, Berkley Publishing Group, New York, 1982, first paperback edition, p. 323.

1.4.1 Scientific Management

Adam Smith (1776) in his book *The Wealth of Nations* discussed how the division of labor could dramatically increase productivity. He noted that breaking work up into small discrete jobs simultaneously makes the job simpler, enables specialization, and allows workers to become highly skilled and efficient at the job assigned to them. During this same period, the concept of interchangeable parts emerged. Interchangeable parts are enablers for the division of labor. These ideas and changes in technology established an intellectual environment that has influenced work design to this day: work is broken up into simple tasks that are repetitively done by workers.

During the Industrial Revolution (1850-1950), emerged the Scientific Management movement, which strongly influenced thought on how to best organize an industrial organization. *Scientific Management* is the application of the scientific method to management. The primary goal of scientific management was to achieve efficiency of resources. Time and motion studies to improve labor efficiency, use of machinery to increase productivity, as well as division of labor characterize the general concepts of Scientific Management. Scientific Management led to the founding of the academic discipline Industrial Engineering, whose curriculum still strongly bears the mark of Scientific Management theory.[11]

It is important to review the times and environment under which the ideas of scientific management were developed. There was a social migration of people to urban areas from rural agricultural areas of the county, as well as an influx of new immigrants from Europe. Cheap and unskilled labor was plentiful. Manufacturing, the main generator of wealth, required large amounts of labor to manufacture the products they sold. These circumstances strongly influenced the development of scientific management. First, the progenitors of scientific management took labor for granted; if one worker did not perform well, there were others to take his place. Given this attitude it is not surprising that worker satisfaction was not a priority or even considered in the design of work. Second, many of these new workers had little education or training. So the division of labor was partly done to create unskilled production jobs these workers could fill. The simpler the job then the easier it was to fill the job with the available labor. The separation of the planning of the work from doing the work also flowed from these circumstances. So, we need to remember the social and economic environment that influenced the thinking that went into the development of Scientific Management.

The ideas of Scientific Management were developed by Frederick Taylor, Gilbreth, and others. Frederick Taylor was an engineer who became a well-known advocate of scientific management. Taylor wrote,

> And this one best method and best implementation can only be discovered or developed through scientific study and analysis... This involves the gradual substitution of science for "rule of thumb" throughout the mechanical arts.

In Taylorism, as it is sometimes referred, there is a sharp division between physical work and cognitive work. Taylor strongly believed that the successful manager was a manager who controlled every aspect of the production process. To achieve this, managers should uncouple planning and execution – i.e., workers only execute what managers plan. This is probably the most well-known principle of Scientific Management (the other principles are listed in the box below. At a lecture he gave in 1906, Taylor explained, "In our scheme, we do not ask for the initiative of our men. We do not want any initiative. All we want of them is to obey the orders we give them, do what we say, and do it quick" ([17] p. 169).

[11] Although interviews of senior faculty by Bailey and Barley [4] suggest Taylor had less of a role in shaping the Industrial Engineering curriculum than many have thought.

Principles of Scientific Management

1. Time studies
2. Functional supervision
3. Standardization of tools and implements
4. Standardization of work methods
5. Separate planning function
6. Management by exception principle
7. The use of slide-rules and similar time-saving devices
8. Instruction cards for workmen
9. Task allocation and large bonus for successful performance
10. The use of the "differential rate"
11. Mnemonic systems for classifying products and implements
12. A routing system
13. A modern costing system

In Scientific Management, the work is systematically analyzed, it is broken down into its minuscule operations, each operation is assigned to a separate worker, and an elaborate set of procedures is generated to regulate each operation. In this approach, there is a sharp division between the planning work and the actual labor itself. This principle is famously illustrated by the story of when Taylor analyzed shoveling. He determined the proper size shovel for various materials (coal, dirt, rock, and so forth). The worker as a result was no longer allowed to choose which shovel to use. Taylor demonstrated that the productivity of the worker was greatly increased.

Henry Ford expanded on the ideas of scientific management by adding the revolutionary idea of having a moving assembly line. In 1913, the Ford Motor Company reengineered their Highland Park automobile assembly operations and cut production time from 750 minutes to 90 minutes per car through work simplification, division of labor, the moving assembly line, and other means.[12]

Henri Fayol, a contemporary of Taylor, proposed general principles of management. These ideas continue to strongly influence management thought up to the present day [37]. The work of Taylor and Fayol is essentially complementary. They both realized that the problem of human resource management is the key to business success. Both applied the scientific method to the problem of management. Taylor worked primarily on the operative level, from the bottom of the organizational hierarchy upward. Fayol concentrated on the Managing Director (his term) and worked downward. Unlike Taylor, Fayol's work reflects a tension between his recognition that managers are not supermen and yet employees should not be allowed enough autonomy and responsibility to solve second-order problems (problems for which there are no precedents, or previous exemplary solutions). Many of Fayol's principles still make sense today. For example, it is still important that if somebody is held responsible for something, then that person needs the authority to ensure its success. Other principles such as discipline need to be adapted to the expectations of society today.

[12] http://www.ford.com/about-ford/heritage/places/highlandpark/663-highland-park.

Fayol's Principles of Management

Division of work. Management should pursue standardization of work so that it can be divided among workers who specialize in a narrow set of tasks. The objective is to produce more and better with the same effort.

Authority and responsibility. The good manager should have official authority deriving from office and his personal authority. Responsibility is a corollary of authority, it is its natural consequence and essential counterpart, and where authority is exercised responsibility arises.

Discipline. Discipline is obedience, behavior, and respect. Discipline is absolutely essential for the smooth running of business and without discipline no enterprise could prosper.

Unity of command. An employee should receive orders from one superior only.

Unity of direction. One head and one plan for a group of activities having the same objective (centralization of authority).

Subordination of individual interest to general interest. The interest of the home should come before that of its members and the interest of the state should have pride of place over that of one citizen or group of citizens. Constant supervision is needed to ensure that the general interest will not be lost in favor of individual interest.

Remuneration of personnel. Remuneration should be fair. It shall not go beyond reasonable limits.

Centralization. Centralization belongs to the natural order. The degree of centralization must vary according to different cases. If the moral worth of the manager, his strength, intelligence, experience and swiftness of thought allow him to have a wide span of activities, he will be able to carry centralization quite far and reduce his seconds-in-command to mere executive agents.

Scalar chain. Describes the chain of superiors ranging from the ultimate authority to the lowest ranks. In short, it is the line of authority. It is an error to depart needlessly from the line of authority, but it is an even greater one to keep to it when detriment to the business ensues. An employee who cannot obtain guidance from a superior should have the initiative to choose a course of action best suited for the organization. Such a course of action is based on precedence set by the management.

Order. In the case of material things – A place for everything and everything in its place. In case of human order – A place for everyone and everyone in his place.

Equity. For the personnel to be encouraged to carry out duties with all the devotion and loyalty of which it is capable it must be treated with kindliness, and equity results from the combination of kindliness and justice.

Stability of tenure of personnel. Generally, the managerial personnel of prosperous firms is stable, that of unsuccessful ones is unstable. Moreover, stability of workers is beneficial. Time is required for an employee to get used to new work and succeed in doing it well. If when he has got used to it, or before then, he is removed, he will not have time to render worthwhile service.

Initiative. Thinking out a plan and ensuring its success is one of the keenest satisfaction for an intelligent man to experience. It is also one of the most powerful stimulants of human endeavor. Hence, it is essential to encourage and develop this capacity to the fullest.

Esprit de corps. Harmony and a sense of belonging to a group is great strength to the organization. Effort, then, should be made to establish the esprit de corps.

1.4.2 Humanist School

The humanist school of management of the early and mid-1900s tried to shift the focus of organization from processes to people [22]. Mayo, who was trained as a sociologist, viewed man as a social and emotional being, and argued that if you treated the workers with respect and tried to meet their needs, then they would be more productive workers. The experiments he led at Western Electric's Hawthorne Works illustrate the basic ideas of motivation that he developed. These experiments, called the Hawthorne Experiments, studied how changes to work conditions such as hours per week, number of rest breaks, time of lunch, and similar changes affected productivity. In the experiments, they found productivity increased regardless of the interventions they made – productivity even improved when the workers were returned to their original, harder work conditions. These experiments do not support the one best way advocated by Taylor, and point to a more complex view of human motivation and performance. Workers at Hawthorne were simply responding positively to the attention from management.

Scientific Management ignores or minimizes those traits that make humans, human. The Humanist School was to some extent a reaction to the Scientific Management approach that viewed man mechanistically. The main criticism, still echoed today, is that Scientific Management leads to repetitious, tedious, and boring jobs that diminish the workers' self-being. The Humanist School, on the other hand, argued that through job enrichment and promoting worker satisfaction then organizational benefits could be realized.

Maslow [21] categorized human needs in a hierarchy starting at the bottom with physiological needs (food, water, shelter), safety, belonging, self-respect, and what he termed self-actualization. He argued that lower-level needs must be satisfied prior to high-level needs. Under a Taylorist approach in the late 1800s through the early part of the 1900s, an enterprise would satisfy the physiological needs and, later, also safety.[13] Following WWII many corporations satisfied belonging and for some workers self-respect. Self-actualization is less common, but is argued by some authors that enterprises should strive to design work so that all workers can attain a level of self-actualization that will also benefit the enterprise [20].

This early work led to the mostly accepted job characteristics theory articulated by Hackman and Oldham [14, 24]. The job characteristics theory posits that five job characteristics produce a psychological state in the job holder that affects job outcomes including performance. The five job characteristics are skill variety, task identity, task significance, autonomy, and feedback. Skill variety is when the job requires a wide variety of the worker's skills and abilities. Task identity is when the worker develops a sense of ownership and responsibility for a meaningful part of the job. Task significance is the degree to which the job impacts the lives of others. Autonomy is the degree of freedom and independence the worker has in deciding how to carry out the work. Feedback is the degree to which the worker received knowledge of the results of his or her work.

These five job characteristics are theorized to affect the worker's psychological state, which in turn affects four outcomes of internal work motivation, growth satisfaction, general satisfaction, and work effectiveness. How and the degree to which they affect these outcomes is moderated by the workers' growth need strength, defined as the need for personal accomplishment. Workers with a high growth need strength react more favorably to enriched jobs. This model better represents the complexity of the human situation. There are many factors that influence worker satisfaction and the link to job performance. Because each worker is an individual then not all workers will react in the same way.

[13] Concern over industrial safety is a recent development. In the 1870s, during the building of the Brooklyn Bridge between 30 and 40 workers died, including its chief engineer John Roebling. – p. 506 in *The Great Bridge* by David McCullough, Simon & Schuster Publisher, 2001.

The Volvo assembly plant in Uddevalla, Sweden is a prominent example of the human relations school. Volvo put into practice what is called the Human System Design by having small teams of skilled workers assemble complete cars from start to finish. The approach turns the ideas of Smith, Taylor, and Ford on its head. A motivation for Volvo to experiment with alternatives to the assembly line was the difficulty they experienced in recruiting and retaining workers. Sweden, at the time in 1990, had a very low unemployment rate of 1.1%. A more humanist work environment was seen as necessary to attract workers.

Volvo closed the Uddevalla plant in 1993 [32], critics saying it never matched the productivity of other assembly plants, although there are many factors that may have come into play for Volvo's poor performance (poor car design, marketing, etc.). In Volvo's Uddevalla plant, it took 16 months of training for a worker to master the assembly work ([26] p. 90); contrast this with Ford's idea of training workers in simple tasks in a matter of hours. In 1990, it took 50-55 man-hours to assemble a car in the Uddevalla plant ([26] p. 150). The best automotive assembly plants were assembling cars in 13.5 man-hours ([36] p. 120). However, a controversy persists over the performance of the Volvo Uddevalla plant because direct comparisons of plants in different geographic regions is difficult, and there are many other factors that played a role in Volvo's decisions; such as the more important management-stated objective to reduce capacity, rather than to close the plant due to poor efficiency.

1.4.3 General Systems Theory Movement

The Scientific Management approach breaks down the production process into simple tasks and then improves the efficiency of each task. This is called a reductionist approach, and it still widely permeates current engineering practice. A different approach is the systems approach that looks at the whole. Bertalanffy, an important researcher of the systems movement and others contributed to the General Systems Theory that says, "the whole is greater than the sum of its parts." What this statement means is the behavior of the system cannot be explained from the behavior of its constituent parts. The system characteristics emerge due to the interaction of the parts. This lead Bertalanffy to propose the existence of general system laws that apply to any system, irrespective of the particular system, its properties, or the elements involved. Bertalanffy was motivated, in part, by the trend towards even greater reductionism and specialization in science. Against reductionism, he said [5], "It is necessary to study not only parts and processes in isolation, but also to solve the decisive problems found in organization and order unifying them, resulting from dynamic interaction of parts, and making the behavior of the parts different when studied in isolation or within the whole...". And against specialization, "Modern science is characterized by its ever-increasing specialization, necessitated by the enormous amount of data, the complexity of techniques and of theoretical structures within every field. Thus science is split into innumerable disciplines continually generating new sub-disciplines. In consequence, the physicist, the biologist, the psychologist and the social scientist are, so to speak, encapsulated in their private universes, and it is difficult to get word from one cocoon to the other...".

Two important observations can be drawn from the systems science movement. First, the relationships between the parts of the system are important to the overall system behavior. For this reason, a reductionist approach might fail because it does not account sufficiently for these relationships. Second, an open system interacts with its environment, and these interactions need to be included in the analysis.[14]

[14]Some authors define analysis as strictly being reductionist. We do not use the term in that way. Later in this book we define analysis, which can be either reductionist or not.

1.5 Summary

To summarize, enterprise engineering is a discipline that views the enterprise as a product that is purposefully designed. A basic premise of this book is that enterprise engineering is a scientific endeavor in its own right; enterprise engineering has an established theoretical basis with domain-specific knowledge. This knowledge includes:

- An enterprise is an open system.
- An enterprise has a life-cycle describing its evolution from conception to retirement.
- The enterprise design method and complex problem-solving method.
- How to conduct enterprise projects.
- Understanding of how an enterprise interacts with its environment.

Enterprise projects usually only address limited aspects of the enterprise. For this reason, we argue it is important to have an enterprise architecture to ensure all enterprise projects conform to an overall enterprise design. To conduct a project, there is an enterprise design method that describes the phases, techniques, and management of these large projects.

This chapter reviewed the historical development of enterprise engineering, tracing its roots to scientific management, humanist school of management, and general systems theory. Current approaches to enterprise systems build on these foundations. Scientific management ideas on efficiency still have great influence on how we design enterprise systems, but it is tempered by a knowledge that humans are important elements of the system that cannot be treated simply as machines. General systems theory introduced ideas on feedback, open versus closed systems, and other system principles that apply to enterprise systems and are discussed more fully in the next chapter. The concepts of the human relations school did not replace the earlier scientific management school, nor has the more recent systems thinking school replaced the human relations school. The older ideas are not abandoned by incorporated into the new ideas. Division of labor and work simplification do improve productivity. But now we also consider the other aspects of the job as well. Consequently, work designers trade off the benefits of smaller tasks versus job enrichment, and the resulting decision depends on the particular case being considered.

Review Questions

1. Describe how many enterprise systems are designed today.
2. Why do we use the term “enterprise” instead of organization?
3. What makes a virtual enterprise different from other enterprises?
4. List the steps involved in a general problem-solving methodology.
5. What is the high-level design of an enterprise and who creates it?
6. In a project that redesigns the inventory control for a manufacturer, identify who the stakeholders are.
7. Argue whether a project that replaces a manual process to do purchase orders with an on-line, web-based system should be called business process reengineering.
8. What is the primary distinction between an enterprise architect and a system architect?

9. Explain how enterprise engineering is different from systems engineering.
10. Explain why it is important to understand the enterprise's environment.
11. Describe the primary goals of scientific management.
12. Take one of the fourteen principles of scientific management and describe how it is, or is not used today in business.
13. A worker in a fast-food restaurant has the job to assemble hamburgers. Assess this job using Maslow's hierarchy of needs.
14. Choose a business (e.g., fast-food, automotive manufacturing, textiles, televisions) and describe how globalization is affecting that business.
15. Explain the difference between "enterprise system" as used in the book and "enterprise system" as when referring to enterprise resource planning systems.
16. Describe how BPR differs from Continuous Process Improvement.

Bibliography

[1] S. Alter. *The Work System Method.* Work System Press, Larkspur, CA, 2006.

[2] Anonymous. A special report on the rich: More or less equal? *The Economist,* April 2, 2009.

[3] B. Arteta and R.E. Giachetti. A measure of agility as the complexity of the enterprise system. *Robotics and Computer-Integrated Manufacturing,* 20(6):495–503, 2004.

[4] D.E. Bailey and S.R. Barley. Return to work: Toward post-industrial engineering. *IIE Transactions,* 37:737–752, 2005.

[5] L. Bertalanffy. *General Systems Theory: Foundations, Development, Applications.* George Braziller Publishing, New York, 1968.

[6] B.S. Blanchard and W.J. Fabrycky. *Systems Engineering and Analysis.* 3rd edition, Prentice Hall, Upper Saddle River, NJ, 1997.

[7] Cisco. Visual networking index – forecast and methodology 2007-2012. Technical report, Cisco, 1008.

[8] W.R. Cline. Trade and income distribution. Technical report, Institute for International Economics, Washington DC, 1997.

[9] R. Culpeper. Approaches to globalization and inequality within the international system. Technical report, Overarching Concerns Program, Paper Number 6, October 2005, United National Research Institute for Social Development, 2005.

[10] T.H. Davenport. Putting the enterprise into the enterprise system. *Harvard Business Review,* July-August:121–131, 1998.

[11] T.H. Davenport and J.E. Short. The new industrial engineering: Information technology and business process design. *Sloan Management Review,* Summer:11–27, 1990.

[12] R.E. Freeman. *Strategic Management: A Stakeholder Perspective.* Pittman Press, Boston, MA, 1984.

[13] H.T. Goranson. *The Agile Virtual Enterprise: Cases, Metrics, Tools.* Quorum Press, Westport, CT, 1999.

[14] J.R. Hackman and G.R. Oldham. Motivation through the design of work: Test of a theory. *Organizational Behavior and Human Performance*, 16:250–279, 1975.

[15] K. Hafner. *Where Wizards Stay Up Late: The Origins of the Internet.* Simon & Schuster, New York, NY, 1998.

[16] M. Hammer and J. Champy. *Reengineering the corporation: A Manifesto for Business Revolution.* Harper Business, New York, NY, 1993.

[17] R. Kanigel. *The One Best Way.* The Viking Press, New York, NY, 1997.

[18] Y.T. Leung, N. Caswell, and M. Kamath. The case for the business process engineer. Technical report, IBM Research Report, RJ10355(A0507-021), Almaden, CA, July 13, 2005.

[19] D.H. Liles, M.E. Johnson, and L. Meade. The enterprise engineering discipline. In *Proceedings of the Society for Enterprise Engineering*, Orlando, FL, June ,1995.

[20] J. Martin. *The Great Transition: Using the Seven Disciplines of Enterprise Engineering to Align People, Technology, and Strategy.* AMACOM Press, New York, NY, 1995.

[21] A. Maslow. *Motivation and Personality.* Harper & Row, London, UK, 1964.

[22] E. Mayo. *The Social Problems of an Industrialized Civilization.* McMillian Press, New York, NY, 1945.

[23] W. McCumber and C.D. Sloan. Educating systems engineers: Encouraging divergent thinking. In *INCOSE International Symposium*, Las Vegas, NV, July 28-August 1, 2002.

[24] G. Oldham, J. Hackman, and J. Pearce. Conditions under which employees respond positively to enriched work. *Journal of Applied Psychology*, 61(4):395–403, 1976.

[25] M. E. Porter. *Competitive Advantage: Creating and Sustaining Superior Performance.* The Free Press, New York, NY, 1985.

[26] H.D. Pruijt. *The fight against Taylorism in Europe.* PhD thesis, Erasmus Universiteit, Rotterdam, NL, 1996.

[27] G. Ritzer. *The Globalization of Nothing.* Sage Publication, Beverly Hills, CA, 2004.

[28] W.B. Rouse. Embracing the enterprise. *Industrial Engineer*, 36(3):31–35, 2004.

[29] W.B. Rouse. A theory of enterprise transformation. *Systems Engineering*, 8(4):279–295, 2005.

[30] O. Sáenz, C. Chen, M.A. Centeno, and R.E. Giachetti. Defining enterprise systems engineering. *International Journal of Industrial & Systems Engineering*, 4(5):483–501, 2009.

[31] A.P. Sage and J.E. Armstrong. *Introduction to Systems Engineering.* John Wiley & Sons, New York, NY, 2000.

[32] A. Sandberg. Volvo human-centred work organization – the end of the road? *New Technology, Work and Employment*, 8(2):83–87, 2007.

[33] J.E. Stiglitz. *Globalization and Its Discontents.* W.W. Norton & Company, New York, NY, 2003.

[34] D.R. Towill. Successful business systems engineering part i: The systems approach to business processes. *Engineering Management Journal*, 7(1):55–64, 1997.

[35] T.J. Williams. *Handbook of Life Cycle Engineering*, chapter The Purdue enterprise reference architecture (PERA), pages 289–330. Springer, New York, NY, 1998.

[36] J.P. Womack, D.T. Jones, and D. Roos. *The Machine that Changed the World.* Harper Perennial, 1990.

[37] D. Wren. *The Evolution of Management Thought.* John Wiley & Sons, New York, NY, 2004.

2

Systems Theory

"In the past the man has been first; in the future the system must be first." – Frederick W. Taylor (1856-1915), originator of scientific management.

Systems theory is a way to view the world, and has influenced scientific, engineering, management, and other fields of inquiry. Systems theory is used in this book as the paradigm for explaining how enterprise systems work, and prescribing designs for how they should work. Ackoff [1] quotes Albert Einstein, " Without changing our patterns of thought, we will not be able to solve the problems we created with our current patterns of thought." This chapter is an attempt to change the reader's pattern of thought. In this chapter, we explain in greater depth what it means to say an enterprise is a system. The chapter presents the principles of systems theory and how they apply to enterprise systems. Then we contrast traditional reductionist thinking with systems thinking. This chapter describes systems dynamics, which provides tools to help apply the systems thinking principles. Using a systems approach we intend to influence how we understand the enterprise, and how to improve the way we do analysis and design of enterprise systems.

After completing this chapter, you should be able to:

- Describe how an enterprise fits the definition of a system.
- List and describe the properties of a system.
- Identify the emergent properties of a system.
- Identify the feedback components and how they function to provide feedback in a system.
- Contrast an open system from a closed system.
- Define system complexity and identify systems that are complex.
- Contrast reductionist thinking from systems thinking.
- Apply systems thinking to understand and solve problems.
- Create a causal loop diagram to model a system.
- Create a stock and flow diagram to model a system.
- Describe the limitations of system dynamics.

2.1 Definition of a System

A *system* is a set of discernible, interacting parts or subsystems that form an integrated whole that acts with a single goal or purpose. We can draw a boundary around the system, and everything inside the boundary is part of the system, while everything outside of the boundary is part of the external environment. Discernible means we can distinguish between each part or subsystem. The parts of the system have their individual performance and behavior, but in the system they are interrelated with other parts, which gives rise to behavior that can only be attributed to the system as a whole. Some of the relationships are designed into the system to obtain a desired system behavior. Other relationships, called side effects, are not formally designed into the system, and many times these side effects are undesirable. A system has one or more defining functions or properties that make it a system. If no function or property can be attributed to the system as a whole, then it is just an aggregation of parts.

The above definition of a system is for any system. In Chapter 1, we had defined an enterprise as a special type of system. We said an enterprise is a complex, socio-technical system that comprises interdependent resources of people, information, and technology that interact with each other and their environment in support of a common mission. In order to understand what it means for an enterprise to be a system, we explore the properties of enterprise systems. Appreciation of these properties will help the analysis and the design of enterprise systems.

2.1.1 Enterprise Boundaries

A system boundary delineates what is part of the system and what is not. The boundary of an enterprise is somewhat arbitrary because it depends on the intentions and aims of the observer [11]. Conventionally, the boundaries of an enterprise are defined where its control over resources ends. However, our observation of modern enterprises questions the appropriateness of this definition. Does the enterprise include its partially controlled distributor? Does the enterprise include temporary workers? or its outsourced call center? Whether these and other objects are included in or out of the system depends on the intention of the observer. In these cases, drawing the boundaries based on ownership may not make sense from a system's viewpoint. The enterprise may not be able to control these entities, but it can strongly influence their behavior. Exerting influence is not command and control; it involves negotiation, suggestion, and persuasion. An enterprise might be able to strongly influence the behavior of its distributor, if say, the enterprise accounts for most of the distributor's business. But, what if a small company sells their product through Walmart, can they hope to influence how Walmart does business? Probably not. Thus, the definition of an enterprise boundary is contextual, and may be different for each project.

2.1.2 Enterprise Subsystems

Subsystems are systems in their own right, but they are also part of the enterprise system. The subsystems can be identified individually and exhibit all the properties of a system. How the subsystems are identified depends on the viewpoint of the observer. A functional viewpoint of the enterprise might define the subsystems of an enterprise as marketing, sales, and manufacturing. A geographical viewpoint of the enterprise might define the subsystem as North American Operations, European Operations, and Asian Operations. Within these subsystems there are more subsystems. Manufacturing has an Inventory Control system, a Production Control system, and others. There are other ways to divide an enterprise into subsystems, and then divide those subsystems into more subsystems. One of the main themes of enterprise design is finding optimal ways to structure the enterprise into subsystems.

Not only is an enterprise composed of subsystems, but the enterprise can be part of a larger system. For example, many enterprises are part of a supply chain, which is also a system. Consequently, from a system's perspective, given any system, we can think of it as being a subsystem of some larger system, and simultaneously, we can think of it as being composed of subsystems, each of which is composed of subsystems, and so forth.

2.1.3 Holism

"The whole is more than the sum of its parts" is a commonly heard system-thinking phrase. What this phrase means is that the whole system exhibits *emergent properties* that are meaningful only when they are attributed to the whole system, and not to any of the individual parts of the system. Peter Checkland, an authority on systems thinking, considers emergent properties to be fundamental to systems science [4]. Emergent properties cannot be deduced from the part's properties. We call this concept *holism*, the idea that a system exhibits properties and behavior that cannot be attributed to any one of its parts.

Enterprises have the property of holism. We can see it in the ability of enterprises to be innovative and develop new products and services or in the observed financial performance of enterprises (revenues, costs, and profits) that cannot be attributed to the parts of the enterprise (e.g., marketing, sales, manufacturing, etc.).

The existence of emergent properties defeats attempts at a traditional reductionist approach because if you decompose the system into its constituent parts, and then analyze each part individually, you will miss the emergent properties. What holism directs us to focus on are the relationships between subsystems because the emergent properties can only be understood by understanding how the parts interact to provide that property. It also directs us to investigate how changes to one part of the system will affect other parts of the system. As a fundamental property of systems, understanding holism means studying the entire system and not just its parts.

2.1.4 Open versus Closed

Enterprise systems affect, and are affected by, their environment. They exchange materials, information, and people with their environment. A system with these characteristics is called an *open system*. The opposite of an open system is a *closed system* that does not have any interaction with its environment. As a result of being an open system, you can only understand an enterprise in the context of its environment [10].

Open systems require certain environmental conditions in order to perform their defining function. For example, Amazon.com cannot exist as an enterprise without the Internet that is part of its external environment. Other environmental conditions that can affect enterprises are the regulatory, market, labor, political, and social environments. Because enterprises are open systems they need to scan their environment and then adapt to the environment.

Enterprise systems can exhibit steady-state behavior, defined as the dynamic adjustment of system components and attributes in reaction to environmental changes so that the system maintains a constant (steady) state. This is called self-regulatory or self-adaptive behavior. As purposeful systems, enterprises can change their environment, which is not unheard of. For example, a company could lobby the government for special treatment, laws, or policies that would be favorable to it.

Viewing an enterprise as an open system also directs us to consider the inputs and outputs of the enterprise. The inputs and outputs cross the boundaries of the system. Essentially, we can view the enterprise as a single, large process that converts inputs into outputs. The inputs are labor, materials, capital, and knowledge. The outputs are money, products, services, and waste. The "process perspective" is a useful way to view an enterprise and forms the basis of the process view described in a later chapter.

2.1.5 Purposefulness

Enterprises have purpose, which is defined in the mission statement of the enterprise. Furthermore, the people in the enterprise have a purpose, their individual goals, and these individual goals may or may not be aligned with the enterprise purpose. Ackoff [1] termed these purposeful systems. Because the enterprise is purposeful there is a rationale that explains its actions. Understanding the purpose of the enterprise system and the people in it is a matter of uncovering this rationale. Rationale depends on the environment – people are influenced by both the business culture and the social culture. An analysis needs to consider who the people are, what motivates them, and how this motivation underlies their actions.

It is useful to describe the other types of systems so that purposefulness is put in perspective. Gharajedaghi [10] distinguishes between state-maintaining systems, goal-seeking systems, and purposeful systems. A state-maintaining system can react to its environment but only in a predefined way. It exhibits no learning and cannot adapt its behavior. The cruise control in a car is a state-maintaining system. A goal-seeking system can respond differently to different events but always in pursuit of the particular outcome (goal). A goal-seeking system cannot change its goal. Goal-seeking systems have all the capabilities of state-maintaining systems. Purposeful systems have all the capabilities of goal-seeking and state-maintaining systems with the addition that they also exhibit free-will, which is the ability to also change their goals. Hence, purposeful systems have choice over what goal to pursue and what means to use to work towards the goal [1].

2.1.6 Feedback and Control

The control of a system is the subject of *cybernetics*, which conceptualizes the mechanisms of feedback and self-regulation whether by mechanical, electrical or other means. Understanding feedback control is fundamental to systems engineering.

The following discussion of feedback control refers to Figure 2.1. The system being controlled transforms inputs into outputs. The system being controlled is subject to unpredictable disturbances from its environment. These disturbances may cause the system to deviate from its desired state. The objective of control is to reduce the deviation between a desired state of the system and the actual state of the system. In order to maintain the desired state, feedback control is implemented as follows:

1. A system parameter indicative of the desired system state is measured and monitored.
2. A reference level that describes the desired value for the system parameter is set.
3. A decision-making function compares the current state of the system to the reference level. The deviation between these two states is the error signal.
4. An actuator alters the system state so as to reach the desired state. The actuator's behavior is usually directly proportional to the magnitude of the deviation.

The conventional example of feedback control is maintaining room temperature with a thermostat. The system parameter being controlled is the room, the controlled parameter is the room temperature, a reference temperature is set on the thermostat, the thermostat measures the difference between the two temperatures, the thermostat turns on/off the actuator, which is the furnace or other heat-providing source. In this way, the room temperature is maintained within some tolerance range of temperatures to the set temperature.

For control to work, the controller must have a resolution greater than what is being controlled. Using the thermostat to illustrate the concept, if the thermostat can only be set in increments of 10 degrees (60°F, 70°F, 80°F, etc.), then it is impossible to control the

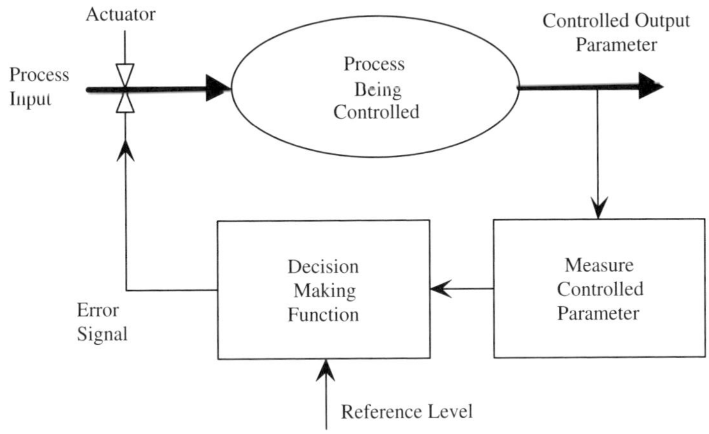

FIGURE 2.1
Feedback in a system.

room at a temperature of 68°F. This concept is due to Ashby's [3] *law of requisite variety* that says the variety of controller must be greater than the variety of the environment. It is the environment that creates the perturbations that affect the system. The controller needs to be able to create as many counteractions as there are perturbations; in other words, the controller must have at least the same variety as the environment. The law can be interpreted as the controlling system must have at least as many states (variety) as the system being controlled. To control room temperature to 68°F, the thermostat must have gradations of 1 degree or less.

In enterprises the controlled system is generally not an electro-mechanical system like the thermostat. Nevertheless, the cybernetic concepts of control still apply. Enterprise control is performed by the management structure, whose primary role is to control the systems they are responsible for. In the model of Figure 2.1, the manager fulfills the role of the decision-making function. The measured control parameter could simply be the manager's observation of the system performance, but it is better if the manager has a performance measurement system to measure the state of the system. The actuator is the set of actions the manager can take to correct any deviations between the system's desired performance and its actual performance.

To illustrate the human control of a system, consider a call center that receives incoming calls from clients and provides information to them. One of several measures of performance for call centers is the throughput rate, defined as the number of calls per hour the center can handle. An automated telephony system measures the throughput rate (among many other performance parameters), and reports the throughput rate to the manager. The manager receives the performance reports with which he can compare the actual performance against the desired performance (the reference level). If actual performance is less than desired performance, then the manager has to decide what course of action to take. The manager usually has several courses of action he can take in order to bring actual performance to desired performance. If the actual throughput is less than the desired throughput, then some courses of action are: (1) add additional call agents to field calls, (2) have the agents spend less time on the phone with each client, or (3) implement technology to automatically answer some calls instead of human agents.

The call center example illustrates the feedback control principle in an organizational context. While the basic principles of feedback control hold for all systems, there are important differences when the system is controlled by humans. First, in many organizational systems we are limited to the precision that we can know the state of the system. If the manager is not sure what the state of the system is, then how can the manager know if the state is different than the desired state? While a college student, the author spent one summer working for Macy's Department store, and remembers the store would shut down early twice a year to physically count store inventory. They did this because after six months they did not have an accurate count of the inventory, which made inventory control near impossible. Good control is predicated on a performance measurement system that lets the managers know the state of the system.

A second limitation is that in most enterprise scenarios there is, not a single parameter like temperature that you want to control, but many parameters. In the call center example, in addition to throughput we would want to control the client waiting time, the quality of the responses to customer calls, and the cost of staffing the call center. Oftentimes, these goals are in conflict. Actions to improve quality of the responses might also simultaneously lead to unwanted increases in client waiting-time or staffing costs. Oliva and Sterman [15] provide an enlightening discussion explaining why service quality often deteriorates in many enterprises due to the management actions taken to improve performance. Unless management understands the relationships between all the goals and how they react to the control actions, then controlling the system is very difficult if not impossible in very complex systems.

A third issue is that the manager does not have precise knowledge of how the chosen course of action taken will affect the system's performance. Frequently, there are no mathematical equations available to predict how an action will affect the system. Moreover, feedback in enterprise systems is often delayed in time and is nonlinear – both characteristics that make predicting the effects of a control action much more difficult. For example, in a software engineering team, if the project starts to fall behind schedule, then the manager might decide to use more overtime. But this course of action will put into play other feedback loops. More overtime can eventually lead to burn-out, or experienced software engineers leaving. These effects will then make the project late again. In the section on system dynamics, we present causal loop diagrams that can help the analyst better understand the many feedback loops that exist in systems.

Collectively, these issues make control by humans of systems much more difficult than the simple thermostat example would suggest. As enterprise designers these principles need to be kept in mind when designing systems.

2.1.7 Complexity

Enterprise systems are usually described as being complex. Yet, the question, "*what makes a system complex?*" is not an easily answered question. Most people have an intuitive sense of complexity, but in order to have an engineering understanding of enterprise complexity, we need to formalize the definition. At a very basic level, we say a system is complex because it is difficult for us to understand how the system functions, and it is difficult for us to predict the behavior of the system. So the question becomes what makes a system difficult to understand and predict?

A widely held viewpoint is that complexity is due to the large number of interacting parts [16]. The more parts a system has, then the more complex that system is. Many enterprises are large, consisting of many different parts, so an enterprise would be considered complex using this definition. However, having many parts may be a common characteristic of complex systems, but in itself, it is an insufficient condition and sometimes not even a necessary condition. To illustrate the shortcoming of defining complexity in terms of the number of interacting parts, consider a Swiss chronograph (watch) that has a large number of interacting parts. Many technically oriented people can open the watch and reverse engineer how the parts work and go together. Moreover, the watch's behavior is well defined and predictable. Because the observed behavior is predictable, then it suggests the watch does not fit our definition of being a complex system. Moreover, some people, such as watch-makers, understand exactly how a watch operates, can name all the parts, and tell how they go together. Instead of saying the watch is complex, we say the watch is *complicated.* The large number of parts makes it difficult to understand, but it is understandable, especially to the watch-maker. So, the number of parts does contribute to complexity, but there are other system characteristics that must be present for us to call the system complex.

Complexity arises from not only the number of parts in the system, but also from the interrelationships of the system parts and the emergent behavior that cannot be predicted from the individual system parts [20]. This is why the watch is not complex – the relationships between the watch components are all determined by mechanics that are well understood. Contrast this with systems that have nonlinear relationships, feedback relationships, and system-wide behavior that emerges as a result of many component interactions. Nonlinear relationships are such that small changes in inputs result in large changes in outputs, no change in outputs, or delayed changes in outputs. Similarly, a large change in input might lead to no change in the output or a small change in the output. Feedback relationships can be negative or positive. Negative feedback slows down the response to the input and positive feedback increases the response to the input. Relationships might be difficult to understand because there are spatial and temporal disconnects between when something happens – the *cause*, and when the system reacts – the *effect.* For example, an enterprise might change its marketing strategy but not see the results of this change for a year or more – indicating a large time delay between an event and its effect on the system. Similarly, an enterprise might experience market losses in one region that affect market conditions in another region – indicating a disconnect in the location of the event and its effect.

We have identified two characteristics that make systems complex: the number of parts, and the network of relationships between the parts. A system with many parts is at least complicated to understand and may also be complex. For it to be complex, the relationships between the parts must be such that system behavior becomes difficult to understand and predict. In general, most enterprises fulfill this criteria; they have both a large number of parts and the parts are related in ways that make it difficult to understand how the enterprise operates and to predict the behavior of the enterprise.

The definition of complexity described depends on the presence of an observer who ascribes the property of complexity to the system. Some researchers conceptualize complexity as an inherent property of the system independent of any observer. If we accept that complexity is to some extent based on the observer, then this means that by taking a different view of the same system we might be able to reduce the complexity of the system because complexity is dependent on the observer. In fact, this is a best practice of systems engineering, to have multiple views or models of a system to gain a better understanding of that system. It also underpins scientific and engineering research that seeks to make the complex understandable.

In many engineering situations it would help to be able to measure the complexity of different system designs. Entropy measures derived from information theory provide a measurement of system complexity [17]. Entropy when applied to thermodynamics measures the disorganization of the system. When applied to system complexity, it measures the amount of information necessary to specify the state of the system [6, 9, 19]. A system that has many parts and relationships, many of which are nonlinear, would require more information to specify its state than a system with few parts and relationships, mostly easily understood. The *information entropy measure* captures both the idea that a complex system is difficult to understand and that its behavior is difficult to predict.

It is worthwhile to avoid confusion with the various other definitions of complexity that are evolving in the literature. In the mathematical and computer sciences they often speak of computational complexity or the complexity of a problem. Here complexity is whether there is an algorithm available to solve a problem in polynomial time. If not, then the problem is complex. Some problems are very simple to state such as the Traveling Salesman Problem, but are very difficult to solve for large instances of the problem. Another term is complex adaptive systems that was coined to describe systems composed of independent, intelligent agents who have separate goals and behaviors [8]. When an agent comes into contact with another agent it may lead to conflict or competition over resources. As a result, the agents learn how to adapt their behaviors. This adaptive behavior leads to self-organization and the emergence of system behavior patterns that cannot be deduced from the individual agents. The theory of complex adaptive systems can help explain why Canadian geese flock in "V" formations, how financial markets work, and how crowds exit stadiums.

To summarize, by saying an enterprise is complex, we are saying the enterprise exhibits properties that make understanding cause/effect relationships difficult, makes predicting system behavior difficult, and makes describing the system difficult. Having better theories and models of the enterprise can help reduce the perceived complexity, but inherently enterprises have this property and it can never be fully eliminated.

2.1.8 Dynamic

An enterprise changes over time; it grows, shrinks, adds parts, removes parts, and changes its parts and the relationships between those parts. Enterprises seem to go through stages of continuous change, characterized by slow, small-scale improvements to limited parts of the enterprise that are interrupted by larger more dramatic reengineering projects cutting across several functions of the enterprise. An example of the many changes, some transformational, is exhibited by the history of IBM. IBM was not conceived as a single enterprise, it resulted as the merger and acquisition of several smaller independent companies in the year 1924. Throughout the years, IBM added new manufacturing plants, research centers, and business centers in the U.S. and other countries. Each new plant, division, or country branch of IBM is a new enterprise. IBM also closed plants, divisions, and country branches throughout its history. In 1983, it discontinued its biomedical products business for example. It established subsidiaries such as IBM Credit Corporation in 1980 that are outside of its core competencies. IBM changed its strategy at times too; in 1986 it revised its corporate goals. IBM changed its organizational structure several times; in 1988 it went through a series of restructuring initiatives to react to market changes. In 1993, it hired its first-ever, outside CEO, Louis Gerstner Jr., who embarked on large-scale changes to IBM's organization and business culture. In 2004, IBM's consulting services accounted for almost 50% of revenue, this in a company that was known for its manufacturing of computer hardware. Also in 2004, it sold its PC division to a Chinese company, Lenovo.[1] The history of IBM is similar to most companies that manage to survive through the years. Enterprises are open systems and must change and adapt to their environment in order to survive. The early leaders of IBM could not have foreseen that IBM would become such a large service-provider. Nor can IBM's current management predict how IBM will look 50 years from now because they cannot predict how the world will change.

[1]Source: IBM archives found at http://www-03.ibm.com/ibm/history/index.html.

2.1.9 Equifinality

Enterprises exhibit the property of equifinality, which means the enterprise can accomplish its objectives with different inputs and with different internal processes to reach the enterprise's goal. Given two enterprises, both can achieve the same outcome by using different inputs and/or different processes. This is part of the reason for the diversity in the marketplace. All automotive companies have similar goals to make profits by producing cars, yet all the automotive companies go about reaching this goal in different ways. Equifinality suggests, contrary to Taylor's maxim, there is no single, best way to reach a goal. Equifinality is an important property to remember when analyzing enterprises because how an enterprise chooses to pursue its goals will be influenced by the culture it operates within. The enterprise analyst needs to be aware that a best practice in one organization might not be transferable to another organization because of the mismatch between the practice and the organization's culture. However, both organizations can achieve the same goals but through different paths.

2.2 System Dynamics

System Dynamics is a theory that says systems should be viewed as interrelated groups of parts, and by modeling the structure of the system we can understand and even predict the system behavior. System Dynamics was started at MIT by Forrester [7], and is derived from ideas in computer simulation, control theory, and decision-making. System Dynamics provides two tools that are useful in the study of enterprise systems: causal loop diagrams and stock and flow simulation models.

2.2.1 Causal Loop Diagrams

A causal loop diagram depicts a system as an inter-related system of variables. The variables are system elements of interest that are connected by arcs denoting relationships. The purpose of the diagram is to depict beliefs about the causal structure of the system. Figure 2.2 shows the notation for three types of relationships. Each node, shown as text, is a system variable. The arcs show relationships between variables. Relationships are either positive (shown as +) or negative (shown as -). The definitions of the relationships are:

- Positive Relationships: If A increases, then B also increases. If A decreases, then B also decreases. (They change in the *same* direction.)
- Negative Relationship: If C increases, then D decreases. If C decreases, then C increases. (They change in *opposite* directions.)
- Delays: The delay may be positive or negative. In a negative delay, if E increases, then F decreases after a period of time. Delays are denoted by two diagonal lines drawn through the arc as shown.

A delay is an interruption between an action and its consequences. Delays occur frequently in dynamic systems. This often results in overshooting a desired outcome.

Figure 2.3 shows a causal loop diagram of a thermostat controlling a room's temperature. In the diagram, the arc from `Heat` to `Room Temperature` is positive. This means that as the `Heat` increases, the `Room Temperature` increases. The arc between `Room Temperature`

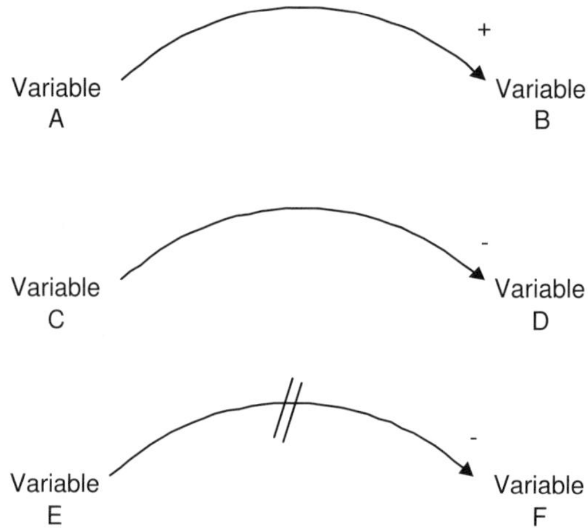

FIGURE 2.2
Causal loop diagram notation.

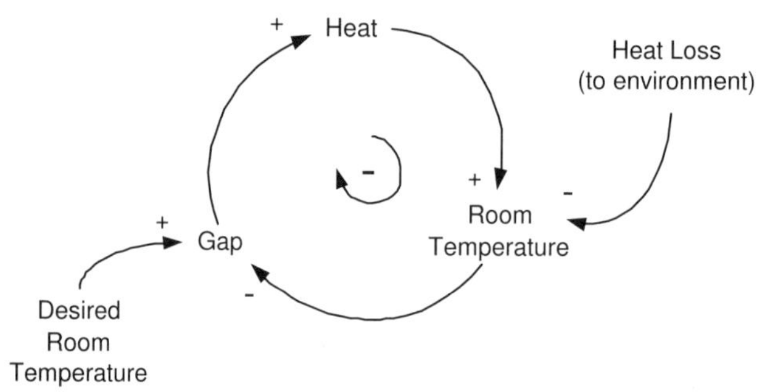

FIGURE 2.3
System dynamics model of feedback.

and `Gap` is negative. This means that as the `Room Temperature` increases the `Gap` decreases. Why is this? The `Desired Room Temperature` is higher than the actual ***Room Temperature***. So as `Room Temperature` increases, the `Gap` (`Desired Room Temperature -Room Temperature`) will decrease. It is left to the reader to verify the signs of the remaining arcs.

A closed loop created by two or more arcs shows feedback in the system. The loop created between `Heat, Room Temperature,` and `Gap` represents a negative feedback loop. It is denoted by a curved arrow with a negative sign in it. Similarly, a positive feedback loop is denoted by a curved arrow with a positive sign in it. In more complex diagrams, it helps the model's readability if the feedback loops are named. The next example does this and names the feedback loops.

To determine a loop's polarity, a quick method is to count the number of negative signs in the loop. If it is an odd number, then it is a negative feedback loop; if it is an even number, then it is a positive feedback loop. The other way to determine the polarity is to select a variable and trace the effect of a small increase as it propagates around the loop. If the feedback reinforces the original increase, then it is a positive feedback loop; otherwise, it is a negative feedback loop.

A central theme in systems dynamics is the feedback occurs in complex loops in most systems. System dynamics has shown how most systems do not have simple feedback loops like in the thermostat, but feedback loops often contain many variables and delays. A more complicated (and complex) causal loop diagram is shown in Figure 2.4 for the operations of a healthcare clinic. Figure 2.4 is of a causal loop diagram showing the relationships between patient demand, patient behavior, clinic capacity, and clinic policies. To understand what the causal loop diagram is showing we will trace through the relationships.

Appointment demand consists of three separate types of appointment demand: follow-up demand, patient-initiated demand, and referrals. As the appointment demand increases, then the appointment delay increases. Appointment delay is the time between requesting an appointment and actually receiving the appointment. Appointment delay depends on the appointment demand and the service rate (number of patients per day) of the clinic to satisfy that demand.

The waiting loop says that as appointment delay increases the fractional no-show rate increases. What this means is that as appointment delay increases, the patient is more likely to forget about their appointment and miss (not show up) the appointment. As the fractional no-show rate increases, the actual no-show rate increases. With increased no-shows, clinical management reacts by overbooking the schedule to compensate for the no-shows. Increased overbooking puts more appointments into each day, assuming some patients will not show up. The effect is to reduce the appointment delay.

The overtime loop says that as appointment delay increases the pressure on management to reduce the delay increases. As the pressure on management builds, they will eventually use overtime to increase capacity and work down the backlog of patient appointments. The increased capacity means they serve more patients each day, and they will decrease the appointment delay.

The walk-in capacity loop says that as appointment delay increases some patients, frustrated with their ability to obtain an appointment, will simply walk in to the clinic. The walk-in rate will increase and after some time the clinic will reserve more slots for expected walk-ins. This reduces the available capacity for patient appointments, which in turn reduces the service rate, and increases the appointment delay.

The patient satisfaction loop says that as the no-show rate increases, then the amount of overbooking increases. Greater overbooking increases in-office patient waiting time. Patients do not like to wait, so as the waiting time increases the patient satisfaction decreases. Lower patient satisfaction tends to increase the no-show rate.

The causal loop diagram is advantageous for communicating about systems of cause/effect relationships. It is useful for problem analysis by encouraging a team of analysts and other stakeholders to think about the factors affecting system performance and how they are inter-related. Alone, the causal loop diagram cannot be used for quantitative analysis – it only shows the existence and direction of the relationship; it does not indicate the magnitude of the relationship. In order to do analysis, you need to supplement the causal loop diagram with a stock and flow diagram for simulation.

2.2.2 Stock and Flow Diagrams

A stock and flow diagram views the system as consisting of stocks that accumulate things and flows that show the movement of things in the system. The "things" that flow and are accumulated can be:

- Materials. This can be any physical object such as products, raw materials, or inventory.
- Money. This can be money in any of its forms such as a savings account, income from the sales of a product, or taxes.

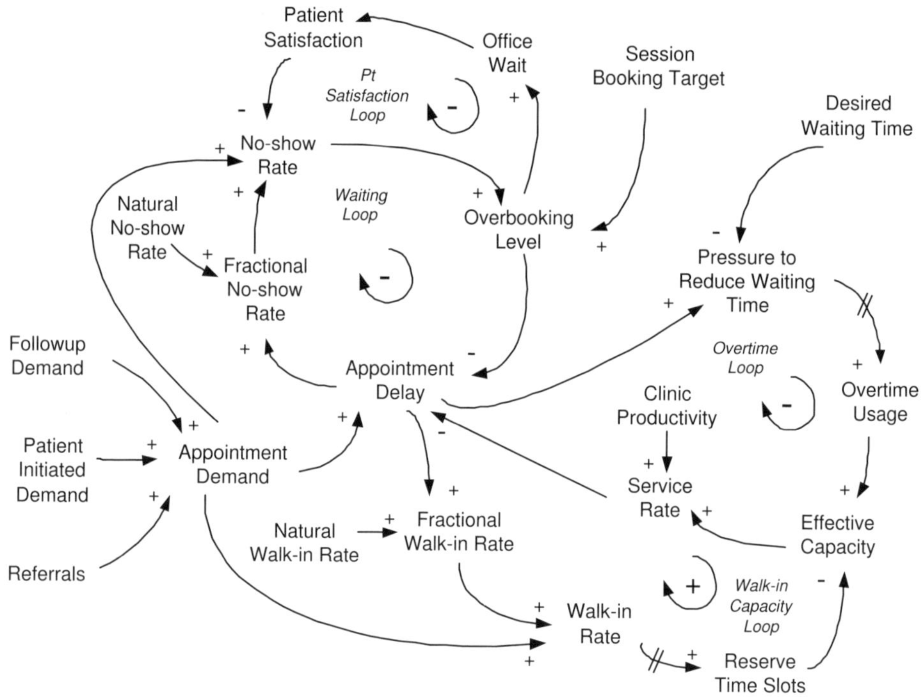

FIGURE 2.4
Causal loop of healthcare clinic.

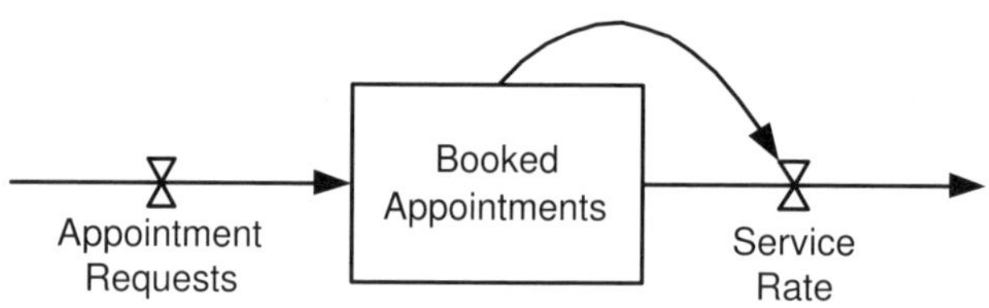

FIGURE 2.5
Patient appointment booking.

- Information Objects. This can be any non-physical object that flows in the system such as customer orders, reservation requests, or purchase orders.
- People. This can be any group of people such as customers, employees, or patients.

Stock and flow diagrams have three types of variables: rate variables, stock variables, and information variables. A stock variable represents the accumulation or depletion of an object, a flow variable represents the rate of change of a stock, and an information variable represents information that can influence either stocks or flows.

In order to simulate the stock and flow model, equations are used to model the relationships between the variables. Stock and flow models are based on the mathematics of differential calculus. The level of a stock is represented as the integration of the flows in minus the flows out. The rate of a flow is represented with a differential equation.

Each stock is the integration of the flows in and out of it, and each flow is a differential equation describe the rate of flow.

Figure 2.5 is used to illustrate a stock and flow model. The model represents a healthcare clinic that receives patient `Appointment Requests` that are stored in the schedule as `Booked Appointments`. `Appointment Requests` is a flow into the stock `Booked Appointments`, representing that as more appointment requests are received, the number of booked appointments increases. The clinic sees patients at the `Service Rate` (number of patients per day). As patients are served, their appointments are removed from `Booked Appointments`. `Service Rate` is a flow out of the stock `Booked Appointments`, representing a decrease in the number of booked appointments as they are served.

The equation to determine the level of a stock variable at any time t is the initial level of the stock plus what flows in up to time t minus what flows out up to time t. In Figure 2.5, let's assume the stock, `Booked Appointments`, initially contains 300 appointments. Let's assume the time unit is days. The equation for the stock is

$$\text{Booked Appointments}(t) = 300 + \int_0^t \text{Appointment Request}(t)dt - \int_0^t \text{Service Rate}(t)dt$$

There are many options to model the rate equations. A rate can be constant or a function of another variable(s). For illustrative purposes, let's define the appointment requests flow as normally distributed with a mean of 55 requests per day and a standard deviation of 7. The equation is

$$\text{Appointment Request}(t) = NORM(55, 7)$$

The patient's `Service Rate` is a function of the `Booked Appointments`. The equation is

$$\text{ServiceRate}(t) = \begin{cases} 60 & \text{if Booked Appointments}(t) > 500 \\ 50 & \text{otherwise} \end{cases}$$

The logic underlying the equation is that if the `Booked Appointments` grows to more than 500 this implies a 10-day wait for patients. To reduce the wait the clinic will work overtime to serve an additional 10 patients on those days. The line linking `Booked Appointments` with the flow `Service Rate` signifies that information about the level of the `Booked Appointments` is used to set the `Service Rate`.

In a stock and flow diagram, you model the system by defining the behavior and properties of interest, representing them as stocks and flows, and including the information links between them. Once the model is created then using software you can simulate the system. The simulation is a continuous simulation: in other words, an individual flow unit cannot be distinguished from the other flow units. The flow through the system is like the flow of water in a river, we cannot trace the path of an individual water molecule but we can observe the overall flow. In many cases, treating, for example, customer orders in this way is sufficient. Only if you want to understand the behavior of individual customer orders would a stock and flow model be insufficient.

A simulation software, such as Vensim or iThink simulates the model and produces graphs showing the system behavior over time. In a small model, it is possible that the equations could be solved by hand. However, in models of actual systems, the number of stocks and flows can be quite numerous so the only practical means to calculate the variable values is by computer simulation. In this example, we would expect oscillatory behavior because the inflow rate (mean of 55) is greater than the outflow rate (constant of 50) when the number of `Booked Appointments` is less than 500. When the level of `Booked Appointments` rises above 500, then the clinic goes into overtime until it drops below 500.

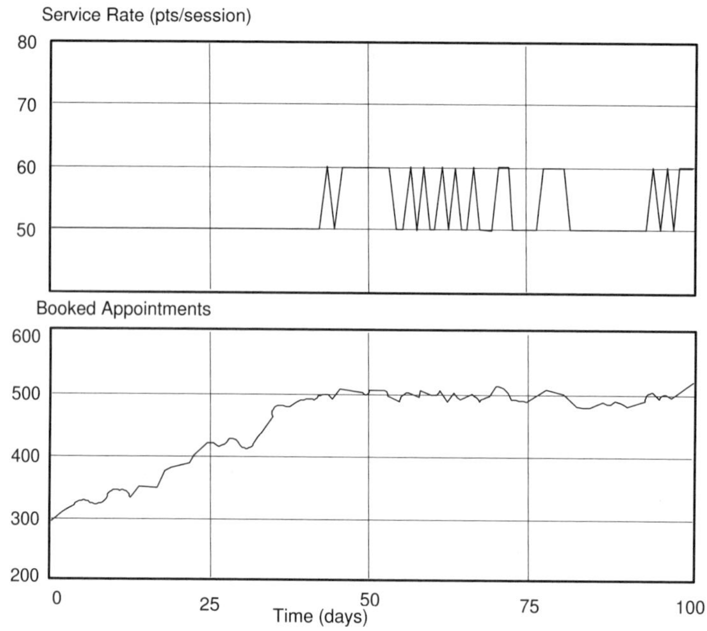

FIGURE 2.6
Dynamic behavior.

Figure 2.6 shows that `Booked Appointments` starts from the initial level of 300 and then increases until day 41 when it reaches 500 booked appointments and triggers the use of overtime. From this time onward, the clinic uses overtime periodically to keep the number of appointments under 500.

2.2.3 Critique of System Dynamics

Perhaps the greatest value of system dynamics is it provides a tool to apply systems thinking. Most alternate systems thinking approaches are long on advice but short on tools that a person can apply. System dynamics provides the causal loop diagram that is very useful when groups of system stakeholders get together to discuss the system behavior. The causal loop diagram can facilitate communication and understanding of the system relationships. To obtain a deeper understanding, a stock and flow model can be created, data collected, and simulations conducted. The simulation provides output of the system behavior and the analyst can now conduct what-if analysis to investigate how different designs or interventions will affect the system behavior.

System dynamics has been criticized that the results are unsuitable for dealing with the "hard" aspects of system analysis [2, 18]. Systems dynamic modelers in their attempt to understand the whole system by including difficult-to-measure influences on system behavior are criticized as being imprecise. The models are criticized as not being valid representations of the actual system, and not following the scientific method. The modeler can create elaborate models on paper, but then how does he obtain data and quantify such variables as customer expectations or perceived value? For the models to work, these types of qualitative variables often appear in the models, but then to simulate the model you eventually need to quantify it. Jackson [12] (p. 80) says, "If system dynamics has no accurate grasp on initial conditions or of the exact impact different relevant variables have on one another, then the claim to make accurate predictions must appear as preposterous." A second potential problem is the model represents all flows as continuous, but oftentimes the systems we analyze have discrete flows, which are more difficult to model in system dynamics.

Forrester [7] and Coyle [5] responded to these criticisms by stating the system dynamics model is concerned with the structural relationships between levels, rates, and their dynamic behavior. Social system models should not be expected to produce precise predictions. Lane [13] points out, "the aim is to work with managers to support debate regarding long-term policy." Forrester goes on to argue that the value is not in the models themselves, but the value is in the learning the analysis team obtains by building the model.

Systems dynamics is a useful way to view systems – it highlights the need to better grasp complex interrelationships of cause and effect, to understand feedback, and to understand nonlinear system responses. However, like any tool it cannot be used in every study of an enterprise system. Causal loop diagrams are useful in early phases of a project to create a picture of the system. For some projects, the stock and flow simulation models are an appropriate means to study system behavior. It is with experience in using the tools that the analysis obtains intuition of the scenarios when they are most useful.

2.3 Systems Thinking

This book takes the position that an enterprise is a system and is best understood, analyzed, and designed when viewed as a system. To do this, the designer must adopt *systems thinking*, a world-view that sees the world as being a system composed of more systems. When applying systems thinking, all problems are systems and the analyst considers the aforementioned system properties in their analysis. Perhaps the best way to describe systems thinking is to first contrast it with the traditional reductionist view that dominates engineering, management, and western culture in general.

2.3.1 Reductionist Perspective

The traditional means to understand a system is the *reductionist approach*. In this approach, a complex problem is broken up into smaller problems, each smaller problem is solved individually, and then the solutions to these smaller problems are recombined to solve the original problem. The reductionist approach has been broadly successful, and continues to be the dominant paradigm underlying engineering work in many domains. The reductionist approach is successful under the conditions when the interactions between the system parts are minimal. When one part of the system has little interaction on other parts of the system, and the other parts have little interaction on that part, then treating the part as an isolated system works well. In this way the engineer reduces the complexity of the entire system, and is able to assemble the components into increasingly complex configurations. The reductionist approach has worked well for automobiles, computers, bridges, and many other highly technological systems.

The reductionist approach does not perform well when the system components have many and/or significant interactions between them. Analysis of supply chains have revealed the limitations of a reductionist approach. In a supply chain, suppliers produce materials and ship them to manufacturers, who in turn produce assemblies that are shipped to distributors who distribute the goods to the retail level, which are then finally purchased by the final customer. A typical approach is that each level in the supply chain (supplier, manufacturing, distributor, retailer) will analyze their inventory situation and optimize it based on their

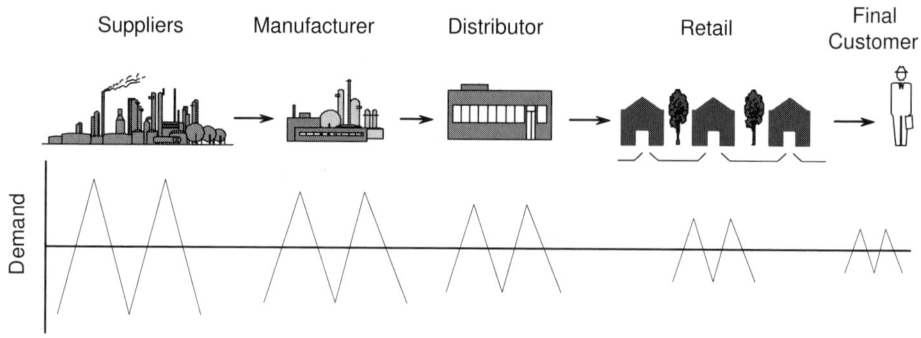

FIGURE 2.7
Bull-whip effect.

local considerations. Invariably this leads to a suboptimal overall policy from the perspective of the entire system (the supply chain). One repercussion is what is called the bull-whip effect. Small changes in demand by the final customer are propagated through the supply chain and increase in intensity such that the suppliers see very large variations in demand. This is called the bull-whip effect in supply chains [14] and causes of the bull-whip effect have been traced to misconceptions of feedback and time delays, batching of orders, and lead-time variability, and price fluctuations.

The reason for the bull-whip effect is there are strong and important interactions between the parts of the supply chain. When each enterprise in the supply chain is treated individually, these interactions are not adequately understood or analyzed. Moreover, in a reductionist approach, important interactions can be ignored such as the information flow in a supply chain is ignored in the traditional reductionist approach.

Too often reductionist thinking conceptualizes problems as straightforward cause and effect. We perceive a problem, the effect, and we seek to determine the cause of the problem and solve it. Many of our tools encourage this simplified thinking. For example, a popular tool used in production is the fishbone or Ishikawa diagram. The Ishikawa diagram highlights a problem and then attempts to list all the causes of the problem. The underlying assumption is there is a direct link between the causes to the effect (problem), or in other words, a linear cause and effect relationship. Not only does reductionist thinking promote linear thinking, but it also usually considers the problem in isolation. For example, an enterprise might have a "manufacturing problem" that is considered unrelated to other functions in the enterprise. Governments attempt to solve the energy problem, the transportation problem, or the economic problems as if these problems are all somehow isolated. Yet, in reality, changes to the availability of energy will have significant effects on the economy and transportation.

2.3.2 Systems Perspective

Systems thinking describes a world-view, or way to approach problems by considering the problem as being part of a system and trying to discern the underlying structure of the system that leads to the observed problems. A systems thinking approach considers the whole system in its entirety and includes an analysis of the interactions between the system parts. The premise is that only a holistic analysis of the system can reveal the complex interrelationships that produce system-wide behavior. This seemingly small change in perspective can have a profound influence on how we address problems. Now, to solve the problem, we need to identify how the different parts of the system are interrelated and how these relationships generate the observed system behavior. Once the system is understood, we can investigate various interventions to solve the perceived problem.

Additionally, the systems approach says you need to analyze open systems in their environment because they interact with the environment. If you separate the system from its environment, then you may miss important feedbacks and other system behaviors. Part of the systems approach is first identifying the larger system that the system under study is a part of. After all, all systems are sub-systems of some other system. Doing this, you try to identify the relationships and flows (material, information, money, etc.) between the system and its environment. Only by doing this can you understand the system behavior.

The systems thinking approach realizes and emphasizes the importance of a multi-disciplinary approach because each discipline brings a different perspective to the problem. The different perspectives help to better achieve a holistic understanding of the system.

Another way in which systems thinking is distinguished from reductionist thinking is that systems thinking places as much emphasis on understanding the problem as on solving the problem. Systems thinking writers describe problems as "messes" [1] or "problem situations" [4]. A mess is a system of problems. The essence of the mess is the systemic nature of the situation, not an aggregate representing the sum of the parts. The elements of a mess are highly interrelated. No part can be touched without touching the other parts. As such, it is an emergent phenomenon produced by the interactions among the parts. Formulation of the mess therefore requires understanding the essence of the behavioral characteristics of social phenomena.

To use systems thinking is difficult and requires practice. As Gharajedaghi [10] observes, most works on systems declare systems thinking is about the whole, but they do not provide any operational methodology to apply systems thinking. The tools available such as causal loop diagrams help immensely. Another way to improve systems thinking is to refer back to the definition of a system and the system properties. Identify all those properties in the system you are studying. In systems thinking, we look for intervention points to solve the problem because we realize actions we take will induce feedback in the system. The problems have multiple viewpoints from each of the stakeholders, it is near to impossible to identify a single objective for the system. Moreover, implementing a solution is more than simply stating here's the solution and go do it. Systems thinking treats the implementation as part of the problem.

Illustration of Systems Thinking

A company, located in Florida, manufactures stairs in a factory and then installs them into what are mostly new homes. The majority of their customers are new-home builders who would develop large tracts of land with up to a hundred homes on them. To serve these customers, the company pursues a successful strategy of providing high service to the home construction industry by delivering and installing the stairs on time. In this way, they differentiate themselves in terms of timeliness, and avoid stairs, a rather low-tech product, from becoming a commodity in which they would have to compete on price alone.

One of the operations of the company is to schedule the delivery and installation of the stairs. The scheduling operation was performed manually. Manual scheduling worked fine when the company was small, but as the company grew, manual scheduling became unmanageable. The schedulers could no longer generate good schedules manually, and most of their day was taken up by scheduling. The operations manager wanted to reengineer the installation scheduling process, and hired us as consultants to study the problem and develop an optimization-based scheduling system.

The company had a pool of approximately 30 independent contractors who did the installations. Each day the scheduler would review the jobs that were scheduled for the next day, review the availability of the installers, and make a job assignment that assigned one or more jobs to each available installer. Constraints governing the feasibility of the job assignment included considering the job skills of each installer compared to the requirements of the installation and the estimated time for each job compared to the available time of each installer. The primary objective was that no jobs would be late. The secondary objectives of the assignment were to minimize non-productive time by minimizing the distance the installers had to travel between jobs. For example, you would prefer to assign two jobs in the same development to a single installer rather than having that installer travel, sometimes hours between jobs on opposite sides of the county. Additionally, an objective was to fairly and evenly distribute the work between the installers. The inputs to the scheduling problem were the list of available installers and the jobs that were due. The problem was first to determine whether a feasible solution existed such that no jobs would be late, and then to present an assignment that minimized non-productive time and evenly distributed the work. If no feasible assignment was possible, the scheduler would prioritize jobs and the system would find an assignment that minimized the weighted sum of late jobs.

We analyzed their scheduling problem following traditional operations research approach. We identified the decision variables, constraints, and objectives and duly developed a heuristic for clustering jobs and then an integer program to optimally assign those jobs to installers. However, in the process of completing the project it became apparent that our system boundaries were far too small to have a large impact. The job assignment was a daily task. On some days, the amount of work exceeded available capacity. Figure 2.8 shows the planned work for three months. The available capacity was approximately 300 hours (30 subcontractors at 10 hours per day). On those days in which the planned work exceeded the 300-hour capacity, there is no feasible solution – some jobs will be late. Yet, the graph shows that on most days the available work is far less than the available capacity. The bigger problem was the variability in work on each day of the week. Clearly, if the jobs could be spread out more evenly over all the days then no jobs would be late because the average capacity exceeds the average demand.

The shortcoming of the thinking done by the operations manager who hired us, and our initial thinking, is that we focused on the more narrow, traditional engineering problem of maximizing utilization of resources. A systems thinking approach, would examine the problem as a system. This implies understanding that the scheduling system is part of a larger system that we can call the house construction system. An investigation of the housing industry and how the due dates are set revealed that far greater improvement could be made without any investment in additional capacity. The project schedule for a house consists of 12 phases and is approximately 6 months in duration. Stairs, which are installed in phase 7 are scheduled approximately 3-4 months in advance. The date set for installation is very uncertain, and also somewhat arbitrary because it does not take into account delays due to weather, materials, or other factors. Moreover, being a day or two late on the stairs installation may not have much of an impact because there is much slack in the schedule between activities. If you ever witness the construction of a home, you will notice that it usually involves a few days of intense activity interspersed with many days of no activity. So the crew that lays the foundation will show up and build the foundation. When they are done it might be a few days or even a week before the crew that frames the house shows up. So, in the six-month duration there might be only two months of actual work being performed on the house and the remaining four months of idle time between work.

Consequently, the stairs installation is almost never on the critical path for home construction and the project manager has great flexibility in moving the stair's installation date forward or back a day or two without any effect on the project completion date. This shifts the scheduling problems; instead of focusing on the daily job assignment, the schedulers

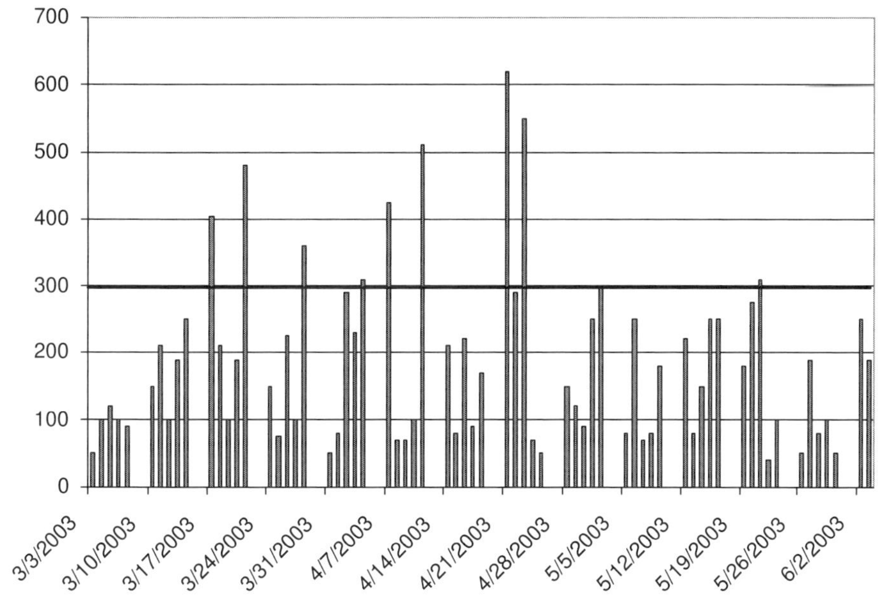

FIGURE 2.8
Planned jobs.

should focus on job planning. Job planning would involve looking at the work scheduled for the coming weeks, and contacting the project managers to negotiate slight changes in the due dates to avoid exceeding capacity on any day.

Several lessons can be learned that illustrate the system thinking concepts presented in this chapter. First, to optimize the stairs installation in isolation is an example of sub-optimization. Without consideration of the interrelationships between the subsystem, defined by stairs installation, with the larger system, or its environment leads to a job assignment that would have little impact on the overall performance of the company. Second, the case study illustrates the difficulty in defining a system boundary. One possibility is that, in trying to capture all relevant behaviors, the analyst draws the boundary increasingly larger and larger. The problem is the analyst might then be overwhelmed by the size of the resulting system definition. So, there is some trade-off in how large of a boundary should be drawn.

Further thought about how we approached the problem revealed our inherent bias towards hard systems thinking. Our initial interviews with the company president, operations manager, and schedulers revealed a complex mess using Checkland's term. We removed much of this mess to formulate a problem in the operations research sense: it had an objective, decision variables, and constraints. Reviewing what was done and thinking about how we could have approached the problem we realize that a systems thinking approach would not be so quick to identify a single objective function and constraints. A systems thinking approach would seek to understand first the system, its relationships, and the structure that gives rise to its behavior. The systems thinking approach recognizes a system is purposeful and would understand how each participant in the system is motivated. In this case, the motivation of the customer is important to understanding how to solve the larger problem.

Furthermore, a systems thinking approach would try to understand the installer's motivation since they are purposeful agents. This would reveal if there is any leverage to improve the system performance by better aligning the installers' objectives with the system's objectives.

2.4 How To Think Like a Systems Thinker

The systems thinking principles derive from the definition of a system, its characteristics, and the idea of focusing on the structure and relationships of the entire system and not just the part you are supposed to solve. The principles are:

- Define the enterprise boundaries that help best solve the problem. Generally, draw the boundaries based on what the stakeholders can control or strongly influence, and allocate all other parts to the enterprise's environment.
- Understand the system structure, identify its subsystems and parts.
- Determine the stakeholders, their perspectives and their goals. If the enterprise system is to achieve its goal, then all the system components must behave consistently with respect to the established goal.
- Remember there are many ways to achieve the goals (equifinality).
- Understand the interaction with the environment (enterprises are open systems).
- An enterprise may exhibit emergent properties, which is behavior that can only be observed at the enterprise level and cannot be deduced by analyzing the enterprise's subsystems in isolation. To try to understand how this behavior emerges look for feedback, non-linearity, delays, and purposeful action by the human actors in the enterprise.
- To maintain stability in a changing environment the enterprise must adapt, which requires feedback loops from the environment.
- Enterprises are purposeful systems because they can both choose their goals and their means to attain those goals. Additionally, the people in the enterprise are purposeful.
- The enterprise is hierarchical, it is composed of lower-level subsystems, and the enterprise is part of higher-level systems (e.g., supply chains). There are multiple ways to define the subsystems depending on the observer's views (e.g., organizational subsystems, process subsystems, etc.).

2.4.1 Implications for Enterprise System Design

We had argued that an enterprise is a system, and this chapter presented systems theory and the tools to study systems. Understanding the individual components of an enterprise is insufficient to understand the entire enterprise. In a system, results do not come from the sum of individual efforts, but from coordinated activities. The only efforts that make sense are those that help achieve the goal. As this is the case, it makes no sense to reward people for their local, individual efforts, but only for how much they have contributed to achieving the goal of the organization. Otherwise we only undermine the common and shared vision we established before.

An important conclusion to be drawn from this diagram is that if a manufacturing enterprise is to succeed, there can be no basic difference in viewpoints, values, and goals among its constituent groups. It is clear that the areas of responsibility are not neatly separated from one another but overlap to an important degree; financial and accounting systems, for example, have a major impact on operations and engineering. Perceived or artificially created boundaries between organizational units, such as those between marketing and engineering, production and purchasing, production control and marketing, or employees and management, both restrict and complicate communication and cooperation. The performance of the system suffers. The challenge to management is to find ways to take advantage of the strengths of the various unit operations and functional groups while discouraging any tendencies to work at cross-purposes or toward conflicting goals. Achieving true involvement among the various activities requires, of course, more than simply reducing the barriers between groups.

The complexity of the enterprise system arises from many directions: the interdependence of the elements of the system, the influence of external forces on it, the impact that it can have on its environment, and the lack of predictability in the consequences of actions. The complexity and the difficulty in assessing the directions that should be followed can create a sense of frustration and futility for the management.

In a system, results do not come from the sum of individual efforts, but from coordinated activities. The only efforts that make sense are those that help achieve the goal. As this is the case, it makes no sense to reward people for their local, individual efforts, but only for how much they have contributed to achieving the goal of the organization. Otherwise we only undermine the common and shared vision we established before.

As an open and purposeful system, we cannot always predict or even control enterprise behavior. The enterprise is subject to perturbations from its environment and the actions of the people in the enterprise. A useful metaphor is to compare the gardener to the watchmaker. A watch is a closed, deterministic system. The watchmaker, as the designer, can specify and build a watch to perform exactly as the watchmaker intends. A garden on the other hand is an open, stochastic system. The gardener, as the designer, cannot specify exactly how many tomatoes will grow or whether they will even grow at all. Instead the gardener specifies how the garden is structured and then over time intervenes in the garden to attempt to attain the desired goal. For example, if it does not rain, the gardener can water the garden. Instead of intervention, the gardener could design the garden for robustness by, for example, planting drought-resistant seeds. As enterprise designers, we are more like the gardener than the watchmaker.

2.5 Summary

This chapter defined systems as consisting of the properties of holism, purposefulness, openness, equifinality, feedback, and complexity. The chapter showed how enterprises have the properties of systems. Traditional reductionist thinking was contrasted with systems thinking. Systems thinking views the enterprise as a whole and assess the system properties to try to understand the system behavior. Systems thinking has strongly influenced our view of enterprises. Many of the current paradigms for enterprise improvement adhere to a system view including lean enterprise systems, total quality management, and supply chain management. To illustrate the nature of systems thinking, it was contrasted with traditional, reductionist thinking. The application of systems thinking is facilitated with the tools of system dynamics: causal loop diagrams and stock and flow diagrams. The chapter demonstrated how to build these models, how they can be used to model, and to analyze systems.

Review Questions

1. Suppose an automatic pool cleaner is not working properly. One way to understand what is wrong is to take apart the pool cleaner and examine each part individually to see if it is broken. Describe how to use a systems thinking approach to troubleshoot for the same automatic pool cleaner.
2. List all the types of items that can cross the boundary of an enterprise. Select one of these items and describe it.
3. Explain what it means to say an enterprise is purposeful.
4. Identify a system and list one or more emergent properties that system exhibits.
5. List and describe examples of a goal-seeking system, a state-maintaining system, and a purposeful system.
6. Name a system property and describe how it applies to an enterprise.
7. Think of a process in an enterprise you are familiar with and explain how the feedback control principles can be applied to that process.

Exercises

1. As the order backlog increases, then there is greater pressure to increase production. So the production manager uses more overtime. As the production increases, then the order backlog decreases. Draw the causal loop diagram for this scenario. Show variables, whether the relationships are positive or negative, and the polarity of the loop.
2. Suppose you are tasked with a project to design a new patient scheduling system for a healthcare clinic. Explain the process of how you would go about defining the boundaries for the project?
3. Present and explain at least three real-life systems, describing the elements that make this system complex. Take into account the definition of complexity from the book in order to describe characteristics of the models. Models must be different from those presented in the book.
4. Construct a causal loop diagram for a bank savings account. You have an account balance that is increased by interest earned at the prevailing interest rate. You can make both withdrawals and deposits on the account.
5. Construct a causal loop diagram for a process with the following behaviors. A company receives customer orders at a known Order Rate. The company has limited Capacity to fill those orders at the Production Rate. The Order Backlog is the accumulation of orders that are awaiting fulfillment; it is the difference between the Order Rate and the Production Rate. When the Backlog grows too large, management will increase capacity, but there is a delay before this happens.
6. Construct a causal loop diagram for a process with the following behaviors. A company has a Desired Staff Level. If the Staffing Gap, which is the difference between the Desired Staff Level and the Current Staff Level, is positive, then the company Hires new staff members.

7. Review the list below and identify each variable as either a stock, a flow, or both depending on context. If you select both a stock and a flow, then provide an example of each instance.
   (a) Inventory
   (b) Orders
   (c) Employees
   (d) New Hires
   (e) Net Income
   (f) Revenue
   (g) Assets
   (h) Booked Reservations
   (i) Reservation Requests
   (j) Perceived Quality
   (k) Defect Rate
8. Your client is the CEO and Director of HR for Acme Inc., whose business is a highly labor-intensive production of the ubiquitous widget. The staff includes 1000 people, with a production rate of 6200 tasks/month. The assumption is held that staff increases are made through hiring only.

   From an interview with the CEO:

   "Three months ago, we were producing at around 6200 tasks per month, but needed to ramp up to 7500. Because of the gap between the production rate and our goal for the production rate, we hired new people – about 20% more. After a delay, inexperienced people, or rookies, arrived on the job. These rookies, with a good amount of training and mentoring from the pros, take some time to mature into pros. The more workers we have, the greater production we have. I expected production to climb up to the new level fairly quickly. But now it is three months later and the new people are on staff, but production is *worse*, not better. Maybe we hired the wrong people, or maybe they're just slacking off. I don't know. We can't exactly figure out what has happened, and what we can do to avoid it in the future."

   Hint: If rookies are trained by the Pros, then this means the Pros have less time to do production.

   (a) Construct a causal loop diagram.
   (b) Sketch a graph of the desired production rate.
   (c) Construct a stock and flow diagram.

Bibliography

[1] R.L. Ackoff. *Re-creating the Corporation.* Oxford University Press, New York, NY, 1999.

[2] H.I. Ansoff and D. P. Slevin. An appreciation of industrial dynamics. *Management Science*, 14:383–397, 1968.

[3] W.R. Ashby. *An Introduction to Cybernetics.* Chapman & Hall, 1957.

[4] P. Checkland. *Systems Thinking, Systems Practice.* John Wiley & Sons, Chichester, England, 1982.

[5] R.G. Coyle. Comment on system dynamics and operational research: An appraisal. *European Journal of Operational Research*, 23:403–406, 1986.

[6] A. Deshmukh, J. Talavage, and M. Barash. Complexity in manufacturing systems, part I: Analysis of static complexity. *IIE Transactions*, 30:645–655, 1998.

[7] J. W. Forrester. Industrial dynamics - a response to Ansoff and Slevin. *Management Science*, 14:601–618, 1968.

[8] M. Gell-Mann. *The Quark and the Jaguar.* Henry Holt and Company, New York, NY, 1994.

[9] M. Gell-Mann and S. Lloyd. Information measures, effective complexity and total information. *Complexity*, 1(1):44–52, 1996.

[10] J. Gharajedaghi. *Systems Thinking: Managing Chaos and Complexity, a Platform for Designing Business Architecture.* Butterworth-Heinemann, Burlington, MA, 1999.

[11] A.D. Hall and R.E. Fagen. *General systems: The yearbook of the society for the advancement of general systems theory*, chapter Definition of a system, pages 18–28. 1956.

[12] M.C. Jackson. *Systems Thinking: Creative Holism for Managers.* John Wiley & Sons, Hoboken, NJ, 2003.

[13] D.C. Lane. Rationality in system dynamics: modelling human and organizational decision making. *Systems Research and Behavioral Science*, 21(4):313–317, 2004.

[14] H.L. Lee, V. Padmanabhan, and S. Whang. The bullwhip effect in supply chains. *Sloan Management Review*, Spring:93–103, 1997.

[15] R. Oliva and J.D. Sterman. Cutting corners and working overtime: quality erosion in the service industry. *Management Science*, 47(7):894–914, 2001.

[16] W.B. Rouse. Engineering complex systems: implications for research in systems engineering. *IEEE Transactions on Systems, Man, and Cybernetics - Part C: Applications and Reviews*, 33(2):154–157, 2003.

[17] C.E. Shannon and W. Weaver. *The Mathematical Theory of Communication.* University of Illinois Press, Urbana, IL, 1949.

[18] J.A. Sharp and S. H. R. Prince. System dynamics and operational research: An appraisal. *European Journal of Operational Research*, 16:1–12, 1984.

[19] S. Sivadasan, J. Efstathiou, G. Frizelle, R. Shirazi, and A. Calinescu. An information-theoretic methodology for measuring the operational complexity of supplier-customer systems. *International Journal of Operations Production Management*, 22(1):80–102, 2002.

[20] J.W. Sutherland and W.J. van den Heuvel. Enterprise application integration and complex adaptive systems. *Communications of the ACM*, 45(10):59–64, 2000.

3

Modeling Concepts

"*The map is not the territory.*" – Alfred Korzybski (1879-1950), Polish-American philosopher and scientist who developed the theory of general semantics.

Enterprise engineering is a discipline in which the artifact of study is not physical. You cannot hold an enterprise in your hand. You cannot look at an enterprise as you would look at a house, bridge, or other building. While we cannot directly see an enterprise system, we can perceive parts of the enterprise system, its inputs, outputs, and behavior. In order to understand enterprise systems and to analyze the properties of enterprise systems, we must model them. For this reason, the enterprise engineer needs to have a full appreciation of the central role modeling plays in enterprise design. In this chapter, we discuss modeling in general with particular attention to the types of models used by enterprise systems.

After completing this chapter, you should be able to:

- Understand modeling as a tool to represent aspects of the enterprise under study.
- Describe how model purpose affects the selection of model type and degree of abstraction.
- Describe and apply the modeling principles of abstraction and views.
- Differentiate between non-analytical, analytical, and simulation models and select the appropriate type for a particular purpose.
- Verify and validate models.
- Describe the modeling process.
- Explain why model reuse is a goal.

3.1 Model Definition

A *model* is an abstract representation of a real-world system that emphasizes some aspects of the system while excluding other aspects. What is included or excluded depends on the purpose of the model. The reason we model is to reduce the complexity of understanding or interacting with a system by eliminating the detail that does not influence its relevant behavior. The suppression of irrelevant details is what helps us to master the complexity of the system being modeled. To illustrate, a map is a model of a region's geography. If the intended purpose of a map is to provide directions to get from one city to another city, then there is no need to show the trees or houses along the route because depicting these details does not further the purpose of the map.

It is important to not confuse the model with the system being modeled. The quote introducing the chapter says, "The map is not the territory." The map is the model of the territory. In other words, the map is an abstraction of the territory onto a two-dimensional representation. Maps make use of symbols; for example, dots to represent cities. Clearly, a city is not a dot. The dot is a symbol to convey the concept of a city and show its location relative to either the map coordinates or in relation to some other reference. A map ignores much of the detail of the territory. Trees, houses, road signs, etc. are not shown on most maps. So, the model describes completely some aspects of the system that the model creator deemed relevant to the model's intended purpose while leaving out other details considered irrelevant. Not only does the modeler decide what to include in the model, but the modeler must also decide at what level of detail to represent it? Some maps will show a road as a line, while other maps provide greater detail and show how many lanes are in the road. Again, the level of detail to show depends on the intent of the model.

The benefits that modeling provides are:

- Visualization – Models help people visualize concepts such as the processes, information, and organizational structure of an enterprise. Some things, such as processes, are not physical, so a model is the only way to visualize it.

- Communication – Models help us communicate about enterprise systems, their analysis, and their design.

- Design – Models are used to specify enterprise designs. We use models to specify the process design, the information system design, and many other aspects of the enterprise. In some cases, we can have a seamless link between the model and the system they represent through the use of Computer-Aided Systems Engineering (CASE) tools [10]. For example, information models can be directly implemented as relational databases in any one of several target systems. In this way, the models directly become the system through a transformation, reducing the implementation time and cost.

- Analysis – Models are an economic means to analyze enterprise systems. It is far cheaper to build a model of a process and analyze the model than to experiment on the actual process. Additionally, models can be used to analyze many design alternatives and explore ideas in a way that is just not feasible when building actual systems. Models often deliver answers far faster than prototyping a system to understand how it works. Consequently, the use of models can reduce the duration of enterprise engineering projects.

In summary, models are indispensable to enterprise engineering, and we might add as well, all other engineering disciplines.

3.2 Features of Models

This section describes three features of models: purpose, abstraction level, and viewpoint. The model purpose is often given by the situation it is needed for, such as for analysis or to communicate instructions to employees. The modeler chooses the abstraction level and viewpoint to fulfill the model's purpose.

Purpose

All models are, or should be, developed with a purpose. The purpose of the model determines what aspects of the system's features and behavior need to be represented in the model and what aspects can be omitted. The model purpose is determined by the modeler. Good practice would have the modeler make the model purpose explicit, but oftentimes it is not.

Models can only be interpreted with respect to the model's purpose. Furthermore, there can be many different models of a single system because each model can have a different purpose and therefore it will show different system features and behavior to serve its purpose.

To illustrate the importance of knowing the purpose of a model, consider the Mercator projection of the Earth shown in Figure 3.1. The Mercator projection is a commonly used map in school classrooms. We grow up looking at this map and "learn" that this is what the Earth looks like. For this reason, many school children might have the misconception that Greenland and Africa cover approximately the same surface area of the world. However, in reality the continent Africa is about fourteen times larger than Greenland. How can it be that Mercator's projection shows them to be roughly the same size? Is the Mercator map incorrect?

We cannot say that Mercator's projection is wrong because it inaccurately depicts the size of each continent without first considering what is the purpose of Mercator's map. The Mercator projection was created with the purpose to aid navigation from the Old World (Europe) to the New World (America). To accomplish this purpose, the projection is drawn so that a course bearing used by a ship can be represented as a straight line on the map called the rhumb line, and the course bearing will have a constant angle with every line of latitude and longitude drawn on the map. This feature is only possible when land masses near the poles are stretched, which distorts their areas. So, Mercator's projection emphasizes key aspects of the Earth to aid the navigation of ships, and Mercator simultaneously excludes other details, such as area accuracy, in order to satisfy his primary purpose. Consequently, Mercator's map is correct and very useful with respect to its purpose – to aid navigation.

An alternate model of the Earth is shown in Figure 3.2. This is the Robinson projection and is drawn to find a compromise between preserving several metric properties of the globe without fully satisfying any of them. Notice that in this model Greenland is much smaller than Africa. Yet, even here the size is distorted at the poles, just not as severely as in Mercator's.

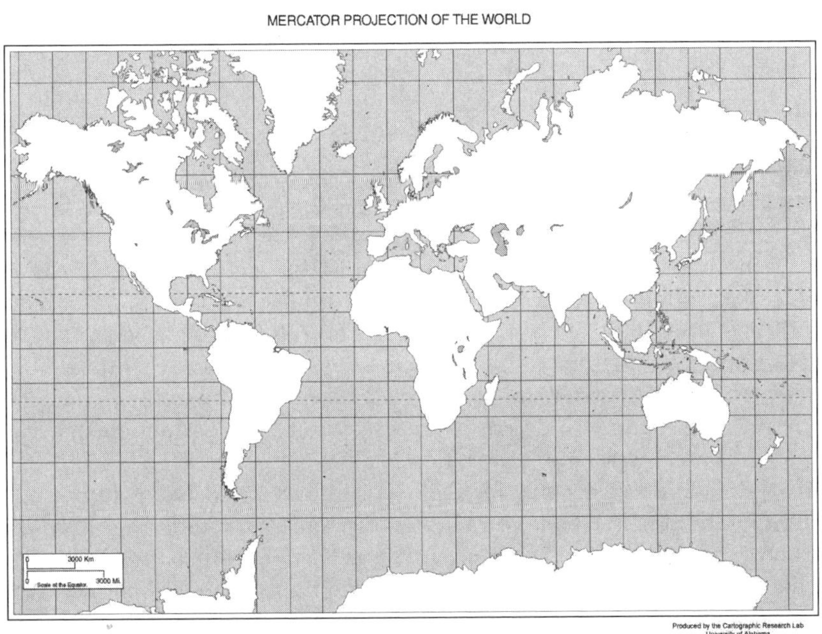

FIGURE 3.1
Mercator Projection (courtesy of the Cartographic Research Laboratory, University of Alabama).

There are more than twenty different projections of the Earth, demonstrating the many ways in which a given physical artifact can be modeled. While there is controversy over the social and political consequences of which map to use, from an engineering perspective, both projections are valid models of the Earth. Each projection serves the intended purpose of the modeler. A key point to remember is a model reveals what its creator believes is important for understanding or predicting the phenomena modeled. The modeler needs to choose what details to represent and how to represent them. Consequently, there is no single correct model of a system. A model can only be valid with respect to its purpose. A model intended for one purpose may not (will likely not) fulfill other purposes.

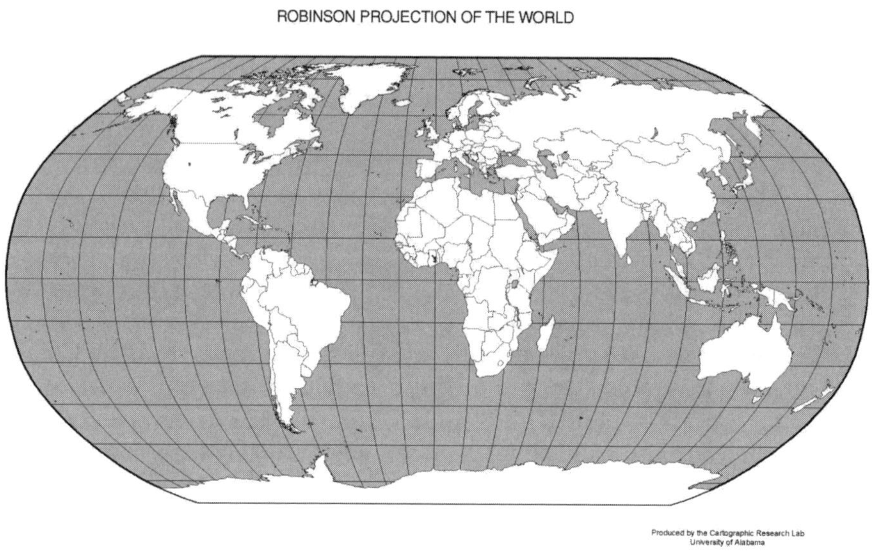

FIGURE 3.2
Robinson Projection (courtesy of the Cartographic Research Laboratory, University of Alabama).

Abstraction

Figure 3.3 shows a series of progressively more abstract sketches of a horse, mimicking Picasso's series of lithographs of the deconstruction of a bull. Each successive representation of the horse removes some detail and emphasizes essential characteristics of the horse, until what is left is basically a stick drawing. The final representation depicts what the artist believes is the essence of a horse: the shape of the horse's back and belly, the shape of the head, the positioning of the legs relative to each other, and the horse's tail. In the same way that an artist creates an abstraction of an object to convey its essential characteristics, an enterprise engineering must also determine how to create an abstraction that captures the essential system characteristics of importance.[1]

The artistic rendition of a horse highlights two interrelated issues facing a modeler in choosing an appropriate abstraction level: what system characteristics to represent in the model, and how should they be represented? A good model emphasizes what is important, recalling that importance is with respect to the model's purpose. The picture shows five depictions of a horse with increasingly more details removed from the model. If the least detailed, most abstract rendition of the horse serves its purpose, then it may be argued that

[1]Checkland [4] says his models describe not "what is" but rather "what I shall (temporarily) take things to be in my analysis" (p. 175).

this is appropriate abstraction level to use. In engineering, a commonly heard term is "back-of-the-envelope calculation" to refer to the practice of an engineer drawing quick sketches and calculations on a scrap piece of paper to explain something to another engineer. Here the purpose is to quickly convey some concept. It makes little sense to invest a lot of time and effort into a detailed drawing if the simpler sketch serves the same purpose. However, if the purpose was a formal design review, then it makes sense that detailed engineering plans and drawings are provided to ensure a thorough review. So, a good rule of thumb is to work at an abstraction level sufficient for the model's purpose, neither more nor less.

In many enterprise modeling techniques, the abstraction level is provided for in the model rules or conventions. This occurs in information modeling using entity-relationship diagrams. By definition an entity represents an object that the enterprise wishes to store data about. Contrast this to process modeling where the selection of an abstraction level is a critical decision left to the modeler. To help make this decision, Chapter 10 describes a process hierarchy and provides indication of when each abstraction level would be appropriate. Many times, it makes sense to show several different abstraction levels in a series of models to serve multiple purposes and/or audiences. Once an abstraction level is chosen, the model must be consistent in its use of the same abstraction level throughout the model. It makes no sense if the detailed drawing of the horse instead had one leg that was simply drawn as a line.

How to represent the details the modeler decides to include is the "art" of modeling. To some extent, the answer is dictated by modeling conventions. Many people are familiar with flowcharts to model the sequence of activities in a process. Flowcharts have defined symbols and rules for how the symbols are connected, what text goes in each symbol, and so forth. Deviation from these definitions and other modeling conventions can likely make the model more difficult to read.

3.3 Model Viewpoint

The enterprise systems we are interested in modeling are too complex for us to represent everything we wish to know about the system in a single model. As was illustrated in the Mercator projection of the Earth, in order to represent some particular characteristic, other characteristics are distorted or removed entirely. A strategy to show everything we wish to show of a single system is to have multiple models of the same system but constructed from different viewpoints. A viewpoint represents only a certain part of the system being modeled. Oftentimes, the views are determined by established conventions adhered to by a discipline. For example, in mechanical drawing we learn to draw three views of the object: the front view, side view, and top view. Individually, none of these views can reveal the three-dimensional shape of the object. For example, Figure 3.4 shows the front view of a physical object. In this view can you tell if surface A is a projection or a slot?

Looking at Figure 3.4, both a projection and a slot are possible. If, for a rather simple mechanical product, we cannot tell all the system features from a single view, then how could it be possible to represent all the features of an enterprise system? It is not possible, and for this reason, there are many possible viewpoints for modeling an enterprise.

Enterprise systems cannot be understood from one perspective alone; multiple perspectives or viewpoints are necessary in order to fully understand the enterprise system. However, for enterprise systems, there is no agreed upon number of views that should be included. In this book, we use three views defined as the process view, the information view, and the organizational view. Some models define views beyond these three. Zachman's framework

FIGURE 3.3
Abstractions of a horse (courtesy of Mario Kim).

defines six views, the three already mentioned as well as a network, time, and motivation view. In software engineering, the Unified Modeling Language (UML) defines more than ten different views of various aspects of software. Each view has a purpose, emphasizes a particular characteristic of the enterprise system, and consequently, ignores other characteristics. Take the above three views of an enterprise. The process view models the input/output relationships and the network of activities in the enterprise. The information view models the information content of the enterprise. The organization view shows the organization hierarchy and which organizational units are responsible for which functions. Collectively, the set of viewpoints must be consistent. For example, if the process view shows an activity generating a purchase order as output, then the information model should show the purchase order and one of the organizational units should have the responsibility for generating the purchase order. In this way, all the views are consistent and show a single enterprise. Notice that showing multiple viewpoints of a system is an alternative means to decompose the problem into several simpler models.

3.4 Modeling Language

Models are expressed in terms of a modeling language. The modeling language defines the modeling constructs and the rules for how the constructs can be used. *Modeling constructs*, sometimes called symbols, are the primitive elements of a modeling language. Constructs

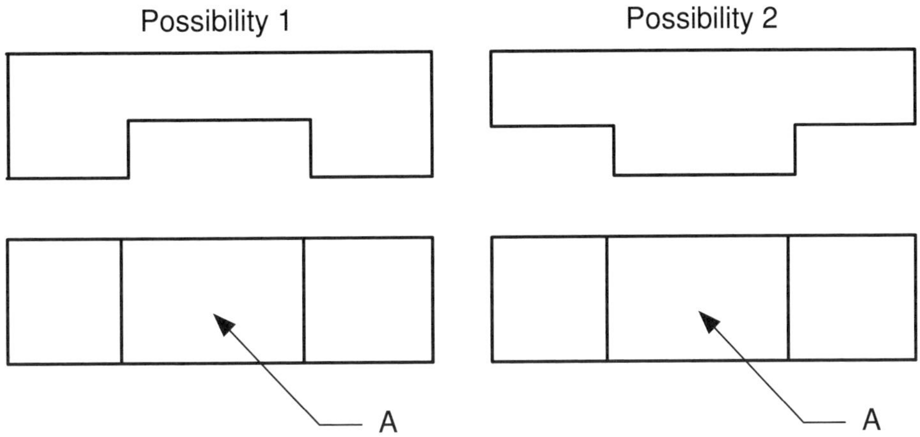

FIGURE 3.4
Two possible top views of the block.

may be graphical or textual. Each modeling construct has a defined syntax for how it can be used in the model and a defined semantics of how to interpret the construct. *Syntax* is used to describe the grammar or the rules of use for the construct. For example, in event process diagrams, all functions can only receive input arrows from events. Event Process Chains can therefore be verified for correctness by checking that every function only receives input from an event. If another model construct has input to a function, then the model violates the modeling language's syntax and is in error. The *semantics* of a model construct define the meaning of the construct. Modeling languages, often through a meta-model, define each construct and its meaning. For example, in an entity-relationship model of the information content of a system, the data dictionary defines each entity and attribute used in the model. Additionally, semantics are embedded in the relationships used in the model. Different relationships convey different intepretations or semantics.

The model constructs are part of the formal language, but models also contain modeling conventions. *Modeling conventions* are general guidelines, not part of the semantics, that improve clarity and aid understanding of the model. For example, IDEF0 has the convention to arrange function boxes going down diagonally from left to right. Adherence to model standards and conventions greatly increases the effectiveness of communication between the engineer and the many stakeholders with whom he deals. A purpose of all models is to provide a representation of the system to help the stakeholders discuss the problem, system, and system design.

There has been a desire in both the research community and industry to develop modeling languages with more formal semantics. Formal semantics help us avoid the problems inherent in the communication of ambiguous representations, and formal semantics enables computers to interpret messages, which allows for computer-to-computer communication without human intervention. To illustrate, consider a company that wishes to share a model with another company. In this case, formal semantics helps ensure the receiving company interprets the model consistent with the intention of the generating company. One such initiative is the use of XML, the eXtendible Markup Language, within many industries for web-based communication between systems. The premise of XML is that an industry group can define an XML schema and then any XML document conforming to the schema can be used in computer-to-computer communication within that industry. For example, in the

TABLE 3.1
Model Classification

Type	Subtype
Non-analytical	
Analytical	Deterministic Stochastic
Computational	Discrete-event Continuous Agent-based

automotive supply chain, companies would be able to send purchase orders via the Web to each other without the need for human intervention.

3.4.1 Types of Models

There are many different types of models, and here we categorize them into three general types shown in Table 3.1. A model can either be analytical, non-analytical, or computational. *Analytical models* are math models in which the relations in the model are described using mathematical terms. Due to their mathematical representation, analytical models can be analyzed, and therefore used to predict the outcome of decisions, design choices, or any other system decision. We can further subdivide analytical models depending on the assumptions they make concerning uncertainty. *Deterministic models* assume there is no uncertainty and represent all variables in the model as known quantities. *Stochastic models* assume uncertainty is important and represent uncertainty as random variables. Another way analytical models can be classified is whether they are generating or evaluating [7]. Generating models provide an optimal solution or policy for the problem. Optimal means the best policy with respect to some objective. Evaluating models let the modeler input data and the model outputs the solution. To find the optimal solution in an evaluating model, the modeler would have to input all possible combinations of input parameters, and then select the best solution.

Non-analytical models are descriptive models that show a static view of the system at a point in time. Examples of non-analytical models are flowcharts, organization charts, or data models. Non-analytical models are generally diagrams supplemented with text descriptions. Descriptive models cannot be simulated or computed, and therefore non-analytical models are not used for quantitative analysis because they include no quantitative values. Qualitative analysis is possible such as to compare different process designs or analyze the flow of materials in a system. These models are best for understanding the enterprise systems and for communication with non-technical stakeholders. Additionally, these models are often input to computational models for analysis.

Computational models exploit the capability of computers to perform calculations rapidly so that the model can represent the time-varying behavior of a system. These are dynamic models and are often called simulation models. The computational models describe the system mathematically as a function of time. Computational models can be either deterministic or stochastic. Deterministic simulations, given the same initial conditions, will replicate the same exact behavior every time the model is solved. Whereas stochastic simulations use random numbers and even with the same initial conditions each time the simulation model is solved it will exhibit different behavior. Computational models are further divided based on the type of simulation. Discrete Event Simulation (DES) is

TABLE 3.2
Comparison of Modeling Approaches

Model Type	Strengths	Limitations
Non-analytical	Widely understood, easy to create, even by non-specialist; relatively quick to create; wide range of model types to model various aspects of enterprise systems.	Limited ability to support analysis of the enterprise (only qualitative analysis possible).
Analytical (deterministic)	Can determine optimal design for a system (limited to model assumptions); can perform sensitivity analysis to understand input/output relationships; relatively low data requirements compared to simulation.	Not well understood by non-technical people; requires skilled engineer to develop; only appropriate in situations where random behavior or variation is not critical to analysis.
Analytical (stochastic)	Can analyze systems where random behavior strongly influences system behavior.	Not well understood by non-technical people; requires skilled engineer to develop.
Computational model	Can analyze systems of high complexity; can perform wide range of analysis.	Requires skilled engineer to develop; validation is important for user acceptance of model; significant data requirements for validation; longer time to create model than either non-analytical or analytical.

a traditional means to model business processes by assuming the system state changes at discrete points in time. Systems Dynamics is a continuous simulation in which the system state changes continuously as a function of time. Underlying continuous simulation is a set of simultaneous differential equations. Agent-based simulation can be either discrete-event or continuous, the difference is in the representation of the system. In agent-based modeling the system behavior is not modeled directly. Instead the behaviors of individual agents in the system are modeled and their interaction gives rise to system-wide behaviors.

Table 3.2 lists the relative strengths and limitations of each of the modeling approaches. Choosing a model type depends on the problem being addressed, the desired results, data available, personnel involved, time constraints, and intended use of model. To illustrate, if the purpose of the model is to understand the organizational structure, then the non-analytical organization chart is an appropriate model. If the purpose is to determine how many bank tellers are needed to provide a desired service level, then an analytical model such as a queueing model or a computational model is needed.

Some readers might draw the conclusion that stochastic models are always superior to deterministic models because the real world is uncertain, and since stochastic models represent the uncertainty, they will better capture desired behavior than deterministic models. This is not always the case; there are many situations in which deterministic models perform well and are preferable to stochastic models. First, in some problems uncertainty is not critical to the system behavior so deterministic models are reasonable. Second, these models are analytical; they are described by math equations. Deterministic models are generally much more tractable than stochastic models. This means that we can usually solve a deterministic

TABLE 3.3
Analytical Versus Simulation Models

Criteria	Analytical	Simulation
Model complexity	Limited, attempt to include only the most relevant system behavior; for example, model workers processing rate as a constant 10 parts/hour.	Extensive, possible to include very complex, detailed behavior. For example, model workers breaks, fatigue, and absenteeism and how it impacts processing rate.
Data requirements	Requires less, e.g., need the mean and standard deviation of process rate.	Requires more, e.g., need the mean and standard deviation of process rate, the work schedule with breaks, mean and standard deviation of absenteeism.
Time to build model	Typically can be built fast, in a matter of hours or days.	Typically requires significant time, can be built in weeks or months.
Optimization	Frequently possible to optimize.	Traditionally not optimized, but evolutionary search techniques, and other meta-heuristics becoming more common.
Time to run	Usually single execution of model to obtain results, in seconds or minutes.	Requires multiple executions to obtain statistically valid output, may take minutes for each run.

model rather easily compared to a stochastic model. For these reasons, deterministic models are widely used.

Computational or simulation modeling is one of the most effective modeling approaches available to the enterprise engineer because systems of almost unlimited complexity can be modeled and analyzed. Since simulation modeling can be used in almost any situation, it is often the first approach considered by many engineers. The limitations of simulation all derive from the greater time, data, and effort needed to build and validate the model.

Table 3.3 compares analytical models against simulation models with regard to important modeling criteria. The complexity of analytical models is often limited by the tractability of the resulting model. What this means is that if the analytical model becomes too large or too complex, then it may not be possible to find a solution to the model in a reasonable amount of time. However, there is a trade-off to consider between whether to use an analytical or a simulation model. Simulation models require significantly more time than analytical models to develop. For example, a queueing model might be more appropriate for a rough-cut capacity analysis than a simulation model because the queueing model can be created, validated, and analyzed in a fraction of the time it would take to do the same analysis with a simulation model.

Models also differ in their ability to accommodate detail. There is a limitation to the amount of detail a mathematical model can have and its tractability. The more detail added, the more likely the model will become unsolvable. Typically, as more detail is added the preferred modeling approach becomes simulation, because simulation can accommodate almost any level of detail (such as bathroom breaks, if you wanted). However, frequently other modeling approaches, such as queueing networks, will provide sufficient answers without the burden of irrelevant details.

3.5 Model Verification and Validation

Model verification and validation (V&V) involves answering two questions. First, does the model match its intended design? Second, does the model sufficiently represent reality to satisfy the purpose of the model? Answering the first question involves verification. *Verification* establishes whether a model correctly represents what was intended by the designer. Answering the second question involves validation. *Validation* determines to what degree the model represents the system behavior of interest. Validation is done so that users of the model can have confidence that the model adequately represents the actual system, and consequently have confidence in the analysis and decisions based on the model. Without this confidence and trust in the model, the model would not be used. So, a primary reason for V&V is to build user confidence in the model.

Model Verification – Establishing that the model is correctly implemented and behaves as intended.

Model Validity – How well a model represents the actual system of interest based on the model purpose.

Two important points to remember when discussing model validity are that model validity is contextual and that model validity measures a degree of confidence one has in the model's representation. Model validity is contextual because it is only valid with respect to the model's purpose. Model validity results in a user having a degree of confidence in the model's results but never absolute faith in all model outputs. The reason is because no single test can absolutely validate a model. Validation is an iterative process such that each successive validation test extends our belief that the model captures essential structural elements and behavior of the actual system to serve the purpose of the model [12]. In the literature, many authors make these two points by saying that there is no such thing as a valid model [8] or that, "all models are wrong" [11].

Collectively, these two points imply that validation in the sense of truthfulness is not possible. Instead models are more (or less) useful given a set of objectives and boundary conditions. In this perspective, model validation is a matter of degree, depending largely on the purpose of the model. An enterprise model must be complete and correct with respect to its purpose. Consequently, to validate a model its purpose is made explicit. Moreover, in the end, there usually is no definitive answer of whether a model is valid or not; rather validity is a continuum and must be traded off against other model criteria such as the cost of building the model or the cost of obtaining further data for validation. An experienced modeler consciously considers the various trade-offs when building a model.

3.5.1 Model Verification and Validation Procedure

To verify a model, the modeler checks whether the model has any errors in logic, and establishes whether the model conforms to the model syntax, semantics, and conventions correctly. It makes sense that verification proceeds validation. After all, there is no reason to attempt to validate a model that might have errors in it or is built contrary to its intended design. Once the modeler feels confident about the model verification, he can then proceed to model validation.

Validation is the iterative process of establishing the degree to which the model describes the actual system for the intended purpose. How the model is validated depends on the type of model. We first describe procedures to validate non-analytical models, which can be used with the other model types. We then describe procedures to validate analytical and computational models.

Non-Analytical Models

If the model is a non-analytical model such as a flowchart, then the only V&V possible is verifying the model syntax is correct, verifying the model represents what was intended, and face validity of the assumptions, structure, and behavior depicted in the model. Because the model is non-analytical there are no mathematical or statistical techniques that can be used in validation. The following V&V can be done for static models:

1. Verify model structure. Check that the model has no language syntax errors. Walk-through all model constructs to verify they are designed as intended and that the logic is correct. Verify the model behaves as designed.

2. Face validity of model. Face validity determines whether the model appears valid to experts who review the model. To obtain face validity, subject matter experts (SMEs) review the model logic, assumptions, inputs, and outputs and provide assessments as to how closely the model represents the actual system for the specified purpose. Face validity is iterative; often many meetings are required before a model is accepted by the SMEs. Face validity is considered a weaker form of validity than statistical validation by many modelers. Yet, it is an important part of validation that should be done for all models.

Analytical and Computational Models

For analytical and computational models, we have available a larger range of statistical validation techniques, and users usually demand such validation. The techniques used depend on the data available. If the model is a process model and you have available historical input data (such as process service times and arrival rates) and you have output data (average waiting time, cycle time, throughput time, and other performance statistics), then you could take the input data, use it in the model, and then compare the model output to the observed output. Statistical validation of computational and analytical models is the strongest form of validation. The V&V procedure for analytical and computational models is to first use the procedure for non-analytical models. Then the following additional validation tests can be performed:

1. Validate toy problems. One way to validate a model is to test simplifications of the model in which the behavior is known. Sometimes these model simplifications are called "toy problems" [12]. A toy problem is a problem reduced in complexity such that the model can be verified and validated by hand. Satisfactory replication of the behavior of these toy problems builds confidence.

2. Validate model at boundary conditions. Another way is to test extreme conditions of the model or test the boundary conditions of the model. Frequently, the system behavior is known at these conditions.
3. Validate model by checking degeneracy. Degeneracy validation can check whether the model degenerates to a known condition. For example, if the arrival rate is greater than the service rate, does the queue grow to infinity? Another type of degeneracy validation is if the model is stochastic you can check the more easily solved deterministic equivalent.
4. Validate model by comparison against a known valid model. Sometimes a model is validated with another model and not the actual system. This is common in queueing network models that are validated against simulation models of the same system.
5. Statistical model validation. If data exist or can be obtained for the actual system, then statistical validation is among the strongest means to validate a model. Statistical validation compares the model to the actual system and calculates a test statistic. To do statistical validation, you state a null hypothesis that the model and the actual system are different. Depending on the assumptions of the underlying distribution and the number of samples available, you choose an appropriate statistical test. Common tests are a student t-test when there are few samples, or a z-test using the normal probability distribution. In hypothesis testing, you assume the null hypothesis is true and then calculate the test statistic. The test statistic is a p-value that defines the probability that you accept the null hypothesis when it is actually false. The smaller the p-value then the more plausible the rejection of the null hypothesis. Statistical validation can provide strong evidence whether a model is a valid representation of the actual system. See Banks et al. [2] for a treatment of statistical validation of simulation models.
6. Check model consistency. Typically, an enterprise project involves generating multiple models from different viewpoints (e.g., process, information, and organization views). The project might have the same view such as the process view, but it contains several models at various levels of decomposition. Given multiple models of the same system from different viewpoints or levels of abstraction, then an important verification step is to check that all the models are consistent. Each individual model must be internally consistent and the set of models must be consistent with each other. An example of poor consistency is if an information flow has different names at different levels of abstraction, whether in the same model or separate models.

3.5.2 Obstacles to Validation

A common validation problem involves validating a proposed system design, or what is commonly called a "to-be" model. In these enterprise design problems, model validation is particularly difficult because validation cannot be done with respect to an actual system. If the model is of a design, then it is not possible to collect data from the actual system because the actual system does not yet exist. Such to-be models are very important and modeling efforts should not be abandoned because the data are difficult to collect. In this case, strategies for validation should focus on face validation, testing boundary conditions, testing "toy problems" with the model, and validating against other models if available. Another strategy is that usually the organization has an existing system and the "to-be" system is a modification of the existing system. One validation strategy is to first model the "as-is" system, validate it with existing data, then make the changes in the model for the "to-be" system. This strategy builds confidence since we validate the model against the "as-is" and assume much of the basic, underlying structure remains the same.

Hypothesis Testing

To test whether a model is a valid representation of the actual system, you do a test for the inference on the difference between two population means. Let $X_1, ..., X_n$ denote the sample values obtained from n experiments with the actual system, and let $Y_1, ..., Y_n$ denote the sample values obtained from n experiments with the model. The null hypothesis that the mean of the model and the mean of the actual system are the same is stated as

$$H_0 : \mu_X - \mu_Y = 0$$

and the alternate hypothesis is stated as

$$H_1 : \mu_X - \mu_Y \neq 0$$

A confidence level α is selected, usually $\alpha = 0.05$, to define the rejection region for the null hypothesis. If a test statistic falls within the rejection region defined by α, then the null hypothesis is rejected. Otherwise, you fail to reject the null hypothesis.

If the sample size is large and the population is normally distributed, then the test statistic, Z_0, is calculated as

$$Z_0 = \frac{\overline{X} - \overline{Y}}{\sqrt{\frac{\sigma_X^2}{n_X} + \frac{\sigma_Y^2}{n_Y}}}$$

If $Z_0 < Z_{\frac{\alpha}{2}}$ or $Z_0 > Z_{\frac{\alpha}{2}}$, then the test statistic falls in the rejection region and the null hypothesis is rejected. If the test statistic falls in the acceptance region, then you fail to reject the null hypothesis. This indicates there is little evidence that the mean of the model and the actual system are different, or in other words, the model is a valid representation of the actual system. Oftentimes, the sample size is too small to justify the Z_0 test statistic and instead the student t-test is appropriate. Additionally, there are other possible alternate hypotheses. See Balci and Sargent [1] for a discussion of simulation model validation.

If the model is of a social system, then it is often difficult to validate because human behavior plays an important role in these systems [5]. Unlike models of physical systems, models of social systems lack well-known physical laws: social systems have many interrelated variables, the relationships are often nonlinear, and there is usually insufficient data for statistical validation. As a result, face validity is the primary validation technique employed in these models.

A threat to validity is uncertainty in model assumptions. Sensitivity analysis addresses this threat. Sensitivity analysis lets users of the model understand the implications of the model assumptions and lets users see the range in which model results are valid. To illustrate how sensitivity analysis can be used to increase user's confidence in the model in the face of uncertainty, we present the work discussed in Kim and Giachetti [6]. This paper presents a stochastic model to overbook doctor's schedules in order to maximize capacity utilization. The authors use patient no-show data from the medical clinics based on their current operations of no overbooking. The problem is whether the no-show data remain the same when overbooking is introduced to the clinics. The new policy might influence patient behavior.

To address this concern (about the uncertainty in future no-show rates) the authors varied the parameters over reasonable ranges to show how different assumptions affect the system performance. In this way, sensitivity analysis is used to increase confidence in the model by showing how modifications to the model assumptions affect the output.

To summarize, the most desirable validation technique is statistical validation with known data, but unfortunately, it frequently is not possible. Regardless, the modelers should be encouraged to avail themselves of all the other aforementioned verification and validation techniques. Collectively, many smaller validation tests can build confidence in the model and demonstrate the model's usefulness.

3.6 Modeling Process

Modeling, defined in the narrow sense, can describe only the process of building a model. However, modeling is not done, or at least should not be done, solely for the sake of building a model. Rather, modeling is done as part of an overall problem-solving process. In this book, we discuss modeling in the context of the entire problem-solving process. The problem-solving process is a systematic methodology to solve complex problems. It consists of identifiable phases of which the model facilitates all phases. The modeling process (shown in Figure 3.5), within the larger problem-solving process, has several observable phases:

1. Project Initiation – Determine what is the purpose of the modeling effort. Obtain a project sponsor, form a project team, and plan the project.
2. Model Construction – Start with a system problem and build a model that is a valid representation of the problem situation. Verify and validate the model.
3. Model Solving – Start with a model and develop a solution to the model which is a set of recommended courses of action.
4. Reporting – Start with a model solution and convert the model solution into a set of recommended courses of action the company can take.
5. Model Implementation – Take the solutions generated and plan and execute their implementation in the enterprise.
6. Model maintenance – As the system changes, make changes to the model so that it represents the actual system at all times.

Foremost, it must be remembered that the modeling process is iterative and the phases overlap.

3.6.1 Modeling Project Initiation

The purpose of project initiation is to establish the objectives of the project, to assemble a project team, and to plan the project. The main activities are:

Formation of modeling team. The model is created by a team of modeling experts who are responsible for the model, but also receive input and contributions from many other people both within and outside of the organization. A good modeling team includes experts in modeling, subject matter experts, intended model users, and a project sponsor. Vernadat [13] recommends that modeling teams of four or five experts seem to be the best trade-off between time to complete the model and accuracy of the model.

A modeling expert is a person who has knowledge and skills in the modeling and analysis

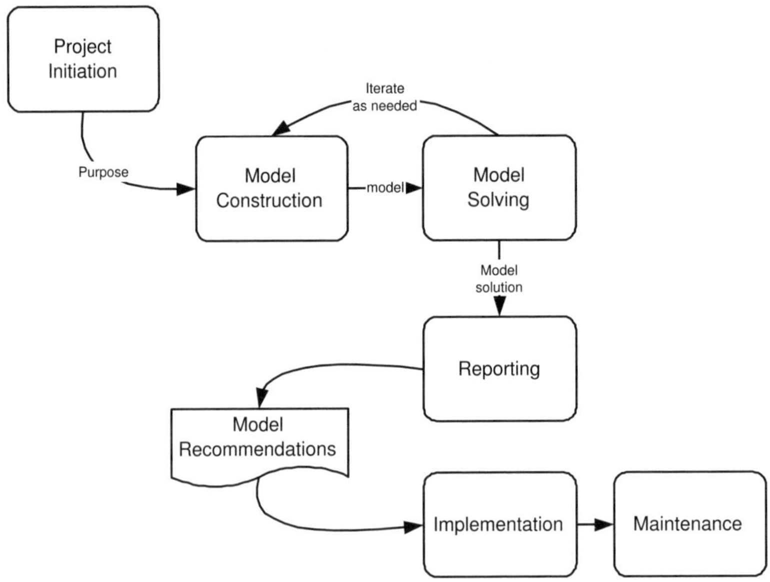

FIGURE 3.5
Modeling process.

techniques but might not be knowledgeable of the problem domain. In industry, such people are industrial engineers, internal business consultants, systems analysts, business systems analysts, or system architects. Frequently, these experts will only be knowledgeable in one or two modeling techniques. A subject matter expert (SME) is a person who has deep knowledge of the business domain being modeled. SMEs are senior employees, not necessarily the managers, of the departments. SMEs can answer questions of why systems operate in a certain way, and SMEs can explain all the details of the system. Domain specialists are IT specialists or other relevant technology domain. These specialists can help determine if proposed model designs are technically feasible; they can provide suggestions based on available technologies (for example use of RFID), etc. The project sponsor is a person in a senior management position who has the authority and power to obtain resources for the project. Generally, the project sponsor is responsible for approving the schedule, budget, and scope of the modeling effort. The project sponsor is usually also the problem owner, the person in charge of the organizational unit, business process, or system that is the focus of the modeling effort.

Generally, as a minimum a modeling team has at least one technical expert and one subject-matter expert. The reason is that a single person rarely has knowledge in both areas, and the knowledge from each complements each other in a way that makes modeling successful. Inclusion of problem owners and other stakeholders in the modeling process can greatly increase the acceptance of the model and the final solutions.

Plan the project. The modeling sub-project needs to develop a modeling plan. The plan would include specification of the modeling activities, a schedule for those activities, identification of milestones and deliverables, and assignment of responsibilities.

3.6.2 Model Construction

In order to build a good model, you must first determine what is the purpose of the modeling effort? As previously argued, all models have a purpose and decisions about abstraction level, viewpoint, and other modeling issues are guided by the model purpose. So, the first task is to define clearly the purpose of the model. We can broadly identify four possible purposes of a model:

1. Understand the enterprise system. The purpose of the model can be to better understand the system structure, key factors, and system behavior. The model can facilitate continuous improvement projects to improve the system. Also, the model can be used to document the system for training or other purposes.
2. Analyze enterprise system problems. The model purpose can be to understand an enterprise system problem and arrive at a solution to better the system. The model can be used to specify requirements for implementation of an information system or other system.
3. Design enterprises. The model purpose can be to create a blueprint of a new proposed system.
4. Manage and control the enterprise. The model purpose can be to understand the system behavior so that it can be managed and controlled.

Determine model type. Once a purpose is defined, the next task is to determine the most suitable model type. The best model type is contingent on the model purpose, data available, system type, users of the model, and developers of the model. All model types can be used to better understand the system. To analyze an enterprise system then usually analytical or computational models are called for. To design an enterprise system all model types are possible choices, likewise for managing and operating the enterprise system.

Determine modeling language. The choice of model language is based on: the expressive power of the language to represent the system being modeled to a sufficient degree, the available tools to support the language, and the availability of expertise to use the language [3].

Determine problem scope. When first building a model, the modeler needs to define the boundaries of the model. You need to define what portions of the system will be included in the model and what will be excluded. Choose the boundaries carefully. Often this is a non-trivial task. In a project to redesign a new process for turning around airplanes between flights, a significant amount of time was spent by the modeling team just to define when the turnaround started and when it ended. Did the turnaround start when the airplane landed? When it was secured at the gate, or when the door on the plane was opened? Did the turnaround process include ticketing? If the boundary is chosen too small, then it is possible that issues contributing to the problem are not examined. If the boundary is too broad and encompassing, then it is difficult to focus on the problem and the modeling team might be overwhelmed with data and issues. Checkland [4] offers that, "there is no principle, no limit to the analyst's freedom to make whatever choice he thinks or feels might lead to insight" (p. 221). The decision of how to define the system boundaries involves decisions on what actors, resources, and activities to include in the analysis and what are only included tangentially to the system.

Fact finding. In Chapter 8, we discuss data gathering and the formation of a data collection strategy. Data collection includes observation, survey, review documentation, interviews, time and motion studies, and other techniques. To build a model you need to understand the system, this is done via fact finding.

Construct the model. Constructing the model is an interactive process. Some of the decisions that must be made during construction are:

1. Define model viewpoint. For example, for a given system you can model it from

an information perspective, a process perspective, or an organization perspective. Each perspective is a separate view of the same system.

2. Abstraction level. Often the level of detail is an economic question. Greater model detail requires greater effort. Greater effort is measured in terms of time, money, number of people involved in modeling, accuracy and comprehension of data collection, etc. This greater effort must be justified by the results of the model. Often there are decreasing marginal benefits such that no model is extremely detailed.

3. Document all assumptions. All models are based on assumptions on how the system works. The modeling team identifies, discusses, documents, and validates all these assumptions. For example, many economic models assume people will act rationally and many queueing models assume people do not balk or renege. Do these assumptions hold for the current situation?

4. Document rationale for decisions made. A model will often exist within the enterprise longer than the persons involved in creating the document. So for example, if a model is to be used again, the new users often need to understand the original rationale for how the model was designed. Unfortunately, this rationale is often not documented.

3.6.3 Model Solving

Analytical and computational models must be solved to obtain the model results. Generally, the solution algorithm is determined by the model type. For example, linear programs are solved by the simplex method, integer programs by the branch and bound algorithm, and so forth. More complicated non-linear models might need specialized algorithms or meta-heuristics to solve.

3.6.4 Reporting

The project objective, problem-solving approach, model, and recommendations need to be reported to the project sponsor. Unless the project team can present the model and solution in a manner clearly understood by the project sponsor, then the effort may have been wasted. Technical writing is an important topic. Some recommendations are to keep the report succinct and to the point. Use simple, direct language. Avoid overly technical jargon. Frequently, avoid most of the equations and math in the main body of the report. These should be left to an appendix. Make judicial use of charts, figures, and graphs to present the data.

3.6.5 Model Implementation

Model implementation is taking the recommendations generated by the model and instituting them in the actual system. Implementation in enterprises often means change management because there are many people in the enterprise that must institute the model recommendations.

3.7 Model Reuse

Given the amount of time, money, and effort put into modeling it would be highly beneficial if models were easy to reuse for the same or other purposes, and if models could interoperate with other models. Model reuse would let us leverage previously done modeling work and hopefully save time and money. Model interoperability would allow one model to share and interact with another model, it would allow the assembly of many subsystem models into a large, enterprise-level model, and it would make reuse much easier. Model interoperability would find application in the formation of virtual enterprises or the coordination of supply chains as two examples.

The benefits of reuse are many, yet in practice reuse occurs infrequently. There are several reasons for the lack of reuse. One obstacle to reuse is that the enterprises we model change over time; likewise, for the models to remain useful they must change as the enterprise changes. However, few managers find they can justify model maintenance for future projects that they do not even know about. Companies develop enterprise models for a particular purpose, usually a project. When the project ends, the project sponsor generally has no more motivation for spending resources to maintain the model. As a result models are not maintained and they become outdated.

There are also technological obstacles to model reuse. To really make model reuse a reality, a search engine and a model repository with meta-data about the model are needed [14]. Without meta-data to describe characteristics of the models in the repository and a search engine to find models, then prospective users of existing models will not be able to find a model to suit their needs.

Finally, reuse is only possible on a large scale if there is model interoperability. In reuse of models, it is unlikely that the entire model would be reused. Rather the modeling team would want to reuse only portions of the existing models or to combine several existing models. For this to happen, the models need to interoperate. This calls for standard modeling languages and a means to capture the model semantics in a formal way. As an example of some of the work trying to develop model interoperability is the Unified Enterprise Modeling Language (UEML) project that is developing standards and technologies to achieve interoperation [9].

Model Reuse

A credit card company operates a mailing center where bills are printed, sorted, and mailed to cardholders. Industrial engineers at this facility constructed a discrete-event simulation model of the facilities' main operations. The model was used to design efficient flow of materials. Five years later, another project was begun and management hoped and even expected that this model could be reused, thus saving tremendous amount of time, effort, and money for the new project. The problem with reuse is that over time the company installed newer and different equipment, processes were modified, policies were changed, and a multitude of other smaller changes were implemented in the facility. As a consequence of these changes, the simulation model was no longer a useful or valid representation of the actual situation in the facility. Models, if they are not maintained, will become outdated because the system they were built to represent changes.

Given the current state-of-practice with model reuse a modeling team needs to approach situations with hesitation where project sponsors claim an existing model can be reused for a new project. Several questions should be asked and investigated to assess the likelihood of reuse. Is the model current? Does it reflect the current structure and behavior of the system we want modeled? Can the model serve our purposes? Is there meta-data or other information about how the model was constructed and what it represents? Negative answers for these questions make reuse improbable.

3.8 Summary

Modeling is the foundation to our understanding of enterprise systems. Developing a proficiency in modeling, as well as understanding the concepts of modeling, is basic knowledge and skills the enterprise engineer must cultivate. The usefulness of models is the fact that they simplify reality, representing the system in a form that we can comprehend. A model that has all the details of the system being modeled would be just as complex as that system and just as inscrutable. The process of modeling helps us clarify our thoughts about the system. For this reason, many modelers find the actual learning they do during the modeling process is as valuable as the model itself.

Too often, students believe complicated models are required to deal with complex systems. But, it is neither necessary nor desirable to build too complicated of a model. What the modeler strives for is a model that makes the complex system seem simple. Do not confuse simple with trivial. The modeler must represent the important relationships, cannot make arbitrary assumptions, and cannot generalize away critical details. What the modeler needs to find is an elegant representation that provides an appropriate level of detail for the system. By attaining simplicity, then the model gains credibility because the users can understand the model. Achieving the right balance is a matter of experience, inspiration, hard work, and sometimes luck.

A model reflects the beliefs of the model builder concerning the system. For this reason, models are useful to ensure that all the stakeholders have a similar conceptualization of the system, its structure, and its behavior. Oftentimes, stakeholders can discuss the same system, but not until they start reviewing actual models do they realize how different their viewpoints can be. A good model can help stakeholders achieve consensus on many issues concerning the system.

The principle of finding the appropriate abstraction level recurs in most modeling tasks. The problem facing the modeler is frequently, as enterprise modelers, the solution is to decompose the system into multiple abstraction levels similar in concept to the multiple abstractions Picasso generated of the bull.

It is often said that modeling is an art. What is meant is that there is no formal, defined process for building a model and determining its worth. Practical modeling experience is indispensable for creating good models. However, as this chapter has attempted to demonstrate, there are general modeling principles that if applied help novice modelers generate good models and avoid pitfalls. The modeling principles that guide modeling decisions are:

- All models should serve a purpose, and the purpose should be documented during the modeling process.

- All models are abstractions; the modeling team needs to determine the appropriate abstraction level to serve the model purpose.

- Enterprise systems and the problems addressed by enterprise engineers are often too complex to represent in a single model. Modelers should consider having multiple models of the system from different viewpoints and/or abstraction levels.

- There are many different types of models. The modeling team needs to choose a model type most appropriate for the problem being addressed.
- Model verification and validation is a process, not an outcome. Validity of a model depends on the model's purpose and is not a yes/no answer.

A goal of the enterprise engineer is to gain experience in building and using models. The more comprehensive our "model base," the more clearly we can think about enterprise systems.

Review Questions

1. Explain a benefit of models.
2. You attend a seminar by a software salesman who makes the statement, "A model shows an absolute view of the system." Argue whether the statement above is correct, incorrect, or partially correct.
3. Explain the difference between verification of a model and validation of a model?
4. List four common purposes of enterprise models.
5. Explain the difference between analytical models and non-analytical models and when each is appropriate to use.
6. Argue whether more detail or less detail is preferable in models.
7. Explain the difference between model syntax and model semantics.
8. What is the difference between different model viewpoints in an enterprise? What are some viewpoints that are appropriate for an enterprise?
9. A company wants to build a model to understand the performance of a proposed information system architecture. They want to know how much capacity in terms of network bandwidth and computer power is needed. What model type: non-analytical, analytical, or simulation is best suited for this purpose? Provide a rationale for your recommendation.
10. A state government organization needs to make their services available to the public. To do this they want to provide a model of the information structure of their services and a process model to depict how it works. What model type: non-analytical, analytical, or simulation is best suited for this purpose? Provide a rationale for your recommendation.
11. What makes "to-be" models especially difficult to validate?
12. Why can statistical validation only be done for analytical and computational models?
13. Explain why it is beneficial to have technical experts and subject-matter experts on a modeling team?
14. What is model reuse and why is it desirable?
15. Under what scenarios are deterministic models not a good choice of model?

Bibliography

[1] O. Balci and R.G. Sargent. A methodology for cost-risk analysis in the statistical validation of simulation models. *Communications of the ACM*, 24(4):190–197, 1981.

[2] J. Banks, J. Carson, B.L. Nelson, and Nicol. D. *Discrete-event System Simulation.* 4th edition, Prentice Hall, NJ, 2004.

[3] P. Bernus. Some thoughts on enterprise modeling. *Production Planning & Control*, 12(2):110–118, 2001.

[4] P. Checkland. *Systems Thinking, Systems Practice.* John Wiley & Sons, Chichester, England, 1982.

[5] S.R. Goerger, M.L. McGinnis, and R.P. Darken. A validation methodology for human behavior representation models. In *The Interservice/Industry Training, Simulation & Education Conference (I/ITSEC)*, Orlando, FL, 2004.

[6] S. Kim and R.E. Giachetti. A stochastic mathematical appointment overbooking model for healthcare providers to improve profits. *IEEE Transactions on Systems, Man, and Cybernetics – Part A*, 36(6):1211–1220, 2006.

[7] G. Koole. Optimization of business processes: An introduction to applied stochastic modeling. Technical report, Department of Mathematics, Vrije Universiteit Amsterdam, NL, 2006.

[8] A.M. Law and W.D. Kelton. *Simulation Modeling and Analysis.* McGraw-Hill, New York, NY, 1991.

[9] H. Panetto, G. Berio, K. Benali, N. Boudjlida, and M. Petit. A unified enterprise modeling language for enhanced interoperability of enterprise models. In *11th IFAC INCOM2004 Symposium, April 5-7, Salvador, Brazil*, 2004.

[10] B. Selic. The pragmatics of model-driven development. *IEEE Software*, 20(5):19–25, 2003.

[11] J.D. Sterman. *Business Dynamics: Systems Thinking and Modeling for a Complex World.* Irwin McGraw-Hill, New York, NY, 2000.

[12] J. Thomsen, R.E. Levitt, J.C. Kunz, C.I. Nass, and D.B. Fridsma. A trajectory for validating computational emulation models of organizations. *Computational & Mathematical Organization Theory*, 5(4):385–401, 1999.

[13] F. Vernadat. *Enterprise Modeling and Integration: Principles and Applications.* Springer, New York, NY, 1996.

[14] L. Whitman and D. Santanu. Enterprise model repository. In *IFAC World Congress Praha 2005, Prague, Czech Republic*, July 3-8, 2005.

4

Enterprise Design Methodology

"Whenever you are asked if you can do a job, tell 'em, 'Certainly I can!' Then get busy and find out how to do it." – Theodore Roosevelt (1858-1919), 26th president of the U.S.

This chapter discusses the nature of engineering design and the methodologies that are used to design enterprise systems. Design is a process whereby a new system is specified to meet a perceived need. To create something new, such as an enterprise, involves not only deciding what to create, but also how to create it. Design problems are really design messes; they are poorly defined, unbounded, vague, and open-ended. Moreover, design teams are always constrained by time, money, and resources. The goal of engineering design methods is to force structure on the design problem so that it can be managed and successfully completed. Moreover, enterprise systems are designed by multidisciplinary teams over a long period of time. Successfully managing projects that involve the work contributions of many different people over a long period of time is a difficult job. Straightforward project management techniques are in general insufficient by themselves. A formal engineering methodology is needed to handle the complexity inherent in any large-scale project of the scope demanded by enterprise system design. In this chapter, we discuss the nature of the design problem faced during an enterprise systems project, and the methodologies, techniques, and the tools that can be brought to bear in completing an enterprise engineering project.

After completing this chapter, you should be able to:

- Describe how enterprise design fits into the problem-solving framework.
- Differentiate between different types of design problems and explain why enterprise design is usually a wicked problem.
- Define the three components of a constrained optimization problem and explain why it is difficult to find an optimal enterprise design.
- Explain why enterprise engineering teams use methodologies.
- Describe and compare the waterfall, spiral, and controlled iteration methodologies.
- List the five CMMI levels and explain why companies are interested in maturing their methodology.

4.1 Design Theory

It is natural to want to design the best system possible, but this raises the question, how do you define "best"? Finding the best solution is called optimization. For simple problems, determining the best design can be relatively straightforward. For example, we might want to buy the laptop computer that minimizes our total cost of ownership (TCO) over the next

TABLE 4.1
Computers and Their Total Cost of Ownership (TCO)

Computer	TCO
Acer Extensa	$1100
Dell Latitude	$1200
HP Pavilion	$1300
Lenovo Y510	$1350
Toshiba Satellite	$1400
Sony Vaio	$1500
IBM ThinkPad	$1550

five years. Suppose we gathered the prices for seven computers. For each laptop computer, we calculate the TCO by choosing a discount rate for the cash flows, and estimating the purchase price, operating cost, maintenance cost, and end-of-life-cycle cost. The data are in Table 4.1.

According to our problem definition (minimize TCO), the Acer Extensa laptop computer will minimize our TCO of $1100. We are able to identify the best laptop computers because we have an objective function that orders the laptops in terms of TCO. The objective function provides a measure of how good a design solution is. Objective functions are equations to either minimize some system attribute or to maximize some system attribute. In the case of enterprise systems, it could be to minimize total cost or to maximize total profits.

Our example of minimizing TCO for the laptop computer misses an important element that is present in all real design problems. The design decision, in this case which desktop computer to buy, is often constrained by other problem concerns. This leads to a constrained optimization problem. The *constrained optimization problem* can be stated as:

> Find the solution that maximizes performance (or minimizes cost) that simultaneously satisfies all the constraints.

In the unconstrained optimization problem above, it is possible the Acer Extensa is missing some important features that we need. The unconstrained problem made no mention of minimum performance requirements on CPU processing speed or random access memory (RAM) that needs to be met. To formulate constraints we make statements such as,

> The CPU processing speed must be greater than X MHz.
> The laptop must have RAM of at least Y Mb.

A computer might have the lowest TCO, but if it does not meet our CPU processing speed requirement or our RAM requirement then it is not a feasible alternative. The set of constraints on our problem define a *feasible region* of designs that satisfy all of our requirements. The objective function tells us which is the best design within this feasible region.

Figure 4.1 visualizes the constrained optimization problem. Each candidate laptop is shown as a point with the price next to it. The constraints are shown as dashed lines with arrows indicating which side of the line satisfies the requirement. The feasible region is the intersection of the two constraints, in this case, at the upper-right where both constraints are simultaneously satisfied. Only those desktop computers that satisfy both constraints are feasible; in other words, they are in the feasible region.

In the constrained optimization problem, we search for the lowest TCO in the feasible region. The optimal solution is the HP Pavilion at $1300. Notice the optimal laptop has

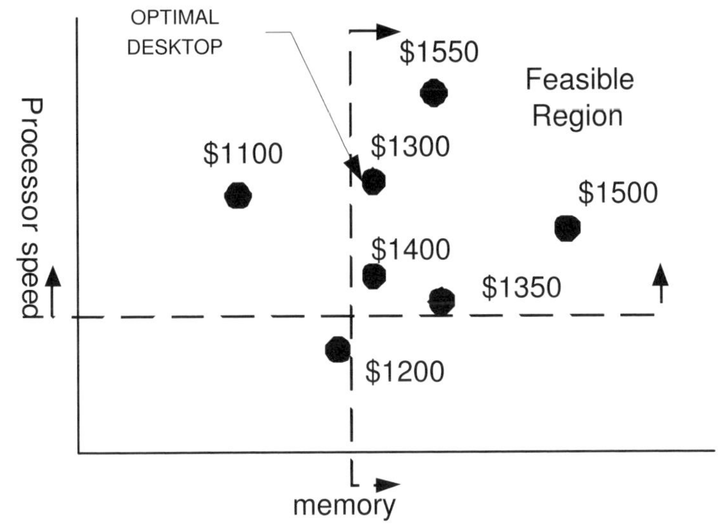

FIGURE 4.1
Desktop computer problem.

neither the highest CPU processing speed nor the largest memory. It does not even have the lowest cost because there is a desktop computer that has a lower cost. What this solution does have is the lowest cost of the feasible solutions, where feasible means the solutions that simultaneously satisfy all the constraints.

The laptop design problem is a constrained optimization problem formulated according to the techniques used in operations research. The field of operations research applies math to solve problems. A cornerstone of operations research is optimization, finding the best solution to a problem. The operations research process is to first formulate the problem by defining the decision variables, objective function, and constraints. In the laptop problem, the decision variable was which laptop to purchase. It was a discrete or integer variable. In many problems the decision variable can be a continuous real number, such as the horsepower of a machine. Once the problem is formulated, then a search algorithm is used to find the optimal solution. The strength of operations research is that large, complex problems can be formulated as mathematical models to identify a feasible solution space, and then that solution space can be systematically searched to find the best solution. The assumption is that the mathematical model is a sufficiently valid representation of the actual system such that the decisions inferred from the model will have the desired outcome in the actual system.

As the system becomes more complex, as the number of stakeholders increases, and as uncertainty about the environment and system increases then formulating a mathematical model becomes an insurmountable task. There are many unknown variables and many unknown unknown-variables.[1] Moreover, the relationships between these variables are only partially understood. Together, the sheer number of variables and the relationships between

[1]As we know, there are known knowns.
There are things we know we know.
We also know, there are known unknowns.
That is to say, we know there are some things we do not know.
But there are also unknown unknowns; the ones we don't know we don't know.
– U.S. Secretary of Defense, Donald Rumsfeld, under President G.W. Bush in a Feb. 12, 2002 Dept. of Defense news briefing.

them can overwhelm and frustrate a design team. Yet, an enterprise designer must arrive at a design for these systems, and clearly some enterprise designs are better than others.

For these reasons, designing enterprise systems is a notoriously difficult endeavor. One of the greatest hurdles facing a design team is simply to understand the problem situation and determine what should be the design goals. In the laptop computer example, the design goal was given to us. But, in most design situations there are many design goals and they can be competing such that the better a solution satisfies one design goal, the worse it satisfies another design goal. Moreover, it is very difficult to quantify all the variables, objectives, and constraints that occur in enterprise design.

Simon ([31], p. 204) says that the key to successful design is

> "the replacement of the goal of maximization with the goal of satisficing, of finding a course of action that is 'good enough.' ... Since the [designer] ... has neither the senses nor the wits to discover an 'optimal' path – even assuming the concept of optimal to be clearly defined – we are concerned only with finding a choice mechanism that will lead it to pursue a 'satisficing' path, a path that will permit satisfaction at some specified level of all its needs."

The concern that Simon and others have voiced is that many problems cannot be mathematically formulated without making gross simplifications or other assumptions that do not carry in the real world. Complicating the problems is the realization that all design problems are constrained by money and time. Gathering more and better information after a point is no longer feasible because it consumes both money and time, both of which are limited resources in any actual design problem.

That is not to say that thinking about design in terms of decision variables, objectives, and constraints is not useful – it is eminently useful. Applying operations research concepts to design provides several benefits. First, it focuses the designers on identifying the main components of the problem (decision variables, constraints, and objectives). Second, it highlights the need to identify the important variables that account for the majority of enterprise behavior. Third, it recognizes that there are multiple, alternative design solutions and part of the design problem is to rank these solutions to identify the best one.

4.1.1 Design as Problem Solving

Design is an elusive term – it can be conceptualized in many different ways. Design is described as a decision-making process [15],[2] as a search process [25], and as a problem-solving process [24]. We think it is useful to think of enterprise design as problem solving, which does not preclude the other viewpoints as will be seen later.

> *Engineering design* is a systematic, intelligent process in which designers generate, evaluate, and specify concepts for systems whose form and function achieve the stakeholder's objectives while satisfying a specified set of constraints [10].

Psychologists have studied how humans solve problems [26]. The following steps (illustrated in Figure 4.2) seem to capture the cognitive process employed.

[2]ABET defines design as, "Engineering design is the process of devising a system, component, or process to meet desired needs. It is a *decision-making process* (often iterative), in which the basic science and mathematics and engineering sciences are applied to convert resources optimally to meet a stated objective."

1. *Scope the Design Problem.* First you need to understand the problem situation [28]. It is the scoping phase that does this. In the scoping phase, you would define the problem boundaries, the objective, system requirements, and constraints on potential problem solutions. Part of scoping would involve creating a model of the design problem. Simon puts it this way, "every problem-solving effort must begin with creating a *representation* for the problem – a problem space in which the search for the solution can take place" ([31], p. 108). Even if not formalized, a problem solver will create a *mental model*, which is an internal representation of the problem. Understanding is influenced by past experience, and for this reason experts outperform novices on many problems. The result of scoping the design problem is an actionable description of the design problem, meaning you can start to specify solutions to the design problem.

2. *Design the Solution.* In the designing phase, you need to create a solution to the design problem. You create design models for a solution that specify the important design features and how they address the design problem. The models might include analytical models, simulation models, small prototypes, or even large-scale pilot systems. The advantage of these models is that they can be quantitatively evaluated.

3. *Evaluate the Solution.* You need to evaluate whether the design solution works as intended. Part of evaluation is identifying where the design can be improved and then making the improvements.

4. *Satisfactory?* Following Simon's idea of a satisficing solution, you determine, along with the stakeholders, if the design solution is satisfactory. If not, then you start again by modifying the scope as needed and continuing through the cycle again. If the current solution is satisfactory, then you end the design process. It helps if you have a definition of what is satisfactory prior to starting the design, but this is not always possible.

Except in the simplest of problems, the problem-solving process shown in Figure 4.2 is iterative. Conklin [8] describes an experiment in which experienced designers were asked to design an elevator and their cognitive processes were studied. They did not follow a linear approach of scoping the problem, designing solutions, and evaluating designs. They were more opportunistic, going from scoping to designing before they completed scoping, and then going back to scoping. The reason seems to be that problem understanding is helped by designing solutions and then evaluating how they would work. Designing solutions and then evaluating them reveals the flaws in their initial assumptions and conceptions. What this teaches us is that a good problem-solving process iterates between scoping the problem, designing solutions, and then evaluating the solutions. In each iteration, the emphasis gradually shifts from spending most of your time and effort on understanding to designing, and then eventually evaluating.

The problem-solving steps described above need to be tailored to a methodology for enterprise engineering and then adapted as the enterprise design project proceeds [24]. Moreover, to manage large design projects a structured methodology is needed [30]. The next section classifies the types of design problems for which a methodology would need to be tailored.

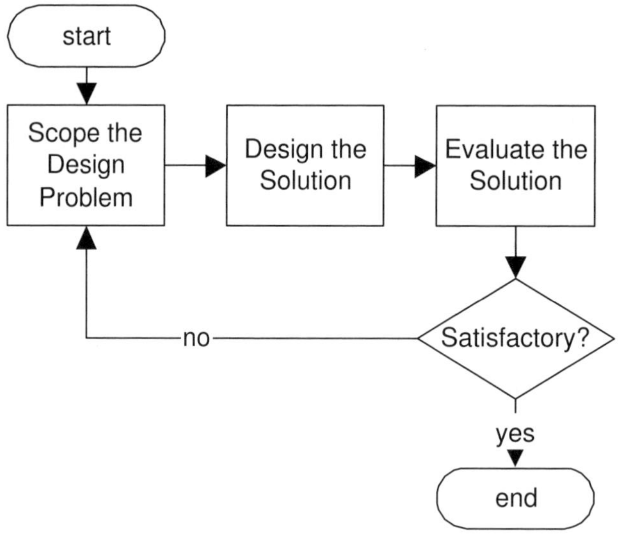

FIGURE 4.2
Design iterations.

4.1.2 The Design Problem

To design a system is to solve a problem, but not all design problems are the same. We draw a distinction between well-defined problems, ill-defined problems, wicked problems, and complex problems.

A *well-defined problem* has readily apparent goals and known constraints. Because the problem is well understood, there is little effort required in scoping the problem. Instead you can proceed rapidly to designing a solution, which involves devising a means to reach the design goal. Oftentimes, known solution methodologies or existing designs are available that can be applied to the problem. This leads to a procedural approach to design. The designer can follow the steps and arrive at a design satisfying the goal and constraints. Solutions to well-defined problems are based on rigorous on-the-job training, formal education, experience, and procedures pre-designed to handle similar situations [22].

An example of a well-defined problem is to design a car's gas tank so the car will have a 200-mile range on a single tank of gas. This is a well-defined problem because the design objective is straightforward and easily formulated, the constraints are known and also easily formulated, and extensive procedural knowledge and rules exist because gas tank design is highly developed. The design procedure is to estimate the miles per gallon based on the car's weight, engine size, and other features, and then calculate the volume required for the gas tank. The designer also needs to determine what material the tank should be made of (metal, fiberglass, or plastic), and what shape the tank should have, which is constrained by the available space. The later design decisions are not trivial, they require engineering, but they are formulaic in that nearly step-by-step procedures are available to generate a design. We can call this "cookbook design." Following an established design procedure is possible because the design problem is a variant of previous design problems (other gas tanks in other cars) [24].

An *ill-defined problem* has undefined, ambiguous goals and unclear means to arrive at the goals. Ill-defined problems usually have never been encountered before and have no pre-set rules and procedures to guide their handling. Such problems are typically less structured, unpredictable, and ambiguous as to "what is wanted." The first task when faced with an ill-defined problem is to understand exactly what is the problem? So, in ill-defined design problems considerably more effort is spent in scoping the design problem than would be in a well-defined problem. Sometimes, it may be possible to turn it into a well-defined problem. The stairs manufacturer example mentioned in Chapter 2 is an example of an ill-defined problem. In that example, the analyst spent time interviewing stakeholders, observing the work, and studying the problem until the problem was well-defined and eventually formulated as a mathematical model.

Horst Rittel (described by Churchman [7]) describes *wicked problems* as, "a class of social system problems which are ill-formulated, where the information is confusing, where there are many clients and decision makers with conflicting values, and where the ramifications in the whole system are thoroughly confusing." A wicked design problem has no single problem formulation: it is indeterminate in that there are many alternative, acceptable formulations [5]. The choice of formulation influences the design, which upon evaluation could lead to further iterations of problem reformulation and redesign that could conceivably continue indefinitely. There are no right answers to a wicked design problem, only good and bad designs. A source of the difficulty is due to what Conklin [8] terms fragmentation: the stakeholders all have different perspectives and are convinced their version of the problem is correct. This results in the social complexity natural in design projects that serve multiple, disparate stakeholders. In a wicked problem, negotiation is necessary to determine what are the priorities in ranking alternatives.

A *complex problem* has an exceptionally large solution space and/or the relationships between variables are nonlinear; both of which complicate the search for a solution ([32], p. 21). A cogent example of a complex problem is the traveling salesman problem (TSP). In the TSP, a salesman needs to travel to n different cities, he needs to visit each city once and only once, and he needs to end in the city he starts at. The problem is to find a circuit through all the cities that minimizes the total distance (or cost) traveled. Conceptually the problem is simple and easy to understand. Yet, to find an optimal solution to even small traveling salesman problems is difficult. The reason is the problem is NP-hard, which means it has high computational complexity, a measure of how the solution space grows exponentially with respect to the problem size. For n cities the number of possible circuits is $1/2\,(n-1)!$. This is a large number. A small TSP problem of 20 cities has $1/2(20-1)! = 6.082 \times 10^{16}$ possible solutions. If it took one second to generate a solution, you could not evaluate all the solutions in a single lifetime.

The TSP is complex only with regard to finding the optimal solution. To find a good solution in a reasonable amount of time, we can use heuristics. One heuristic for the TSP is the nearest neighbor rule. In this heuristic, you select a starting city, and then the next city in the sequence is the closest city (i.e., nearest neighbor). For the third city, you choose the closest to the second city, and continue until all the cities are part of the route. The heuristic is unlikely to find the optimal solution, but it will usually find good solutions that are more than acceptable for most people.

Relating to our current discussion, a complex problem might be difficult to solve optimally, but a complex problem can be well defined as is the case of the TSP. If you had two solutions to the TSP, it is an easy matter to compare their total distance and the shorter circuit is the better solution. This is possible because the problem is unambiguous with a single clear objective. Compare this to a wicked problem where the design team has difficulty just stating the problem, and if given two potential solutions, the team might not agree on the better solution.

Most enterprise engineering design problems are ill-defined problems and many fit the definition of wicked problems. To help identify wicked problems, Rittel [27] described ten characteristics, here paraphrasing the main points are:

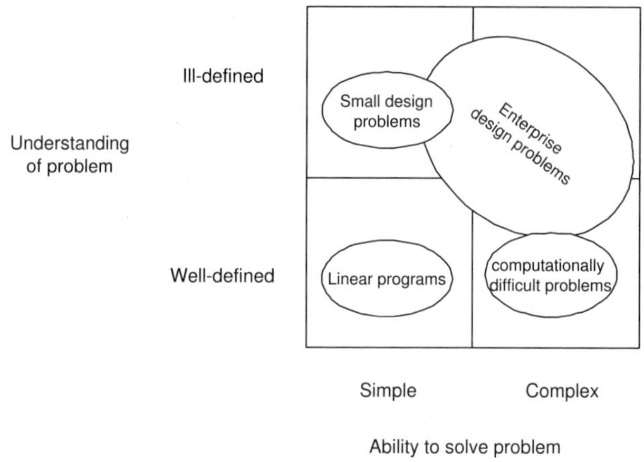

FIGURE 4.3
Problem typology.

1. There is no definitive statement of the problem, each stakeholder has a different perspective of the problem, and all are right.
2. There is no definitive solution, and therefore no "stopping rule," so the design process could go on indefinitely.
3. To understand the problem, the designer creates a solution and therefore gains greater knowledge about the problem and then can generate another solution.
4. There are competing solutions that are neither correct nor wrong, only good or bad.

It is important to remember that the perception of a problem cannot be removed from the person faced with solving the problem. What is a difficult problem for one person could well be a simple problem for another person. What this means is the problem classification is not absolute, but depends on the observer.

Figure 4.3 classifies the types of problems along two dimensions of how well we can understand a problem, and once understood, what is our ability to solve the problem.[3] Well-defined problems can be simple to solve (e.g., a logistics network optimization problem is both easy to formulate and easy to solve with network programming algorithms) or difficult to solve (e.g., the TSP as previously discussed). Ill-defined problems can also be simple to solve (e.g., small design problems that have few alternatives) or difficult to solve (e.g., most enterprise design problems). In this classification, wicked problems are both ill-defined and complex.

A useful observation is that simply gathering more data, after a point, will not help solve a wicked problem. The amount of data that can be collected is un-ending. Moreover, the human mind is limited in how much information it can process (Simon's "bounded rationality"). To address wicked problems, designers must be comfortable with ambiguity and uncertainty. The next section describes design thinking that helps deal with wicked problems.

[3]Dörner [9] presents a similar matrix to describe problem barriers according to the characteristics of the problem (clarity of the goal criteria and familiarity with the means).

4.1.3 Design Thinking

To understand the problem situation and scope the design problem, the design team needs to obtain information. Information is obtained through data collection techniques and involves asking questions of oneself and of the stakeholders. Questions that seek to reveal facts and that have a truth value that can be verified are characteristic of convergent thinking. These types of questions are phrased such that only a single or few specific answers exist. To pose such a question, the designer needs to have sufficient prior knowledge to have narrowed down the design problem. Asking convergent questions is indicative of convergent thinking.

What if you do not know enough about the design problem to ask convergent questions, as would occur in the beginning of a design project? To obtain the knowledge needed to subsequently narrow down the design, you ask divergent questions. A divergent question is phrased such that multiple alternative, known answers exist, as well as multiple unknown, possible answers. The answers have no truth value, you cannot verify if the answer is correct or wrong. The aim of asking divergent questions is to *diverge* from the known facts to the possibilities that can be created from them. Asking divergent questions is indicative of divergent thinking.

Convergent thinking is oriented towards deriving the single best (optimal) answer to a clearly defined problem. It emphasizes logical, rational thinking and focuses on accumulating information, recognizing the current problem as matching a known problem, and reapplying established techniques. Convergent thinking is based on familiarity with what is already known. It is most effective when the problem is structured such that a solution can be generated based on stored knowledge and information through a rational process, logical search, mathematical algorithm, or decision-making strategy.

Divergent thinking, by contrast, involves producing multiple and alternative solutions to the problem. It requires making unexpected combinations, recognizing relationships among objects even if only remotely linked, and transforming information into unanticipated forms. Divergent thinking goes from a problem to multiple potential solutions. Different people will generate different solutions, none of which can be verified as either correct or wrong, only on the continuum from good to poor. Synonyms for divergent thinking are out-of-the-box thinking or creative thinking.

Psychologists view divergent thinking as a major component of creativity. Divergent thinking leads to fluency, defined as the ability to rapidly produce a large number or ideas to a problem; flexibility, defined as the capacity to consider a variety of approaches to a problem simultaneously; originality defined as the tendency to produce ideas different from those of most people, and elaboration, defined as the ability to think through the details of an idea and carry it off.

Designers use cycles of divergent thinking to expand and explore the design space and then convergent thinking to develop a deep understanding of specific aspects of the design. The specific aspects of systems thinking discussed in Chapter 2 – recognizing the systems context, reasoning about uncertainty, making estimates, and performing experiments – might be characterized as desirable habits of mind that also reflect the notion of convergent-divergent thinking discussed here.

Many people fall into thinking ruts that unconsciously place constraints on their thinking that are difficult to break out of. Frank Herbert poses this exercise to discuss thinking.

Divergent Thinking Example

How are the following numbers ordered?

8 5 4 9 1 7 6 3

In school, you probably encountered problems like this. Because they are numbers, most people, especially engineers, will try to find some underlying mathematical relationship. However, the relationship is not mathematical – the numbers are ordered alphabetically. Before dismissing this problem as a mind game, you should know that many recent approaches devised to promote divergent thinking involve similar problems (e.g., see [23]). The idea is to encourage people to look at a problem differently and reach a different solution that may not normally even be considered. Divergent thinking is aided by brainstorming and other techniques – all of which can be learned.

One form of divergent thinking embedded in most design methodologies is to generate multiple alternatives. Unfortunately, in practice this step is frequently glossed over, or the alternatives are all variations of the same design concept (e.g., all different types of enterprise resource planning (ERP) systems). One way to improve the design process is to seek out truly alternative solutions to the design problem.

Convergent versus Divergent Thinking

Divergent Thinking – Given a problem think of all the ways the problem can be solved.

Convergent Thinking – Given a problem determine a solution and develop a detailed solution plan.

The ideas about design thinking, convergent thinking, and divergent thinking inform our enterprise design methodologies. Issues that need to be considered throughout the design process are:

1. In the early phases, the design team is open to understanding the problem in many different ways.
2. The design team considers what are the side effects of their design decisions.
3. The design team avoids phrasing design goals as solutions. For example, never say, "The goal of the system is to have a Web interface to improve customer service." This is a solution; the goal is better stated as, "the goal is to improve customer service." A Web interface is but one possible solution; there might be other better solutions.
4. The design team actively generates as many alternative designs as possible. This helps them better understand the problem and fully explore the design solution space.

4.1.4 Design for Change

The designers of enterprise systems must contend with the inevitable change the enterprise will experience during its life cycle. The ability of systems to cope with change is mentioned in various forms throughout the academic literature, industry trade publications, and surveys. The nature and extent of what changes will be experienced cannot be foreseen; nevertheless, we know some of the types of changes that can occur. Changes originate in the enterprise's environment and include market changes, economic changes, political changes, social changes, or technological changes.

To design for change, one option is called robust design. The idea is to design the enterprise so that it is robust, or insensitive, to changes in its environment. To do this,

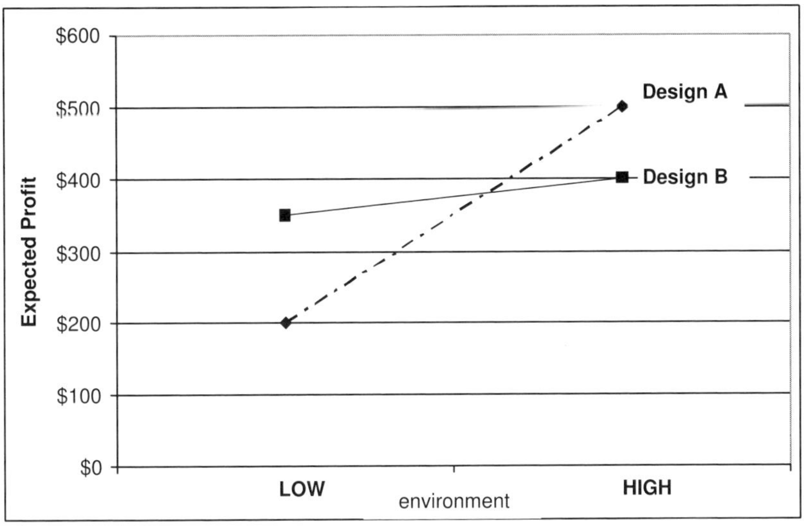

FIGURE 4.4
Robust design example.

the design team analyzes the uncertainty in the environment, quantifies the probabilities of these environments, and then analyzes how different designs will perform in each of the environments. The team can then select the more robust design alternative, defined as the one that works well over a range of uncertain future environments.

To illustrate robust design, consider two design alternatives A and B. The environment has two possible outcomes, low and high, and let's assume each outcome is equally likely. We have some prediction of how each system will perform for each environmental outcome as shown in Figure 4.4. If the goal is to maximize expected profits, and the vertical axis represents expected profits, then Design A will have higher profits in the high environment than Design B. But if the environment happens to be low, then Design A generates less expected profits than Design B. Robust design says we do not know what the environmental outcome will be, so we should select the alternative that will perform best over all the possible environmental outcomes. In this contrived example, Design B performs best with little variation over the possible environmental outcomes – it is the more robust design.

Robust design requires us to identify the possible environments a priori, and then estimate how different enterprise designs would perform in each of those environments. This might be an obstacle in highly dynamic environments where we cannot even identify possible future market, political, or social environments. An alternate approach to handle environmental uncertainty is to design the enterprise so that it can quickly adapt to unexpected environmental changes. This is called agility. Characteristics that make an enterprise agile are:

- Close to the market (or environment)
- Flexible, quick, and efficient information flows
- Decentralized decision-making
- High employee autonomy
- Willingness to innovate

Being close to the market is necessary for an enterprise to know what changes are occurring in its environment and trying to forecast what the environment will be like in the future. Knowing that the environment is changing is necessary prior to reacting to any change, yet alone being proactive about change. Having flexible and efficient information flows ensures this information gets to the decision makers in the enterprise. A decentralized organization structure and high autonomy for employees throughout the organization can help enterprises adapt faster to environmental changes. A willingness to innovate can be with the enterprise's product or its organization structure. There may be times when the environment has changed such that the current organizational structure no longer serves well, then the enterprise must be willing to change the way it is organized. Also, agile enterprises must be willing to develop new capabilities and/or products in anticipation of future changes.

Related to agility are other enterprise characteristics such as flexibility and scalability [13]. Flexibility describes the enterprise's ability to respond to pre-planned, anticipated changes. For example, if the enterprise is planning to enter foreign markets, then design the systems to work in any one of the foreign markets is a matter of enterprise flexibility.

Scalability is the ease with which the enterprise's systems can grow in size over time. Scalability is very important to information systems. For an e-commerce company, a scalable system would be a design that allows it to smoothly grow its systems to handle more and more customers as the business grows. If the company has to redesign its information systems and system architecture in order to grow, then it is not scalable.

Given that our political, economic, social, and technical environment is very dynamic, all enterprises must be designed to accommodate change in their environment. The different terms: agility, flexibility, and scalability describe different types of change and how the enterprise reacts to them.

4.2 Design Methodologies

In this section, we review design methodologies. Prior to describing the methodologies, we first define terminology. A *methodology* is a standardized development process governed by a set of principles and common philosophy that defines a set of activities, methods, best practices, deliverables, and automated tools. System developers and project managers use methodologies to develop and continuously improve an enterprise system. Methodologies divide the project into *phases* that are defined by specific activities carried out in the phase and the deliverables generated in the phase. An *activity* describes a coherent work process that generates one or more deliverables. Some activities are wholly within a single phase and other activities cross over two or more phases (a cross life-cycle activity). Methodologies might dictate the use of specific techniques, where a *technique* is a set of precisely described procedures to accomplish a task (e.g., a technique is use-case modeling to document requirements). Each phase completion is usually marked by a milestone. *Milestones* are critical project events, such as completion of the data model or delivery of the requirements document.

The activities produce one or more outputs. These outputs are called deliverables. The *deliverables* are tangible or demonstrable products created during the project. Example deliverables are requirements specification, information models, and documentation such as the project initiation plan. Sometimes a deliverable coincides with a milestone. Deliverables and milestones help maintain the distinctness of the phases.

According to these definitions, we can say a methodology is divided into one or more phases, each phase has one or more activities, and each activity is supported by one or more techniques. The methodology describes a process for a team of engineers to work together in the design of an enterprise. It defines the roles of what people must do, what deliverables they produce, the activities, and how the activities are organized. The final output of a methodology is the enterprise system.

The project will be completed by a team of individuals including a project manager, systems analysts, systems designers, subject matter experts, and others. The enterprise engineering project can take anywhere from a few months to complete or several years depending on the scope of the project. Some of the individuals will be assigned to the project part-time or for only a particular phase or activity. Thus, people will be joining and leaving the project during its execution.

To manage such large projects (large in terms of project scope, project team size, and project duration) the team will employ tools. Common, even in very large projects, are tools such as templates, checklists, spreadsheets, and databases. Many of these tools are made available in common software products such as MS Word and Excel. For example, a project team might have a spreadsheet to document requirements in MS Excel. One or more of the activities might be supported by software packages designed especially for those activities. These tools are called *Computer-Aided Systems Engineering (CASE)* tools. CASE tools have become common in systems engineering as well as software engineering (where the "S" designates Software). CASE tools are available for activities such as requirements management, configuration control, and process modeling and design. The advantages of CASE tools are many. CASE tools can increase the productivity of the team. They help enforce project policies (e.g., consistent phrasing of requirements). Most case tools have a central repository that all project team members would have access to. This way all team members are working on the same documents and models. CASE tools will also enforce consistency between different enterprise views (e.g., that a process information output matches data attributes that can be found in the data model). Finally, some CASE tools, especially for software, can automatically generate some of the design. For example, a CASE tool might automatically generate the database schema from the data model.

Best Practices... ?

A best practice is a method or technique that experience demonstrates will consistently produce superior results. Many consultants, CASE tool vendors, and software vendors all claim to use best practices – and they all differ from one another! If we ignore unsupportable claims, and marketing hype, this is still possible because every design project is unique and the best practices must be intelligently adapted to the situation and applied to the problem at hand. Tiwana and Keil say, "It is not the chosen methodology *per se* that drives project risk, but how well it fits a given project."

The One-Minute Risk Assessment Tool, A. Tiwana and M. Keil
Communications of the ACM, Nov. 2004, (47)11, 73-77.

Why use formal methods to conduct an enterprise engineering project? Enterprise systems are too complex to design in an ad hoc manner. The project team is usually too large to manage through informal methods. Moreover, there is high uncertainty in early phases of the design problem. For these reasons, enterprises use formal methodologies to do enterprise engineering projects. Formal methods establish control points and plans that help track the progress and budget; they define and produce consistent documentation that reduces lifetime costs of the enterprise system; they promote quality through proven techniques; they provide direction on how to integrate many individual efforts and align them towards the project goal; and they provide a road map to communicate project progress to all the stakeholders. The methodologies help the design teams gradually remove the uncertainty about the problem as the project progresses toward a solution. Studies have shown that projects that follow a methodology perform better than projects that do not. Performance is defined in terms of delivering required functionality, being on schedule, and being on budget.

The following subsections review three common development models: the waterfall model, the spiral model, and the controlled iteration model.

4.2.1 Waterfall Design Model

The waterfall model describes a methodology in which the phases are conducted sequentially, using the metaphor of water falling down steps. The waterfall model (shown in Figure 4.5) defines the phases in terms of their activities; each phase has a single activity. In the waterfall model, the team completes phase 1, then starts phase 2, and so on until the project is completed. While this rigid interpretation was not present in the original description provided by Royce [29], nor in the extended model provided by Boehm [1], this is how most project managers interpret and implement the phases. One of the main arguments in favor of the waterfall model is that changes made early in the life cycle are less expensive than later in the life cycle. Or, stated in another way, errors found in early phases are less costly than if the error is not discovered until a later phase.

The waterfall model makes strong and restrictive assumptions on the nature of requirements. The waterfall model assumes:

1. Requirements are known in advance of implementation.
2. The requirements have no unresolved high-risk implications.
3. The nature of requirements will not change, or be changed minimally during development and system evolution.
4. The right architecture for implementing the requirements is well understood.
5. The requirements are compatible with all the key system stakeholder expectations.
6. There is sufficient time to conduct the project sequentially [3].

In most enterprise design projects, these assumptions are erroneous. If, as was argued in the previous section, most enterprise design problems are wicked problems, then it is impossible to know all the requirements in advance of any design or implementation work. Doing so is inconsistent with the problem-solving methodology. Moreover, if the environment changes rapidly the sequential approach does not work well.

The waterfall model has largely fallen out of favor in most domains. In software engineering, problems arise due to the poor communication inherent in the functional teams associated with waterfall methods, inability to specify all requirements prior to starting any code, and poor integration of the multiple components in a system [6]. In enterprise engineering, it is very difficult, if not impossible, to derive a complete requirements specification prior to any analysis or design. In the automotive industry, the waterfall model is associated with long lead-times to bring a new car to market. To bring cars to market faster, companies use more parallel development methods.

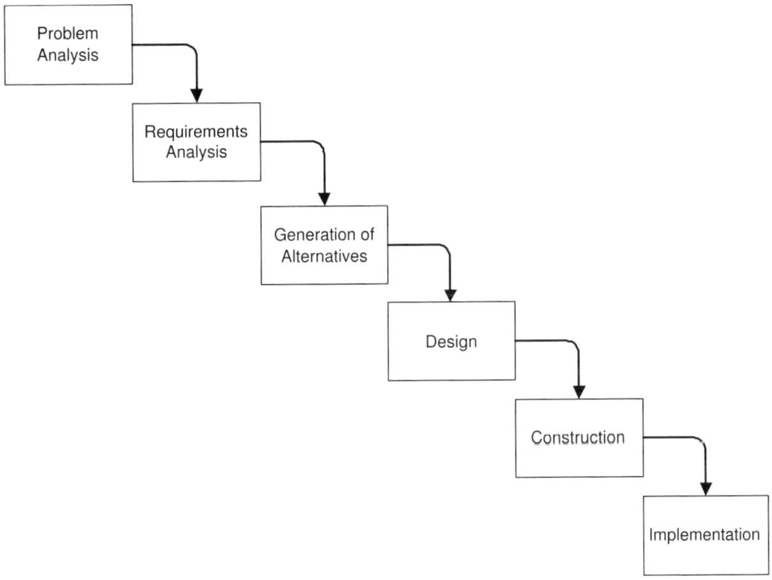

FIGURE 4.5
The waterfall methodology.

Waterfall models treat activities and phases as the same concept. A potential management problem with the waterfall method is that if you are 80% done, then this means you have nothing to deliver to the customer because everything is 80% completed. Later, we will contrast this with an iterative approach in which if the project team is 80% done, then at least it has something ready to deliver to the customer.

One of the main reasons the waterfall methodology performs poorly is due to the nature of dependencies in many projects. The waterfall method assumes that all project dependencies are sequential. However, in many projects the dependencies are reciprocal, such that a decision made in one activity depends on a decision made in another activity, which depends on the decision made in the first activity. Reciprocal dependencies in which activity A depends on activity B, which depends on activity A are poorly served by the sequential, waterfall model.

The waterfall model continues to be used, even if it is not the most appropriate model for the project at hand. This is likely because of the seeming logic of the sequential process and the ease of managing the waterfall methodology. As the first formal method popularized, it is the standard by which all others are compared, consequently the waterfall model is much maligned in the literature.

4.2.2 Spiral Design Model

The spiral model, shown in Figure 4.6, is an iterative system development model. It is used in both software development and systems engineering. In the spiral model, development proceeds through all the phases in each iteration [2]. At the end of each iteration, a prototype is delivered, which reduces risk because the stakeholders can evaluate the prototypes and determine whether the project is progressing to the desired final product. To illustrate the spiral, we describe a variation of the spiral model called Rapid Application Development (RAD). RAD is used to develop software and involves significant prototyping. A RAD method to develop an e-commerce system would identify a system function, analyze it,

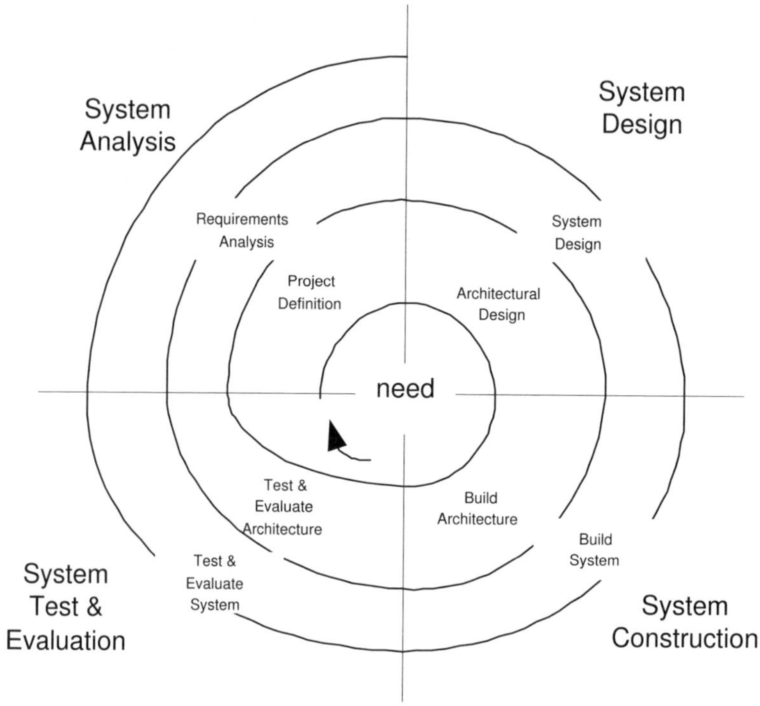

FIGURE 4.6
The spiral model.

design it, build it, test it, and then deliver it to the customer. Then a new iteration begins, going through all the phases, to deliver a second system function.

It is difficult to envision how the spiral approach could be applied to enterprise systems because prototyping of enterprises is not as straightforward as it is for software. Also, it is difficult to manage the many interactions that may be required [6]. What the spiral methodology teaches us is the idea of multiple iterations of the activities as a means to handle risk.

4.2.3 Controlled Iteration Design Model

The controlled iteration model takes the spiral model but adds more structure to the iterations. It also makes a distinction, missing from the waterfall and other models, between phases in the development of the project and activities that span multiple phases.

The controlled iteration methodology is best exemplified by the Rational Unified Process (RUP) [6] intended for software development. The RUP is an iterative and incremental methodology that uses object-oriented modeling to support the activities. RUP distinguishes between phases and workflows [18]. Phases divide a project over time, and activities describe work that occurs, to varying levels of effort, through multiple phases. Figure 4.7 represents the two dimensions and illustrates the idea that activities overlap phases and have different intensities (the vertical heights) as the project progresses. For example, requirements engineering can happen continuously throughout all phases, although more intensively during inception. Similarly, other activities including analysis, design, implementation, and test also span multiple phases. The support of the problem-solving phases is evident in the def-

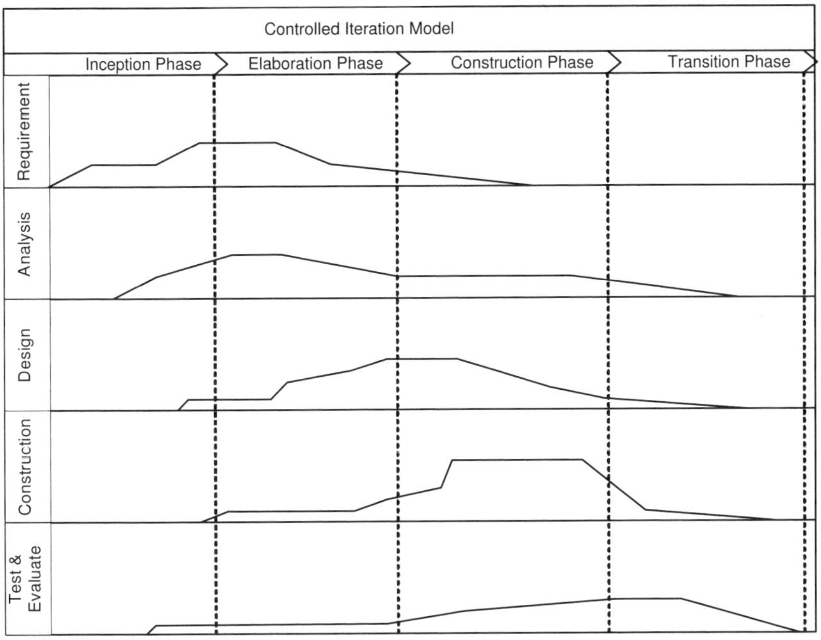

FIGURE 4.7
Rational unified process.

inition of workflows in the RUP (requirements and analysis map to scoping, design and implementation map to design, and test maps to evaluation).

The four phases of RUP are:

1. Inception – The development team works to understand what to design and build. The project team creates a project vision, develops a business case, and specifies high-level requirements.
2. Elaboration – The project team works to understand how to build the system. The team designs a baseline architecture and further elaborates on requirements.
3. Construction – The project team builds the product and conducts tests.
4. Transition – The project team validates the product and obtains stakeholder acceptance.

In RUP, a project is divided into a number of iterations and for each iteration the project goes through all four phases. In this way it is similar to the spiral methodology. RUP uses the unified modeling language (UML) and the CASE Tool Rational Rose.

The RUP methodology is frequently used in the software industry; about 10,000 companies are applying it [17]. The RUP methodology is strongly tied to using object-oriented modeling; the methodology might not work as well with other modeling techniques.

The controlled iteration methodology is premised on a few fundamental concepts. First, the methodology assumes that in most projects the requirements will change as the project progresses. Changing requirements are seen as a major cause of late projects, projects going over budget, and dissatisfied customers. Second, the methodology says that integration cannot be left to the end of the project. To achieve integration of all the system components, the team needs to continuously work on integration. Having multiple iterations throughout development accomplishes this. Third, in an iterative approach it is easier to identify and mitigate against risks earlier in the development process. Fourth, the definition of phases, major milestones, and activities makes management easier than a pure spiral model. In summary, the iterations introduce feedback into the development process earlier than the waterfall methodology and throughout the development. The feedback can be used by the development team to design and develop a better system.

4.2.4 Methods in Practice

There are many other methodologies not reviewed here. A problem faced by an enterprise engineering team is which method to adopt? Once a method is adopted, does the team use it "out-of-the-box" or does it tailor it to their needs? Evidence suggests that in many cases methodologies are not strictly followed, and in some cases, methodologies hardly reflect how the design actually progresses.

Truex et al. [33] critique the general assumption that systems development is a rational, goal-driven and managed process that is facilitated by a methodology. They coin the term amethodological development that views system development as a unique, individual, negotiated and opportunistic process, driven by accident as much as by plan. Various case studies support this view. Nandhakumar and Avison [21] discuss a case study in which the methodology is described as a useful fiction to suggest a controlled process when in fact it was not closely followed. Field studies find there is a wide difference between the formalized phases and activities prescribed by a methodology and what is actually done during the development project [11].

What we learn from this line of research is that in practice teams view methodologies pragmatically, and they borrow and use parts of a methodology that they deem appropriate for their project. Given that the methodologies tend to be generic so that many different enterprises can use them, it seems appropriate that they be tailored in practice. It is difficult to have a universal methodology that any enterprise could use unadulterated. Moreover, since the enterprise environment changes with time the methodology should adapt to the changes as needed. Brinkkemper [4] calls this methods engineering – the customization of a methodology according to the situational environment. The issue facing a project team is how to tailor the methodology? Glass [14] presents this as an outstanding issue facing researchers and the users of methodologies. He phrases it as the need for advice on how to assemble a project-specific methodology from method ideas borrowed and adopted from many reference methodologies.

4.3 Methodology Principles

There are principles, sometimes called best practices, that have emerged from years of experience over many projects. Many of these principles are incorporated into the design methodologies outright; either explicitly in the techniques and tools employed, or as part of a guiding philosophy. In this section, we discuss some of the methodology principles that should be incorporated into every enterprise engineering project.

1. Have stakeholders participate in analysis and design – Getting all the stakeholders involved in all phases of the project improves the quality of the final system and greatly mitigates user resistance to the new system. It is especially important to have the users participate in the system design. The quality of the final system will be better because the users have better knowledge of the business processes

than the implementation team. Their knowledge will help in identifying problems and generating requirements. User participation also speeds up the problem analysis and requirements analysis. In addition to helping in the analysis and design of the system, user participation can greatly reduce problems often associated with system deployment. The reason is the users will start to understand the magnitude of the changes involved much earlier. They will have more time to prepare, consider, and finally accept the change. From the user's perspective, the change will be a smoother transition than the possibly abrupt change if the users were excluded from involvement with the project. In a nutshell, user involvement leads to user "buy-in." The users have a stake in the new system and will more likely want it to succeed. The users will erect fewer obstacles to the system – making deployment that much easier.

2. Use an iterative, problem-solving approach – The problem-solving approach described is iterative, and the design methodology should emulate the problem-solving iterations. Iteration benefits the design methodology in two ways: First, iteration enables constant monitoring of progress since stakeholders will see actual deliverables throughout the design process. Second, iteration also helps reduce risk by identifying problems early. Engineering design is conducted with imperfect models, incomplete information, and often with ambiguous objectives as well (due to being a wicked problem). Errors are unavoidable because design is a matter of making many assumptions in often problematic situations. Waiting to gather more and more information to reduce the risk of less than perfect decision-making often does not work because the project must move on. In industry, experts refer to this as "analysis paralysis." In system design, many decisions must be made with less than perfect knowledge. Through iteration, any errors can be identified early in the project when they are less costly to deal with.

3. Use an enterprise architecture – An enterprise architecture provides a holistic, unifying view of the enterprise. It describes the long-term, stable portions of the enterprise and how they should be designed. All enterprise engineering projects should be consistent with the enterprise architecture. The enterprise architecture can help project communications, ensure the project deliverables conform with other parts of the enterprise, and ensure achievement of the enterprise mission and vision.

4. Provide for good project communication – The entire project team can be quite large, making it difficult for individual team members to appreciate how their activity contributes to the entire project and the impact their decisions and actions have on other aspects of the project. A major role of a project manager is the coordination of all the project activities to ensure progress towards the overall project goal. Clear communication of project status, impending milestones, potential risks, and other information will help the project. One of the primary reasons communications improves the project performance is team members are more likely to coordinate their activities by mutual adjustment (informal communication) rather than more expensive forms of coordination (formal meetings, design reviews, etc.) [20].

5. Make extensive use of modeling – Creating and using models provides many benefits to design methodologies. Models help team members and stakeholders envision and think through abstract concepts. Models form an important part of the project's documentation. Models are necessary for analysis (e.g., queueing models to analyze process performance) as well as design (e.g., to specify the

information content of the system). The project team should make models part of the deliverables and track the progress of model development.

6. Document throughout project – During the course of an enterprise engineering project, many decisions will be made by different team members. In later stages of the project, or sometimes after the project is completed, problems can emerge where somebody wants to understand why a system was designed in a certain way. Without documentation it is often impossible to recreate the design intent that went into the system. This is but one reason why documentation is important. Another reason is that the project team is large, team members come and go from the project. To promote good communication and consistency of purpose without excessive rework, then the team needs good documentation. Finally, documentation also forces formalization of the design decisions made, and documentation opens up those decisions to critique from other team members. This is, in general, beneficial to the team. For these reasons, it is important to document all major decisions throughout the project as those decisions are made. A good methodology defines hard deliverables that include documentation.

7. Build prototypes when possible – Prototyping is when the design team builds a mock-up, scaled-down version, or other partially functional system. The prototype can help the team explore the problem space. Everyone involved gets to try out their ideas and validate their understanding of the problem. Prototyping also provides an ideal mechanism for stakeholder discussion and feedback because it provides a physical manifestation of the system that stakeholders can relate to more easily than abstract ideas presented in documentation and models.

8. Establish standards – The project team can be large and the project duration can be many years. To manage large projects the team should establish standards for: training, communication, documentation, modeling, terminology, and anything else that is done in multiple activities. Standards reduce errors and project duration because all team members developed shared expectations for documentation and models.

9. Continuously verify quality – Testing that the design concepts should be done throughout the design process. All intermediary and final design deliverables should be tested, and they should be tested at different times throughout the design process. Early tests of concepts can be done with computer models such as simulation, the design team can visit other enterprises that have similar designs to the one proposed, requirements can be tested, etc. Early testing avoids wasted effort and surprises late in the design cycle. It is also an explicit phase of the problem-solving methodology.

10. Design system for growth and change – Enterprises as open systems constantly interact with their environment. As a result, the enterprise will change: change can be in growth, priorities, markets, organizational structure, and so forth. Accordingly, the enterprise design team should anticipate the need for the enterprise to change over time and design flexibility into the system. Maier and Rechtin [19] present this idea as an architecture heuristic to build the system "with good bones." Meaning that if the system is designed with a good structure, it can be changed in the future to accommodate changing requirements.

4.4 Capability Maturity Model Integration

The Capability Maturity Model Integration (CMMI) is directed at improving the quality of the system by improving the quality of the methodology producing the system. The premise is that formally managed methods will lead to projects that develop quality projects, on time, and within budget. The CMMI is published by the Software Engineering Institute (SEI) in Pittsburgh, PA. The CMM (the original version of the CMMI) was commissioned by the U.S. Department of Defense to help them qualify software vendors' capabilities. From there, it quickly evolved into a powerful tool to guide process improvement initiatives, not only for Software Development but for many related fields such as Systems Engineering, Product Acquisition, Team Management, Research and Development, etc. CMMI was developed to consolidate the various CMM models into a single model applicable to any development project. CMMI integrates three previous CMM models from the separate domains of software engineering (SW-CMM), systems engineering (SECM), and integrated product design and development (IPD-CMM).

CMMI provides a choice of either a continuous representation or a staged representation. The staged representation is similiar to the SW-CMM and is commonly used. It defines stages of progress such that each stage lays the foundation for improvement to the next stage. The continuous representation treats each process separately so that a company could choose to improve different processes at different rates. Here we discuss the staged model because of its established history of use. The staged model defines five maturity levels (shown in Figure 4.8) that describe an evolutionary path recommended for an organization to improve the processes it uses to develop and maintain its products and services.

Maturity Level 1: Initial

At maturity level 1, the project team does not have a stable environment, the project follows no consistent process, and the project is characterized by chaos. A project team at maturity level 1 may develop systems that work, but it does this over budget or behind schedule – thus it is technically a project failure. Success is usually predicated on the ability of the team members and not on having well-defined processes. Maturity level 1 project teams are more likely to over-commit, abandon processes in a time of crisis, and are unable to repeat their successes.

Maturity Level 2: Managed

At maturity level 2, the project manager tracks the processes against the project plan with respect to making major milestones, staying on schedule, and on budget. At this level, the project team has established the foundation of project management practices. The project team defines team roles and fills them with qualified people. The team members are provided with adequate resources to produce the defined activity outputs. Figure 4.8 shows the processes the project team masters at maturity level 2. The project team works with the project stakeholders to ensure the project is working towards the goals as planned. The work products and services satisfy their specified process descriptions, standards, and procedures.

Maturity Level 3: Defined

At maturity level 3, the project team uses a defined methodology that defines project activities and describes them in standards, procedures, tools, and methods. Figure 4.8 shows

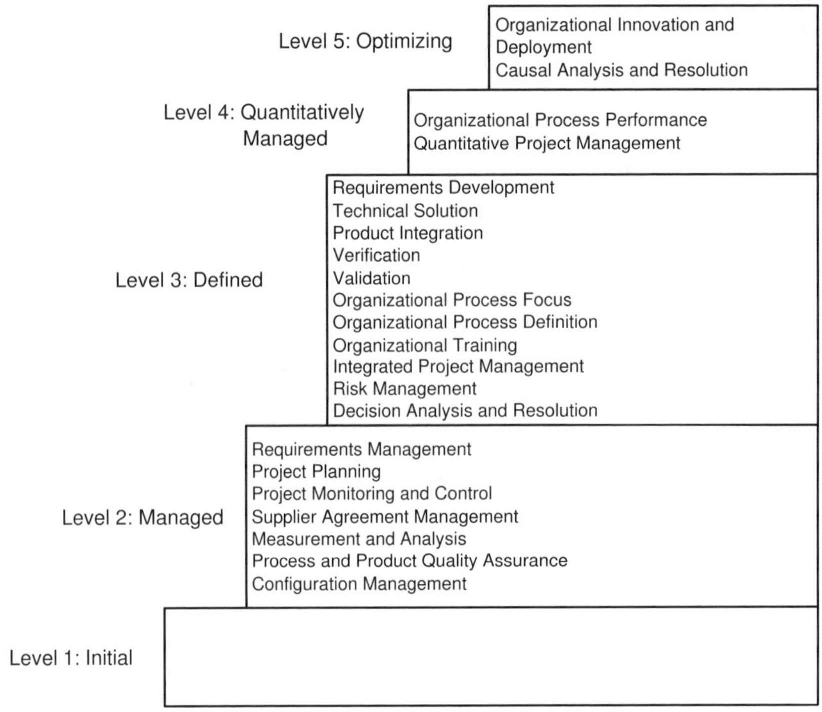

FIGURE 4.8
CMMI levels.

the processes the project team masters at level 3, in addition to those from level 2. The project's processes are stable, predictable, and consistent.

Maturity Level 4: Quantitatively Managed

At maturity level 4, the project team establishes quantitative objectives to measure the project's quality and performance. The performance measures are used to manage the project's processes. Quality and process performance is understood in statistical terms and is managed throughout the life of the processes. The project team establishes statistical quality control by defining process measures, statistically analyzing them, and then using the results to guide management intervention to improve the process. The main difference with maturity level 3 is the use of process measurement to improve the processes.

Maturity Level 5: Optimizing

At maturity level 5, the project team continuously monitors the project processes and improves them according to the statistical data. The project team understands how to differentiate common cause variation from special cause variation. The emphasis of maturity level 5 is the continuous improvement of the project's processes through both incremental and radical improvements.

To be a particular maturity level, all the process areas for that maturity level must be that level or higher. A process area is a cluster of related practices in an area that, when implemented collectively, satisfy a set of goals considered important for making improvement in that area. CMMI defines 22 process areas that are assigned to maturity levels as shown in Figure 4.8. To achieve maturity level 2, all process areas assigned to maturity level

SG1 Develop Customer Requirements

SP1.1 Elicit Needs

SP1.2 Develop the Customer Requirements

SG2 Develop Product Requirements

SP2.1 Establish Product and Product Component Requirements

SP2.2 Allocate Product Component Requirements

SP2.3 Identify Interface Requirements

SG3 Analyze and Validate Requirements

SP3.1 Establish Operational Concepts and Scenarios

SP3.2 Establish a Definition of Required Functionality

SP3.3 Analyze Requirements

SP3.4 Analyze Requirements to Achieve Balance

SP3.5 Validate Requirements

FIGURE 4.9
Goals and specific practices for requirements development.

2 must achieve capability level 2 or higher. For example, shown in Figure 4.9 is the requirements determination process with its subprocesses required for maturity level 2. The CMMI specification goes into greater detail to describe exactly what each goal and sub-practice entails.

The idea of the CMMI is that as organizations improve their processes, they are more likely to succeed in project objectives. CMMI is methodology neutral in that a waterfall model or a spiral model could equally be at the same maturity level. The CMMI specification is quite long, and achieving even maturity level 2 involves significant investment by an organization. CMMI by trying to be applicable to all projects (software, systems engineering, integrated product design and development) might have process areas and sub-practices that are not relevant to enterprise engineering.

The CMMI model is based on the premise that improving process maturity will result in better project performance and product quality. Evidence has been accumulated, mostly through case studies of companies that implemented CMMI, and the results show significant improvements in product quality, projects meeting schedule, and projects meeting budgets [16]. In general, data show that companies achieving CMMI levels 2 and 3 improve in most measures of performance and receive positive returns on their investments [12]. These results strongly support the CMMI model and the use of formal design methods.

4.5 Summary

This chapter described the nature of engineering design and how it applies to the design of enterprise systems. The goal of any design project is to find the best solution to the design problem. In terms of enterprise systems, we argued the design problem is a wicked problem for which there is no "best" solution but only satisfactory solutions. To do enterprise design, the design team uses a methodology. A methodology provides a structured approach for a team to design enterprise systems. Not only does it help the team solve the design problem, but is also helps the team define the problem (i.e., the design goals). Many experts cite cases where the design team was solving the wrong problem – a methodology helps a design team avoid this pitfall.

There are many methodologies, techniques, and tools. The best methodology to use depends on the context of the enterprise project. Regardless of the methodology, there are best practices that all methodologies should incorporate into their process. An important tool for evaluating how good an organization manages design projects is the CMMI model. The CMMI model assesses the maturity of the organization's project management processes.

Review Questions

1. Compare and contrast phase, techniques, and activities.
2. List the three components of an optimization problem.
3. Argue whether it is possible to have an optimal enterprise design.
4. Explain why it is difficult to define an objective function for an enterprise design problem.
5. Select an enterprise system you are familiar with and describe in words what would be an objective function, decision variables, and constraints for the design of that enterprise system.
6. Explain the difference between an optimizing and a satisficing solution.
7. Compare and contrast an ill-defined problem from a wicked problem.
8. Describe how the four problem solving steps can be applied to design a new inventory control system for a manufacturer.
9. Explain why the problem-solving technique is iterative. Why would a single iteration usually be insufficient?
10. Contrast convergent thinking and divergent thinking.
11. Explain why the waterfall methodology has largely fallen out of favor with most companies?
12. Compare a traditional waterfall design methodology to an iterative methodology. What are the advantages/disadvantages of one approach versus the other?
13. What is the purpose of the CMMI model and what does it tell us about a company's methodology?
14. Describe a cross life-cycle activity that must be done throughout the enterprise project.
15. Select a phase of the enterprise development methodology and explain its main inputs and outputs.
16. Judge how important it is to get stakeholders involved in a development project.
17. List and describe one reason why is it important to get all the stakeholders involved in an enterprise engineering project.

18. What project or system characteristics would you need in order for the waterfall method to be successful?
19. Explain how project risk is expected to change as a company's CMMI level increases.

Bibliography

[1] B. Boehm. *Software Engineering Economics.* Prentice Hall, Englewood Cliffs, NJ, 1981.

[2] B. Boehm. A spiral model of software development and enhancement. *IEEE Computer*, May:61–67, 1988.

[3] B. Boehm and D. Port. Escaping the software tar pit: Model clashes and how to avoid them. *Software Engineering Notes*, 24(1):36–49, 1999.

[4] S. Brinkkemper. Method engineering: Engineering of information systems development methods and tools. *Information and Software Technology*, 38(4):275–280, 1996.

[5] R. Buchanan. Wicked problems in design thinking. *Design Issues*, 8(2):5–21, 1992.

[6] M.R. Cantor. *Object-oriented project management with UML.* John Wiley & Sons, New York, NY, 1998.

[7] C.W. Churchman. Wicked problems. *Management Science*, 4(14):B141–B142, 1967.

[8] J. Conklin. *Dialogue Mapping: Building Shared Understanding of Wicked Problems.* John Wiley & Sons, Upper Saddle River, NJ, 2006.

[9] D. Dörner. *Problemlösen als Informationsverarbeitung.* 3rd edition, Kohlhammer, Stuttgart, Germany, 1987.

[10] C.L. Dym, A.M. Agogino, O. Eris, D.D. Frey, and L.J. Leifer. Engineering design thinking, teaching, and learning. *Journal of Engineering Education*, January:103–121, 2005.

[11] B. Fitzgerald. The use of systems development methodologies in practice: a field study. *The Information Systems Journal*, 7:201–212, 1997.

[12] D. Galen and M. Avrahami. Are CMM program investments beneficial? analyzing past studies. *IEEE Software*, 23(6):81–87, 2006.

[13] Martinez L.D. Sáenz O.A. Giachetti, R.E. and C. Chen. Analysis of the structural measures of flexibility and agility using a measurement theoretical framework. *International Journal of Production Economics*, 86(1):47–62, 2003.

[14] R. Glass. Matching methodology to problem domain. *Communications of the ACM*, 47(5):19–21, 2004.

[15] G.A. Hazelrigg. A framework for decision-based engineering design. *Journal of Mechanical Design*, 120(4):653–659, 1999.

[16] H. Krasner. Accumulating the body of evidence for the payoff of software process improvement. *IEEE Software Process Improvement*, pages 519–539, 2001.

[17] P. Kroll and P. Kruchten. *The Rational Unified Process Made Easy: A Practitioner's Guide to the RUP.* Addison-Wesley Professional, New York, NY, 2003.

[18] P. Kruchten. *The Rational Unified Process An Introduction.* Addison-Wesley, New York, NY, 2000.

[19] M. Maier and E. Rechtin. *The Art of Systems Architecting.* CRC Press, Boca Raton, FL, 2002.

[20] H. Mintzberg. *Structure in Fives: Designing Effective Organizations.* Prentice Hall, Englewood Cliffs, NJ, 1992.

[21] J. Nandhakumar and D.E. Avison. The fiction of methodological development: a field study of information systems development. *Information Technology & People*, 12(2):176–191, 1999.

[22] A. Newell and H.A. Simon. Computer simulation of human thinking: A theory of problem solving expressed as a computer program permits simulation of thinking processes. *Science*, 134(3495):2011–2017, 1961.

[23] S.B. Niku. *Creative Design of Products and Systems.* John Wiley & Sons, Hoboken, NJ, 2009.

[24] G. Pahl and W. Beitz. *Engineering Design: A Systematic Approach.* Springer, London, England, 1996.

[25] I.C. Parmee. *Evolutionary and Adaptive Computing in Engineering Design.* Springer, New York, NY, 2001.

[26] J.E. Pretz, A.J. Naples, and R.J. Sternberg. *The Psychology of Problem Solving*, chapter Recognizing, defining, and representing problems, pages 3–30. Cambridge University Press, Cambridge, UK, 2003.

[27] H.W.J. Rittel and M.M. Webber. Dilemmas in a general theory of planning. *Policy Sciences*, 4:155–169, 1973.

[28] S.I. Robertson. *Problem Solving.* Psychology Press, Philadelphia, PA, 2001.

[29] W.W. Royce. Managing the development of large software systems. In *Proceedings of the IEEE WESCON*, pages 1–9, 1970.

[30] A.P. Sage and J.E. Armstrong. *Introduction to Systems Engineering.* John Wiley & Sons, New York, NY, 2000.

[31] H. Simon. *The Sciences of the Artificial.* 3rd ed., MIT Press, Cambridge, MA, 1996.

[32] J.D. Sterman. *Business Dynamics: Systems Thinking and Modeling for a Complex World.* Irwin McGraw-Hill, New York, NY, 2000.

[33] D.P. Truex, R. Baskerville, and J. Travis. Amethodical systems development: the deferred meaning of systems development methods. *Accounting Management & Information Technologies*, 10(1):53–79, 2000.

5

Enterprise Architecture

"Architecture aims at eternity." – Christopher Wren (1632-1723), English architect of Saint Paul's Cathedral.

This chapter presents the concept of enterprise architecture to support the analysis and design of enterprise systems. An enterprise architecture is important to deal with the complexity inherent in enterprise systems and to align the many subsystems so that they all work harmoniously towards the enterprise's goals. Enterprise architecture provides the foundation on which enterprise integration is accomplished. Among the first decisions that are made in an enterprise engineering project is what enterprise architecture to adopt.

After completing this chapter, you should be able to:

- Describe the reasons industry is increasingly turning to enterprise architectures.
- State the purpose of an enterprise architecture.
- Explain the relationship between a reference architecture and an enterprise architecture.
- Describe the components of an enterprise architecture.
- Describe the differences between reference enterprise architectures.
- Describe the views and perspectives of Zachman's framework.
- Describe the components of TOGAF's architecture.

5.1 Introduction

It is a rare occasion that a project involves the wholesale design of an entire enterprise. Rather, projects are done to design a new system, to reengineer a business process, to institute a continuous improvement program, or to complete some other project that alters the current business operations, hopefully for the better. It is largely through the multitude of projects conducted over a period of time that we contribute to the evolving design of a business. After all, few of us are ever in the position where we can design an entire business from the bottom up. These projects shape and change parts of the business contributing to the constant evolution of the business's overall design. Would it not be useful if all these decisions that affect the design of an enterprise adhered to some consistent vision of how the enterprise should operate? It is precisely this role that enterprise architecture is intended to fulfill.

An enterprise architecture is a high-level design of the entire business. It describes the structure of the business processes, how they are coordinated with each other, and how technology supports them. It helps us understand the business's complexity by showing how all the different systems are linked together. In this definition, we say a design of the entire business, yet in practice much of the discussion of enterprise architecture takes place within the software engineering community. However, in truth, enterprise architecture should be, and is intended to be for the entire business. The software component is only one part of the enterprise architecture. It is for this reason that industrial engineers, among other professionals, should be aware of and even participate in the development of enterprise architecture. Without widespread participation, it is possible that the resulting enterprise architecture will not address important business needs, or individual projects may even run counter to other projects being done in other parts of the business.

An enterprise architecture provides a high-level design of the entire enterprise that will guide all other enterprise projects. An architecture represents significant, broad design-decisions for the enterprise, from which all other design decisions should be consistent. Architecturally significant decisions are those that the architect (or architecture team) needs to make in order to address the concerns of strategy, structure, and enterprise integration. Architectural decisions include deciding how to decompose the enterprise into views, how to decompose each view into different abstraction levels, policies on technology usage, decisions on business culture to guide organizational design, and decisions on what modeling conventions to use. These are but a few examples of what would be included in the enterprise architecture.

An *Enterprise Architecture* describes the structure of an enterprise, its decomposition into subsystems, the relationships between the subsystems, the relationships with the external environment, the terminology to use, and the guiding principles for the design and evolution of an enterprise.

When discussing enterprise architecture, it is reasonable to draw parallels with the traditional architecture of buildings. An architect creates blueprints for the building that guide its construction. The power of the blueprint is it aligns all the work on the building so that as long as the carpenters, roofers, electricians, plumbers, and other workers follow the blueprint, then they are assured their part will work together with the other parts and lead to the envisioned building. Without the blueprint, there is no insurance the final product will match the vision, nor will it necessarily be done in the most cost-effective manner. Jeanne Ross, Peter Weill, and David Robertson in their book, *Enterprise Architecture as Strategy* [13] describe an amusing example of what can happen when a house is built without architecture as is the Winchester mansion in California. Sarah Winchester, the heiress of the rifle manufacturer, had a mansion under constant construction from 1884 until 1922. There were no plans for the house; construction was based on her sketches done on paper and even her tablecloth. Some of the resulting oddities are stairs that lead into the ceiling, a door that leads to nowhere, a blind chimney that stops short of the ceiling, and dozens of other oddities. Moreover, because there was no master plan, old construction would be ripped up to make way for new construction, walls would be removed and then added, windows relocated, and so forth – greatly adding to the overall cost.

Many businesses evolve without any master plan much like Ms. Winchester's house. To a large extent, in the past, the enterprise was not viewed as a whole system that could be rationally designed. More likely, parts or subsystems of the enterprise were designed in isolation without a holistic perspective of how the parts would work in the entire system. Enterprises would come into being and change not as a result of a conscious, purposeful design effort, but due to ad hoc, sometimes short-term decisions made individually by many different people. Michael Porter in his book, *Competitive Advantage* [12], cites a study by Ernst and Young that found in many companies the key business processes were designed long before there was information technology (IT). Even though these companies had new IT systems, they continued to follow the paper-based business processes but with IT. What this study and other observations indicate is that for many enterprises their current structure does not reflect the results of a rationale design, but the result of a multitude of small changes made over time without consideration of the overall enterprise. The result is a hobble of systems that often do not work together very well, poor investment choices, under-utilization of resources, and diminished operation performance.

To illustrate the result of not having an architecture, consider the example taken from a cruise line operating out of South Florida. The company, like many others, grew rapidly over the years including an acquisition of a smaller cruise line. Figure 5.1 shows the resulting system architecture, or lack thereof, that included many separately developed and maintained systems. The overall system has much redundant data, business logic was tightly coupled to the data layer, and the execution of business rules was inconsistent across the systems. Problems experienced include sometimes erroneous data, errors in reporting, expensive maintenance costs, and difficulty in modifying or improving business processes. It is the primary purpose of enterprise architectures to avoid these problems.

Figure 5.1 shows just what can happen when a company builds its systems without an enterprise architecture. The company from the hospitality industry grew each year and added to its systems in a haphazard way. The systems were separately developed and maintained by different organizational units. The result was thousands of separate data files throughout the company. Examining Figure 5.1, when an application needs to access data it is programmed to access data from multiple different files. So, for example, if a reservation is updated with a change in the cabin number, then the new cabin number must be changed in multiple locations. As a result, the company had a large amount of data redundancy and duplicated business rules across multiple systems. When data are added to the system, it must be added to multiple locations. Because the data get populated in so many different locations, the company experienced recurring problems of data in one location not matching data in another location. In addition, there may have been inconsistency in the execution of business rules among the various systems. While this approach works, it is expensive to maintain, it is fragile (changes to the underlying data sources may easily break the application), the number of application operations needed to accomplish a task is excessive, and it is hard to extend the system (additions require rewriting much of the code).

Lastly, consider the system design shown in Figure 5.2. The diagram shows the systems, databases, and information flows in a large multi-national beverage company. The resulting system complexity is beyond what anybody would consciously design. Rather it reflects the many individual decisions, mergers, and acquisitions over a long period of time.

If the cruise company had an enterprise architecture, then the system evolution and development could have gone differently. An enterprise architecture would have highlighted the needed information flow, contained principles for interoperability, and provided a holistic plan to guide individual system-development decisions. One of the primary benefits of an enterprise architecture is that it shows how all the system components fit together. The benefits of having an enterprise architecture are:

- To provide a model that lets all stakeholders understand and communicate the overall business design. Enterprises are complex so models that abstract away this complexity are very valuable for understanding the enterprise systems. Moreover, by providing an integrated set of models, accessible to all stakeholders, then it helps create a shared vision of how the enterprise is designed and how the enterprise operates. Consequently, an enterprise architecture helps stakeholders plan, manage, and effectively utilize the enterprise's systems and resources.

- To provide a high-level, holistic design of the business indicating how all the subsystems

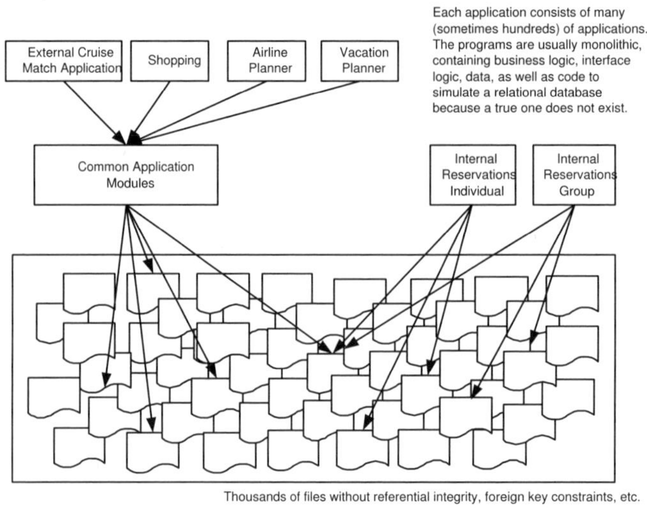

FIGURE 5.1
Results of system evolution at a cruise line.

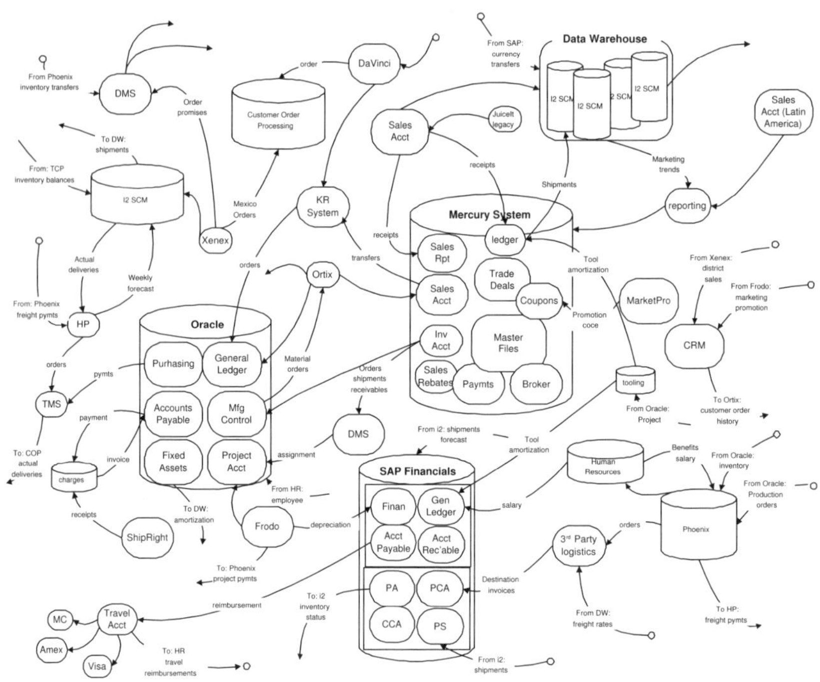

FIGURE 5.2
System diagram of a large, multinational beverage company.

will interoperate and coordinate their work. Essentially, the enterprise architecture helps us avoid sub-optimization. We shift our emphasis from developing the best subsystem (e.g., inventory control system, order fulfillment system, etc.) and instead consider how the subsystems work together to reach overall system goals.

- To express the architectural principles of a long-term vision of the enterprise, and the governing principles of the enterprise to guide all other projects. The statement of these principles promotes effective planning and decision making. It aligns all the projects towards achieving the overall enterprise objectives.
- To ensure legal and regulatory compliance. In the U.S., this is especially important to help organizations comply with the 1996 Clinger-Cohen Act in the U.S.

Collectively, these benefits should reduce the operating costs by making the business more efficient, they should also reduce the costs of infrastructure because all new projects are built to an overall plan, and they should improve the business processes through better system integration and coordination.

5.2 Enterprise Architecture Frameworks

Given the value of having an enterprise architecture the question is, how to develop one? Every enterprise could develop its own architecture as it sees fit. The problem with this approach is that developing an enterprise architecture is an enormous undertaking fraught with the risk of omitting crucial elements, creating inconsistent interfaces between the views, and not finishing the project in a reasonable time period. An alternative approach is, instead of developing an enterprise architecture from scratch, an enterprise can start with a reference enterprise architecture. A *reference enterprise architecture* (or simply reference architecture) is a generic architecture that can be used as the starting point to derive an enterprise's architecture.[1] A reference architecture is a meta-model, in that it is a model of how you should model enterprises. It describes a structured set of models that collectively represent the building blocks of an enterprise system. Reference architectures embody the knowledge gathered, on a large scale, from a multitude of enterprise engineering projects. The idea is that an enterprise by using a reference architecture is more likely to include all relevant views in their architecture, will adopt best practices into their architecture, and will be able to create the architecture much more rapidly than if they did it from scratch. Additionally, the use of a reference architecture provides some insurance that the resulting architecture will be useful and completed on schedule.

A reference architecture must be generic enough to be useful for a wide range of enterprises while at the same time it must be detailed enough that the commonalities are not superficial. This is more or less the quandary faced by all modelers. Figure 5.3 shows how a reference architecture could be used by many different companies to derive enterprise-specific architectures. The three enterprise architectures will all be different (but similar) even though they were all derived from the same reference enterprise architecture.

[1]Terminology is not consistent in the literature. Zachman and a few others call it a framework; in this book, take reference architecture to be equivalent to what they term a framework. Second, some authors differentiate between two types of reference architectures. Williams [18] defines Type I architectures that represent the structure of the enterprise system at a particular point in time, and Type II architectures that represent the structure of a methodology to analyze and design an enterprise system. In this book, we shall only refer to Type I architectures as reference architectures, and Type II architectures will be referred to as methodologies.

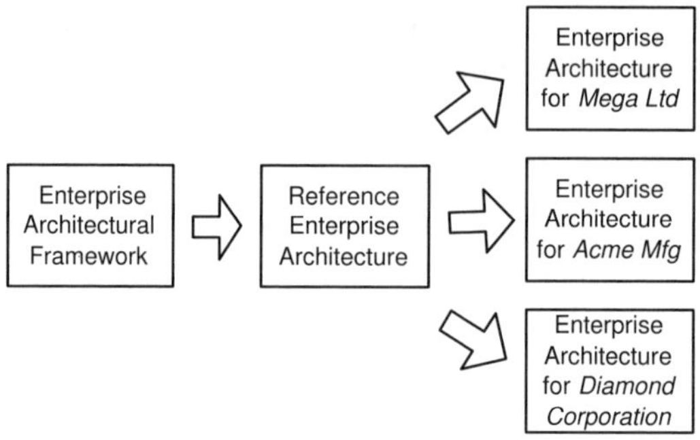

FIGURE 5.3
Using a reference architecture.

A reference architecture prescribes a decomposition that allows engineers to develop solutions to the subsystems relatively independently while ensuring the solutions will integrate well with the other subsystem solutions so that an integrative, effective system solution is obtained. The selection of the appropriate abstraction level, if left entirely to the modeler's discretion, would likely be inconsistent from one model to the next. The reference architecture helps the enterprise modeler by defining appropriate abstraction levels. Moreover, the relationships between the abstraction levels are explicitly defined in the model using the modeling constructions and notation.

The reference architecture:

- Provides a unified, unambiguous definition of terminology.
- Establishes a common means to organize, interpret, and analyze architectural descriptions.
- Identifies architectural concerns, generic stakeholders, viewpoints, and abstraction levels.
- Facilitates communication of enterprise design.
- Helps stakeholders make decisions about enterprise design and operation.
- Is applicable to a wide range of enterprise systems and scenarios.
- Is developed in such a way that it encourages reuse.
- Provides users with some confidence that use of the reference architecture will be successful in the current project.

The idea of having an enterprise architecture has gained wide acceptance. This has led to the proliferation of many different reference architectures. In academia, reference models have been created in Europe (CIMOSA [9] and ARIS [14]), and in the U.S. (PERA). In industry, reference models have been created by IBM (Zachman's Framework) and by industry groups such as The Open Group's TOGAF architecture. Moreover, many large enterprise system vendors provide reference architectures to describe their software and how it will work in the enterprise. For example, SAP has developed over 1000 reference models for business processes in various industries such as automotive, finance, and insurance.

Governments have reference architectures. In the U.S., largely due to the 1996 Clinger-Cohen Act, government agencies are developing enterprise architectures. The Clinger-Cohen Act says each federal agency must have a Chief Information Officer (CIO) who is responsible for developing, maintaining, and implementing sound and integrated information technology systems to achieve the agency's strategic goals. This act has largely been interpreted as requiring U.S. federal agencies to have enterprise architectures.

The U.S. Department of Defense has the DoDAF (Department of Defense Architectural Framework). The CIO Council, formed by the CIOs of major government agencies, developed the Federal Enterprise Architecture Framework (FEAF) [2] for the U.S. federal government. Later the Office of Management and Budget (OMB) assumed responsibility for enterprise architecture and published the Federal Enterprise Architecture (FEA) [3].

National and International standards bodies have published or are working on several different enterprise architectures. IEEE 1471 [4] describes the recommended practices for architectural descriptions of software-intensive systems, and in 2008, it became the ISO 42010. There is an international standard titled Generalized Enterprise Reference Architecture and Methodology (GERAM) to specify what should be part of reference enterprise architectures [1]. GERAM is not per se a reference architecture but a meta-reference architecture.

Surveys of enterprise architectural framework usage show that Zachman's Framework has the largest market share followed by TOGAF [5, 10, 15]. The results may be skewed because most respondents are from the U.S. In addition to the general architecture frameworks, many enterprises use industry-specific frameworks. For example, in one survey [5] eTOM was mentioned, which is the Enhanced Telecom Operations Map. The surveys also suggest that many companies still use proprietary architectures. In the following subsections, we describe the two most commonly used reference architectures: Zachman's Framework and TOGAF. These are reference enterprise architectures – as a reference architecture, you "refer" to the reference enterprise architecture to specify what to model, how to model, and why to model. Understanding Zachman's Framework is important because it strongly influenced many of the U.S. government's reference architectures, so to understand of Zachman's provides prerequisite knowledge to understand FEAF and FEA.

5.2.1 Zachman's Framework

Zachman was an employee of IBM who first published in 1987 a framework [19] that has since evolved into the Zachman Framework for Enterprise Architecture, draws upon the discipline of classical architecture/construction and engineering/manufacturing to establish a common vocabulary and set of perspectives for defining and describing enterprise systems.[2] Zachman's framework is a logical structure for classifying and organizing those elements of an enterprise that are significant to both the management of the enterprise and the development of its information systems. Zachman's Framework is among the most widely used reference architectures; it has also strongly influenced the federal government's FEAF.

The Zachman Framework describes an enterprise architecture along two dimensions of the stakeholders and the perspectives of the enterprise. The framework is a classification scheme that Zachman claims is complete, in that it contains all the cells that exist. There are 36 cells in the framework. In the current form of the architecture, shown in Figure 5.4, the rows represent the various stakeholders. The stakeholders are from top to bottom:

1. The first row, SCOPE, establishes the context for any enterprise system project.

[2]To quote from Zachman, "The Framework as it applies to Enterprises is simply a logical structure for classifying and organizing the descriptive representations of an Enterprise that are significant to the management of the Enterprise as well as to the development of the Enterprise's systems. It was derived from analogous structures that are found in the older disciplines of Architecture/Construction and Engineering/Manufacturing that classify and organize the design artifacts created over the process of designing and producing complex physical products (e.g., buildings or airplanes)." - Zachman, J. *The Framework for Enterprise Architecture: Background, Description and Utility*, from www.zifa.com, accessed on April 10, 2009.

The scope is frequently defined in a Statement of Work (SOW) or other similar document. It defines the enterprise goals, strategy, boundaries, and purpose.

2. The second row, ENTERPRISE MODEL, presents the business owner's view that defines, in business terms, the nature of the business, including its structure, functions, organization, and so forth. This row involves the first formal modeling effort. The SOW provides the starting point for model creation. The integrity of the rules developed for this row must be maintained throughout the remaining rows.
3. The third row, SYSTEM MODEL, presents the designer's view that defines the business similar to the owner's view, but in a more rigorous detailed form. This row is sometimes called the conceptual or *logical view*. For example, row two would define the business functions from the owner's perspective in terms of what activities need to be done. Row three would define it in terms of data transformations, input/output relationships, constraints on the activities, who can do each activity, and so forth.
4. The fourth row, TECHNOLOGY MODEL, presents the builder's view, a technological view that describes how the system functions, information, organization, and so forth are implemented. This row is sometimes called the *physical view*, and prepares the model for implementation. The person performing this activity is proficient in technology. The models associated with this row directly support the future implementation.
5. The fifth row, COMPONENTS, presents the code that implements the application. The person performing this function is the implementer. The implementer is accountable for implementing the provided design.

The columns represent different viewpoints of the enterprise. When it focused on the information system, Zachman's original framework had three views of data, process, and interface. The current version of Zachman's framework re-did the views in terms of the prerogatives: what, how, where, who, when, and why?[3] These map to the columns as follows:

1. The first column, DATA, is the "What" things that are involved in the subject area. The data consist of a list of things. The Entity is a major data item that stands for a class of a business thing. Entities are connected via relationships.
2. The second column, FUNCTION, is the "How" things are done. This consists of a list of processes that are performed.
3. The third column, NETWORK, is the "Where" things are done. This could be a list of locations or network nodes that need to be supported to allow the knowledge to be captured and used within the subject area. The design team should understand if the available network supports that current application. The additional knowledge that is captured to support this conclusion should be documented.
4. The fourth column, PEOPLE, is the "Who" that are part of the enterprise. It contains models about the people and organizations within the business. It describes the rules dealing with who is responsible for a piece of knowledge, who participates in a process, and other work policies.
5. The fifth column, TIME, is the "When" something is done. This could be a fixed time that triggers a process, the time of an event that triggers another process, or a time sequencing that says this process must be performed before a

[3]" I keep six honest serving men (they taught me all I knew): their names are What and Why and When And How and Where and Who" from *The Elephant's Child* by Rudyard Kipling [8].

	Data (What?)	Function (How?)	Network (Where?)	People (Who?)	Time (When?)	Motivation (Why?)
Scope (Planner's Perspective)	List of business data	List of business functions	List of locations for business operations	List of major organization-al units	List of major business events	List of enterprise goals
Enterprise Model (Owner's Perspective)	Semantic Data Model	Business Process Model	Business hierarchy mapped to location	Organization Chart	Strategic Plan & Timeline	Business Plan
System Model (Designer's Perspective)	Logical ER Model	Activity Level Model	Distributed System Architecture	Job roles & responsibility	Business Schedule	Business Rules and Policies
Technology Model (Builder's Perspective)	Physical ER Model	Data Flow Diagram	Information Technology Architecture	People (Who?)	Control Structure	Reward System & Mgmt Control
Detailed Representation (Subcontractor's Perspective)	Data Dictionary	Process Specification & Code	Network Infrastructure	People (Who?)	Timing and Sequencing Definitions	Supplier Contracts & Performance Criteria

FIGURE 5.4
Zachman's framework.

follow-on process. The successful implementation of an application requires this information. The formal requirement that this be addressed within a design will improve the resulting quality of the design. It is generally used in relationship to the Data and Function columns.

6. The sixth column, MOTIVATION, is the "Why" things are done. Business goals and strategies are listed here. These goals and strategies can be expressed as natural language sentences. In many cases, the sentences that measure these goals and strategies will have derivation rules for determining the performance of the company or organization. The derived fact instances are compared with stated goals that are called out in a business objective.

Zachman's framework distinguishes each stakeholder's perspective of the enterprise and shows how to incorporate all the stakeholders' perspectives into the understanding of an enterprise. The idea that different stakeholders can have different perspectives of the same enterprise is illuminating to many analysts, designers, and engineers. In this way, Zachman's framework serves the important need of a reference architecture – to facilitate communication. It helps users of the framework understand how other stakeholders view the system and thus can work better together in the design of an enterprise system. When every cell is populated with appropriate artifacts, there is a sufficient amount of detail to fully describe the system from the perspective of every stakeholder looking at the system from every possible angle (descriptive focus).

Describing the different stakeholder perspectives is not sufficient in itself, but the framework also explains how they fit together. It is similar to the Indian parable of seven blind men each feeling a different part of the elephant, but none can put it together to define the elephant as a whole.[4] Well, Zachman's framework shows how these different perspectives come together; it shows how the different views correspond to each other and, it shows how models are linked to the technology that implements the model concepts.

[4]Parable of *The Blind Men and the Elephant* by John Godfrey Saxe (1816-1887).

A criticism of Zachman's framework is that ideally the viewpoints should be orthogonal. In other words, two different viewpoints should show different views of the enterprise without overlapping. In Zachman's framework, the time column seems to be confounded with the data and function column. It is difficult to see how this column could be modeled separately from other views – a violation of the orthogonality principle. A second criticism is that Zachman does not specify what type of models to populate each cell. So, the DATA view could be populated with entity-relationship models, object-oriented models, or any other model. Critics also say that the 36 cells are too many, and few organizations would ever populate every cell. There is nothing in Zachman's framework that prevents an organization from adapting it to its needs. Lastly, some critics are disappointed that the framework has no associated methodology. It is strictly a taxonomy for the models. Every other reference architecture that is reviewed in this book includes a methodology. What an enterprise contemplating using Zachman's framework would have to do is determine what methodology would integrate with the framework.

What the Zachman framework contributes to the enterprise design problem is clarity of thought. It provides confidence to the design team that if they address all stakeholder levels (planner, owner, designer, developer, translator, and worker), satisfy the basic six interrogative questions about the system (what, how, where, who, when, and why), and have traceability between the stakeholders and across the questions, then the design team created a good enterprise architecture.

Over the years the Zachman's framework has been widely adopted by the architecture community and it is incorporated into and influence of it can be found from many other enterprise architecture frameworks [16], e.g., Federal Enterprise Architecture Framework (FEAF) [2], Treasury Enterprise Architecture Framework (TEAF), and The Open Group Architecture Framework (TOGAF) [17].

5.2.2 TOGAF

TOGAF is the acronym for The Open Group Architecture Framework that is developed by The Open Group. TOGAF Version 8.1 Enterprise Edition is an industry standard architecture framework that can be freely used by any enterprise developing enterprise architecture for its own use [17]. TOGAF evolved from earlier work by the U.S. Department of Defense on the Technical Architecture Framework for Information Management (TAFIM).

TOGAF describes four architectural views:

1. Business architecture – Describes the processes the business uses to meet its goals. It links strategy formulation to strategy implementation.
2. Application architecture – Describes how specific applications are designed and how they interact with each other.
3. Data architecture – Describes the enterprise's logical and physical data resources and how the data are managed.
4. Technical architecture – Describes the hardware and software infrastructure that supports the business processes, applications, and their interactions.

In addition to these views, TOGAF allows for the definition of other views, but it does not say what these views should be in the same way that Zachman's Framework does.

TOGAF consists of three main parts: the Architecture Development Methodology (ADM), the Enterprise Continuum, and the Resource Base. The ADM is a method for developing an enterprise architecture to meet the business and information technology needs of an enterprise (see Figure 5.5). According to the TOGAF specification, each iteration of the ADM must include decisions regarding: (1) the breadth of coverage of the enterprise

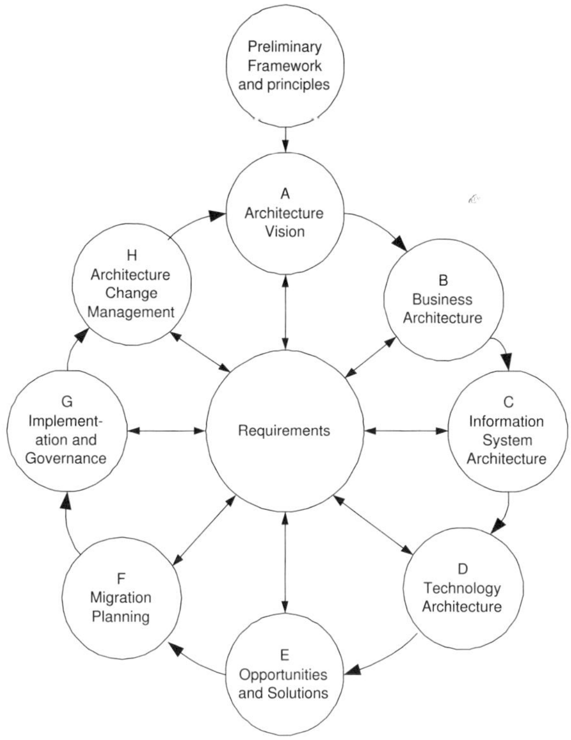

FIGURE 5.5
TOGAF architecture development method.

to be specified, (2) the depth of details to be specified, (3) the extent of the time horizon aimed at, and (4) the architectural assets to be leveraged in the organization's Enterprise Continuum.

The Enterprise Continuum may be viewed as a "virtual repository" of all the architecture assets available to an organization. The term "assets" refers to architectural models, architectural patterns, architecture descriptions, and other artifacts. These artifacts may exist within the enterprise and also in the IT industry at large. The ADM promotes the idea of reuse, and suggests throughout the methodology when the architectural development team should look for assets both within and outside of the firm. Additionally, TOGAF includes two reference models that may be used:

1. TOGAF Foundation Architecture, which describes an architecture of generic services and functions to build specific enterprise architectures. The two components of the foundation architecture are:
   (a) TOGAF Technical Reference Model (TRM) is a model and taxonomy of generic platform services.
   (b) TOGAF Standards Information Base (SIB) is a database of open industry standards to define the services and other IT components of the architecture.
2. The Integrated Information Infrastructure Reference Model is designed to help the enterprise design boundary-less information flow.

The Resource Base is a set of tools (guidelines, templates, etc.) that are available to help the enterprise architect in the use of the ADM.

TOGAF is primarily an IT architecture that looks to align the IT view with the business view. The views confirm this assessment in that three of the four views are all technical in nature (the application, data, and technical views). Unlike Zachman's Framework, TOGAF does not define different abstraction levels corresponding to the stakeholders' views. It does not forbid a company from building these models, but neither does it promote the practice. Since the decomposition of the enterprise into views to deal with complexity is one of the main motivators for enterprise architecture, it is useful to discuss how well the architectures accomplish this. TOGAF could be criticized because the views are not orthogonal: they overlap significantly. The business and data views correspond to columns in Zachman's Framework but the application and technical views correspond to rows. The application view describes the software systems that support the business processes. So, it is a somewhat more technical view of the business process. Moreover, the technical view describes the infrastructure that should be integrating the data view, application view, and business view. This overlapping of views can confuse design teams and other users of the architecture if they are not conversant with TOGAF.

TOGAF's strength is its ADM methodology that describes a detailed approach to generate enterprise architectures. Given the large scope and complexity of building an enterprise architecture, this is a worthwhile contribution to make the architecture design process easier and smoother.

5.2.3 Other Enterprise Architectures

The Computer Integrated Manufacturing Open System Architecture (CIMOSA) is a reference architecture developed by a research consortium in Europe, named AMICE, to describe, as its name implies, manufacturing systems. CIMOSA defines three dimensions of life-cycle phases, views, and implementation stages. The four views it provides of the enterprise are the information, function, resource, and organization views. CIMOSA is still active (see www.cimosa.de), but the architecture is probably not widely used anymore. The CIMOSA Association is focused on enterprise integration and standardization efforts. Many of the CIMOSA researchers contributed to the development of the Generalized Enterprise Reference Architecture and Methodology (GERAM).

The Architecture of Integrated Information Systems (ARIS) was developed by A. Scheer in Germany [14], and has been adapted by SAP, a large vendor of enterprise resource planning (ERP) systems. ARIS defines the House of Business Process Management that describes the enterprise by four inter-related views of data, function, organization, and control. The core of the ARIS architecture is the control view, which integrates the other three views of data, function, and organization. The control view is modeled with Event Process Chains (EPCs) that allow the modeling of the business processes. ARIS is well developed commercially; there is a toolset available to implement ARIS and as mentioned it is used by SAP to describe their ERP system.

GERAM was developed by the IFIP-IFAC Task Force on Architectures for Enterprise Integration [1]. GERAM is not a reference architecture like the others, instead GERAM is meant to assess reference architectures for completeness. GERAM cannot be implemented directly like the other reference architectures because it has no symbols or methodology of its own. GERAM was specified by evaluating three existing enterprise architectures of CIMOSA, GRAI, and PERA; taking the best characteristics of each; and then combining them into GERAM. The GERAM life-cycle phases are identification, concept, requirements, preliminary design, design, detailed design, implementation, operation, and decommission [11].

5.2.4 Summary Enterprise Reference Architectures

An enterprise reference architecture embodies all the knowledge gained from a history of enterprise engineering projects. There are several different enterprise architectures that have gained acceptance; in this chapter, we reviewed the more prominent reference architectures which are Zachman's Framework and TOGAF. A brief mention was made of ARIS, CIMOSA, and GERAM, which have contributed to the development of enterprise architectural ideas. A commonality of these architectures is the idea of representing different views of the enterprise. The data, function, and organization view appear in all of them. The architectures differ in whether they include a resource view and in the representation of process. Some, such as Zachman make no distinction between function and process. Others, such as ARIS model function separately from process, which is the control view of that architecture. Regardless of the views, the concept is that the reference architecture provides a holistic view of the enterprise, through the integration of multiple, orthogonal views.

One of the main benefits of the enterprise architectures is they organize our knowledge about enterprises, they standardize terminology, and they describe proven methods to model, analyze, and design enterprises. Consequently, an enterprise reference architecture helps the members of an enterprise engineering project team to organize integrated models of the enterprise. It helps ensure interoperability of the enterprise designs, and it helps control the costs of the enterprise project. As a result, enterprise reference architectures support design by helping us manage complexity.

5.3 Developing the Enterprise Architecture

To develop an enterprise architecture is a project that requires a cross-functional team led by a facilitator. The facilitator is a person knowledgeable in enterprise architectures. The facilitator convenes a group of stakeholders to discuss the business vision and how it can be realized by the enterprise architecture. One of the first decisions is whether to use a reference architecture, and if so, which reference architecture to use. It is recommended to use a reference architecture because: it will provide a comprehensive view of the enterprise, minimizing the risk of leaving out important enterprise views and/or components; it will minimize the time and cost necessary to build an enterprise architecture; and lastly, it embodies the knowledge gained from many other enterprise architectures that can be reused in the current project.

The decision to adopt a reference architecture as a starting point does not mean it needs to be used verbatim. The reference architecture can be adapted to suit the particular needs of the enterprise. Reference architectures define the main concerns that should be included in the model, but a particular concern may be of low priority for your enterprise. For example, Zachman's framework might be selected, but the enterprise might decide to collapse two views into a single view, or ignore a view completely because it is of lower priority than the other views.

The team decides what to include and what not to include in the enterprise architecture. Only those decisions that are system-wide, high priority, and important to the business are included in the enterprise architecture. Table 5.1 shows a decision matrix to categorize whether decisions should be included in the enterprise architecture or not. Only those decisions that are system-wide and high impact are included in the architecture. Decisions that are local should in general not be included. However, the enterprise architecture might articulate guidelines, principles, or policies to guide those decisions.

TABLE 5.1
Architectural Decisions Are Both System-wide and Have High Impact on the Enterprise

	Low Impact	**High Impact**
System-wide	Not an architectural decision	Architectural decisions
Local	Not an architectural decision	Not an architectural decisions (but might set guidelines and policies to direct decisions)

While developing the architecture, the architectural design team articulates and debates architectural principles. A good enterprise architecture provides a flexible, scalable structure that can be modified over time as needs change. An enterprise architecture describes guiding, architectural principles. Example statements of architectural principles are:

- "The enterprise will provide access to information to authorized users to perform their jobs independent of physical location."
- "New systems shall be able to interoperate with the existing ERP financial's module. They shall be tested to determine they can import and export data to the financial's module."
- "New system development projects shall provide an end-to-end process model, and all organizational units involved in the process shall sign off on the project."

It is easy to see how the architectural principle will guide later projects. Whenever a project is initiated, the project team will have to ensure information availability regardless of physical location. Including this principle in the architecture ensures an enterprise-wide conformance.

Once the views and architectural principles are established, the team focuses on developing baseline models of the current enterprise situation. A strategy to develop the enterprise architecture is to evolve it top-down from the more abstract model levels to more detailed levels. This strategy has several advantages. First, it is difficult to develop an entire enterprise architecture in a short period of time. The team then defines how it wants the enterprise architecture to look.

Once the team completes the to-be models, it then develops an overall plan on how to close the gaps between the as-is model and the to-be model. In many cases, the enterprise architecture is nothing more than a conceptual architecture that gradually emerges as the result of implementing many individual projects [7].

An enterprise architecture includes the components specified in Table 5.2 (adopted from [6]). The enterprise architecture is widely distributed in the organization for feedback, comments, and to help buy-in by the entire organization.

5.4 Enterprise Reference Architecture

This book uses the enterprise reference architecture shown in Figure 5.6. The architecture maps the enterprise design methodology, represented by rows, and the three enterprise views,

TABLE 5.2
The Components that Are Included in the Enterprise Architecture Documents and Their Description

Component	Description
Strategy	Strategy map, goals, corporate mission and vision, business philosophy, corporate culture and values.
Name	Architecture name, description, version, and overview information, date of issue, change history, scope, context, and glossary.
Stakeholders	Identification and description of enterprise stakeholders and their architectural-related concerns.
Views	The enterprise architecture should model multiple views of the enterprise. Moreover, it is possible to model these views at multiple levels as recommended by Zachman. The definition of each architectural viewpoint, its boundaries, and its rationale. The views should be consistent, so that an information object in one view matches the information input shown in the function view.
Models	The models to populate one or more architectural viewpoints.
Principles	Architectural principles and rationale for all architectural choices and decisions that should guide all enterprise projects.
Conventions	Policies and rules.

represented by columns. The three enterprise views are: information, process, and organization. The three enterprise views were selected based on obtaining a complete perspective of the enterprise with a minimum set of views. Additionally, the architecture identifies the main stakeholders corresponding to each methodology phase. Similar to Zachman's framework, each cell is populated with models and other documentation. Collectively, when filled it presents a holistic view of the enterprise project: why the project was done, how it was done, and what was accomplished.

5.5 Summary

This chapter argued and demonstrated that companies need enterprise architecture to avoid problems with enterprise integration. An enterprise architecture was defined as the high-level design of the enterprise. A best practice is to use an architectural framework as the starting point to develop a specific enterprise's architecture. The chapter reviewed two architectural

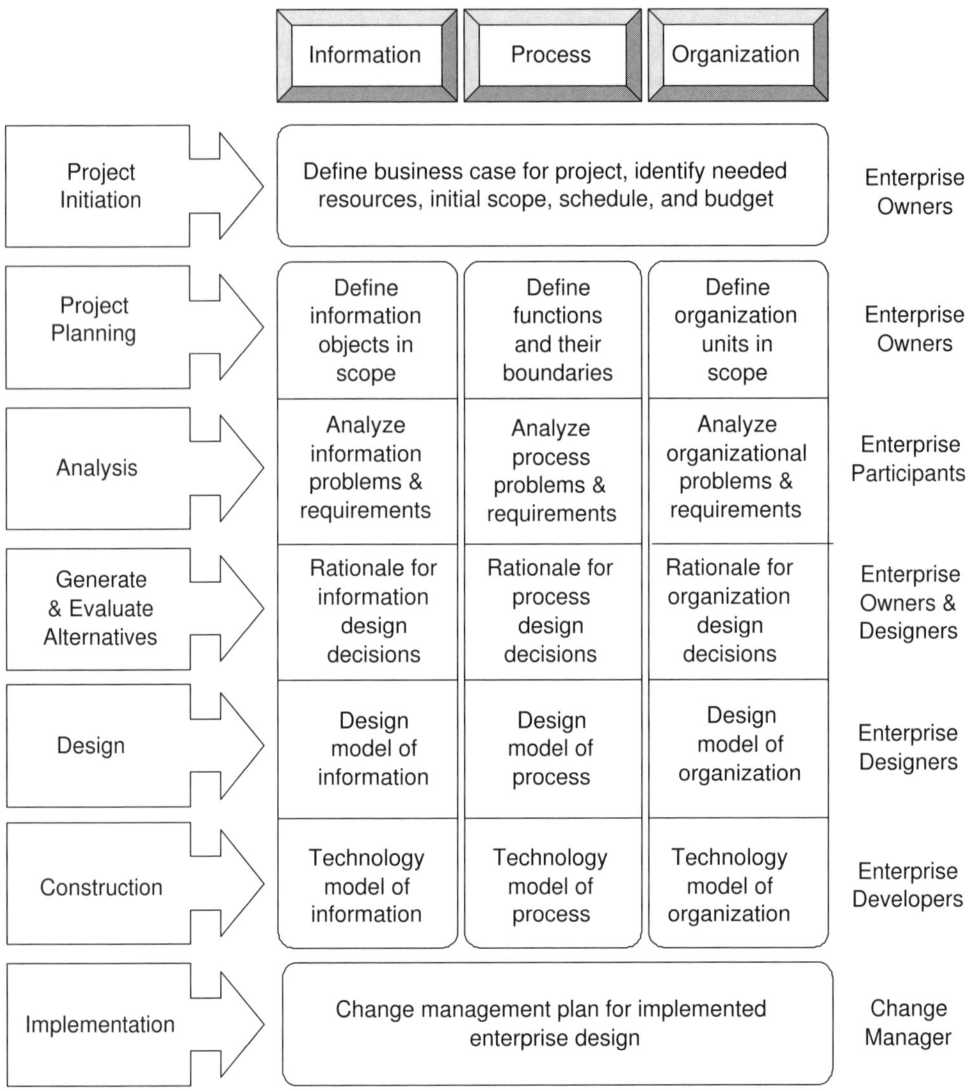

FIGURE 5.6
Enterprise reference architecture.

frameworks: Zachman's and TOGAF. These architecture frameworks showcase the variety of approaches to architecture. Many of these architectures came out of the IT industry and were later extended for coverage of all enterprise aspects. The field of enterprise architecture is ongoing and new standards and developments are underway.

An enterprise architecture accomplishes two goals beneficial to the enterprise engineering project: first, it creates a holistic view of the enterprise showing how all the various perspectives join together to form the enterprise; second, it establishes design standards and answers long-term decisions that all future projects conform with. The enterprise architecture *should* underpin every project the enterprise embarks on.

Review Questions

1. Describe three common views in an enterprise architecture.
2. Describe the relationship between a reference architecture and an enterprise architecture.
3. Describe four benefits derived from having an enterprise architecture.
4. List the components that make up an enterprise architecture.
5. Compare and contrast the architectural approaches taken by Zachman and TOGAF.
6. Describe characteristics of the types of decisions that should be included in enterprise architectures.
7. Explain why many enterprise architectures divide the enterprise into separate views.
8. List some architecturally significant decisions and some decisions that are not architecturally significant.

Bibliography

[1] P. Bernus and L. Nemes. Requirements of the generic enterprise reference architecture and methodology. *Annual Review in Control*, 21:125–136, 1997.

[2] The Chief Information Officers Council. Federal enterprise architecture framework, version 1.1. Technical report, www.cio.gov/documents/fedarch1.pdf, 1999.

[3] FEA. FEA consolidated reference model document version 2.1. Technical report, the Federal Enterprise Architecture Program Management Office, Office of Management of Budget, 2006.

[4] IEEE. IEEE standard 1471-2000: IEEE recommended practice for architectural description of software-intensive systems. Technical report, 2000.

[5] Infosys. Win in the flat world: Infosys enterprise architecture survey 2005. technical report, InfoSys, 2005.

[6] ISO. ISO 42010 - systems and software engineering - architectural description of software-intensive systems. Technical report, 2008.

[7] B. Iyer and R. Gottlieb. The four-domain architecture: An approach to support enterprise architecture design. *IBM Systems Journal*, 43(3):587–598, 2004.

[8] R. Kipling. *Just So Stories*, chapter The Elephant's Child. 1902.

[9] K. Kosanke, F. Vernadat, and M. Zelm. Cimosa: Enterprise engineering and integration. *Computers in Industry*, 40:83–97, 1999.

[10] D. Minoli. *Enterprise Architecture A to Z: Frameworks, Business Process Modeling, SOA, and Infrastructure Technology.* CRC Press, Boca Raton, FL, 2008.

[11] O. Noran. An analysis of the zachman framework for enterprise architecture from the GERAM perspective. *Annual Reviews in Control*, 27(2):163–183, 2003.

[12] M. E. Porter. *Competitive Advantage: Creating and Sustaining Superior Performance.* The Free Press, New York, NY, 1985.

[13] J.W. Ross, P. Weill, and D.C. Robertson. *Enterprise Architecture As Strategy: Creating a Foundation for Business Execution.* Harvard Business School Press, Cambridge, MA, 2006.

[14] A.W. Scheer. *Architecture of Integrated Information Systems: Principles of Enterprise Modeling.* Springer-Verlag, Berlin, Germany, 1992.

[15] J. Schekkerman. Trends in enterprise architecture 2005: How are organizations progressing? Technical report, Institute for Enterprise Architecture Developments, 2005.

[16] A. Tang, J. Han, and P. Chen. A comparative analysis of architecture frameworks. In *Proceedings of the 11th Asia-Pacific Software Engineering Conference*, pages 640–647, 2004.

[17] TOGAF. The open group architectural framework enterprise edition, version 8.1 2003. Technical report, http://www.opengroup.org/architecture/togaf/, 2003.

[18] T.J. Williams. *Handbook of Life Cycle Engineering*, chapter The Purdue enterprise reference architecture (PERA), pages 289–330. Springer, New York, NY, 1998.

[19] J.A. Zachman. A framework for information systems architecture. *IBM Systems Journal*, 26(3):276–292, 1987.

6

Enterprise Analysis and Design Methodology

"How does a project get to be a year behind schedule? One day at a time." – Fred Brooks (1931-), software engineer and author of *The Mythical Man-Month.*

This chapter describes the enterprise design methodology adopted throughout the remainder of the book. The enterprise design methodology is based on the ideas of problem solving and controlled iteration. It is a seven-phase methodology that covers the life-cycle from project initiation to project implementation.

After completing this chapter, you should be able to:

- Name the enterprise design methodology phases and briefly describe each one.
- List and describe the cross life-cycle activities.
- Explain how the enterprise design methodology supports the problem-solving approach.
- Describe the inputs, activities, and outputs of each life-cycle phase.
- Define the major milestones and deliverables for an enterprise engineering project.
- Create a work breakdown structure for a project.
- Estimate activity duration and cost for the project.
- Perform a risk analysis for a project.

6.1 Enterprise Design Methodology

The Enterprise Engineering Methodology (EDM) comprises seven life-cycle phases shown in Figure 6.1. Each phase is delimited by milestones. Each phase comprises one or more activities and generates one or more deliverables. Some of the milestones are also deliverables (e.g., the Project Plan) and other milestones are meetings or events that mark the end of a phase (e.g., Kick-off Meeting marks end of project initiation).

While the phases occur sequentially, this does not mean there is no iteration within phases or overlap of phases. Within each phase the problem-solving iterations take place. In Chapter 4, the problem-solving activities were described as understanding, designing, building, and testing. Figure 6.2 shows the proportional amount of effort spent on each problem-solving activity for each of the phases. In the early phases, the majority of the time is spent understanding the design problem. Gradually, the proportion of effort for understanding decreases. However, the later phases still include some understanding because as you design, build, and test designs you will reveal issues that you did not previously think about. These issues will instigate additional efforts by the team to do some analysis to better understand the problem. Notice that even in the first stages some effort, albeit very little,

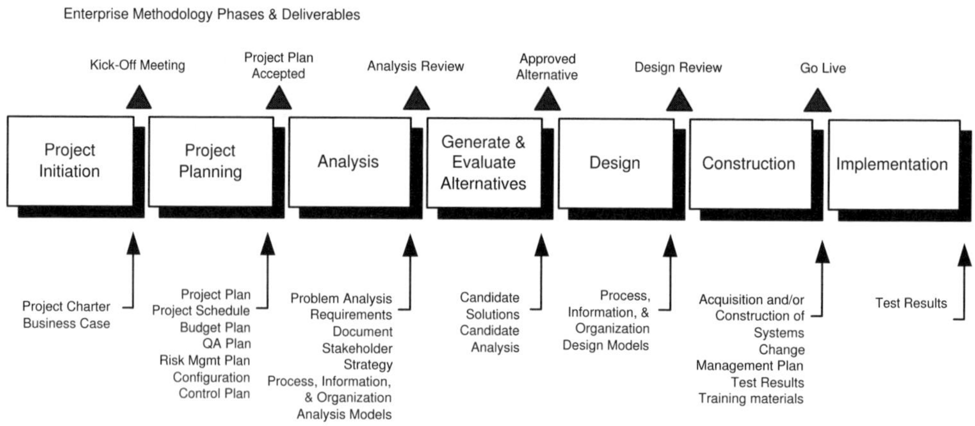

FIGURE 6.1
Enterprise design methodology.

is categorized as design effort. Most projects start with some ideas about how to design the solution. These mental models will influence the way the problems are addressed. What is important is not to be so biased that you cannot abandon the mental model if superior design concepts emerge.

The problem-solving activities of build and test do not really start until analysis, at which time the team would be engaged in making prototypes, analytical models, or simulation models with the intention of increasing their understanding of the problems. Towards the later phases increasingly more effort is dedicated to build and test.

The distribution of effort in the construction phase suggests a sizable amount of design work remains. However, the nature of the design work in the construction phase differs from that during the design phase in that it is at a lower abstraction-level and more detailed. Design during the design phase would involve developing the process models, defining the information flows, and specifying organizational roles to enact the process. Design during the construction phase would involve designing and coding the algorithms for the process, creating the Structured Query Language (SQL) and other commands to enact the information flow, and specifying account privileges for each organization role. By the implementation phase very little design efforts remain.

6.2 Cross Life-Cycle Activities

Cross life-cycle activities are those activities that span multiple life-cycle phases. The enterprise design methodology recognizes five cross life-cycle activities: project management, requirements management, quality assurance, configuration control and management, and risk management. These cross life-cycle activities are described in the following subsections.

6.2.1 Project Management

Project Management is the act of planning, directing, and controlling project resources (people, equipment, and materials) to meet the project objectives. The project plan is developed during the first two life-cycle phases. During these phases, the project manager needs to

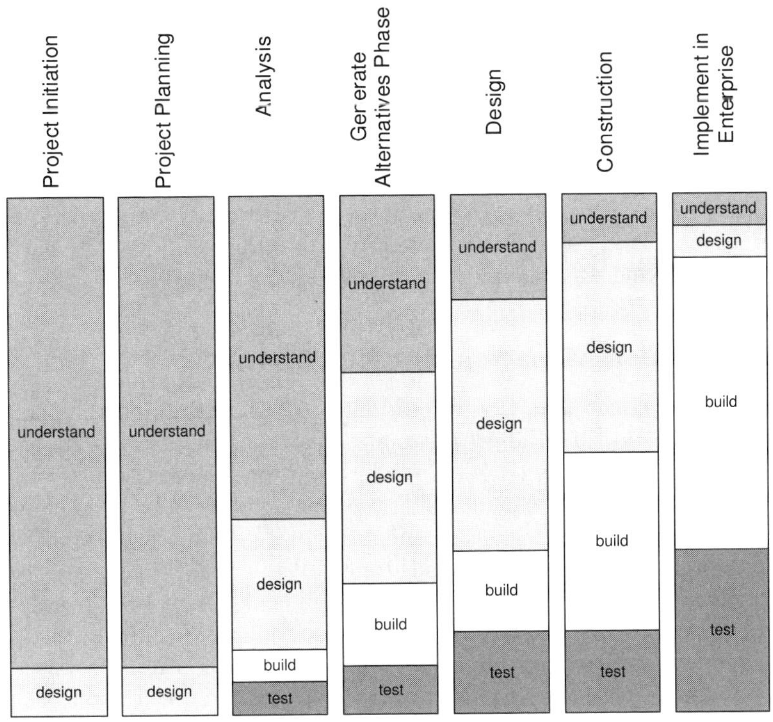

FIGURE 6.2
Problem solving in the life-cycle.

assemble a team, define the project, and develop a project plan. During the first two phases, intensive effort is put into developing the plan. In the remaining phases, project management continues, but the emphasis changes to the day-to-day supervision of the project, coordination of activities, and seeing that the project progresses as planned. Managing the day-to-day execution of the project is a full-time job for most projects and no less intensive than the project planning during the first two phases.

6.2.2 Requirements Management

Requirements Management is the process of first specifying the requirements and then managing change to the requirements. During the analysis phase that requirements are elicited, specified, and documented. Thereafter, the team must manage those requirements. The project team establishes and maintains a plan for performing requirements management. Requirements management also documents the requirements to ensure bi-directional requirements traceability. Additionally, requirements are a configuration item to be tracked and controlled as part of the configuration management process.

6.2.3 Quality Assurance

Quality Assurance (QA) is a planned and systematic approach to provide adequate confidence that an item or product conforms to established standards, procedures, and policies. QA is conducted throughout all life-cycle phases. A QA team is formed independent of the main project team to ensure objectivity. The QA team conducts periodic reviews to verify that the project processes are being used effectively, and if not, the QA team concludes non-compliance and documents its observations with the intention of improving the processes. The QA team also audits the deliverables for adherence to standards and provides feedback to the project team. The QA processes include three main activities: quality planning, quality assurance, and quality control. Quality planning identifies relevant quality standards for the project and how to satisfy them. Quality assurance is the evaluation of the overall project performance to ensure the project will satisfy the quality standards. Quality control is the process of monitoring specific project results to ensure they comply with the quality standards and identifying ways to improve the overall quality.

6.2.4 Configuration Management and Control

Configuration Management and Control establishes and maintains the integrity of the project's products. To do this, the project team creates a configuration management plan (a standard plan for software is IEEE 828, which can be adapted to enterprise systems). The configuration management plan describes the process of identifying configuration items, systematically controlling changes to the configuration items, and maintaining the integrity and traceability of the configuration items throughout the project. Configuration items are project documents, project deliverables, and any item that the project would wish to maintain control of. Example configuration items are: requirements document, test plan, process models, software artifacts, and so forth. For each configuration item, a baseline is established, where a *baseline* describes the formally reviewed and agreed upon basis for further development. Once a configuration item is a baseline, then it can only be changed through the formal change control process.

The project team also establishes a Change Control Board responsible for the change management process. The change management process handles change requests that if approved lead to a new revision of the configuration item. A general change process is:

1. The change is requested by any project team member or enterprise stakeholder.
2. The change request is assessed against project goals.
3. Following the assessment, the change is accepted or rejected.
4. If the change is accepted, the change is assigned to a team member for incorporation into the existing baseline documentation.

Configuration control is an important part of enterprise projects to manage the evolving system and coordinate changes to the system's design. The importance of configuration control is highlighted by its inclusion in Level 2 of the CMMI model.

6.2.5 Risk Management

Risk management is the systematic process of identifying, analyzing, and responding to project risk. Risk is the possibility of suffering diminished project success, measured in terms of project schedule, cost, and scope. Risk has two components: the probability of an undesirable event, and the consequences or impact of that event.

The risk management process involves:

- Risk management planning – the activity to develop an organized, comprehensive process to identify and track risk issues; to develop risk-handling plans; to plan continuous risk assessments; and to assign resources to do risk management. The risk management planning activity develops the Risk Management Plan.

- Risk identification – the activity to identify risk events that can negatively impact the project's schedule, budget, or deliverables. General risk categories are: technical, business, configuration, implementation, organizational, and project management. The project management team reviews the entire project plan, identifies, and documents all potential risks.

- Risk analysis – the activity to assign a probability to the risk event and to assess the consequences of the risk event. When expert opinion is used to assign risk probabilities, then exact, precise probabilities cannot be assigned with any reliability. Instead an ordinal scale as shown in Table 6.1 is preferable.

TABLE 6.1
Risk Probabilities

very unlikely	unlikely	possible	highly likely	almost certain
0.1	0.3	0.5	0.7	0.9
0.05	0.2	0.5	0.8	0.95

The consequences of a risk are measured against the project schedule, cost, and scope. For example, the risk of an unproven technology failing might have the following consequences: the project schedule will be delayed by 6 months, the project cost will increase by $100,000, and the project scope will have reduced features that the unproven technology was to provide. Again, since assigning precise consequences is difficult because we are trying to predict an uncertain future, it makes more sense to use a pre-defined scale (see Table 6.2).

TABLE 6.2
Risk Consequences

Project Objective	**very low**	**low**	**moderate**	**high**	**very high**
costs	insignificant cost increase	< 5% cost increase	5-10% cost increase	10-20% cost increase	> 20% cost increase
schedule	insignificant schedule slippage	schedule slippage < 5%	overall project slippage 5-10%	overall project slippage 10-20%	overall project slippage > 20%
scope	scope decrease barely noticeable	minor areas of scope are affected	major areas of scope are affected	scope reduction unacceptable to client	project deliverables are effectively useless
quality	quality degradation barely noticeable	only very demanding applications are affected	quality reduction requires client approval	quality reduction unacceptable to client	project deliverables are effectively unusable

The risk of an event is the product of its probability and its impact on the project. Figure 6.3 is a risk matrix used to view all the risks simultaneously in terms of their probability and impact. The shaded regions denote higher-level risks. The risk matrix is used by project teams to prioritize risks. In the figure shown, the project team would likely spend the greatest time and effort to respond to Risk D, followed by Risk A, and Risk F. The other risks may be deemed small enough that they do not warrant proactive management but only tracking.

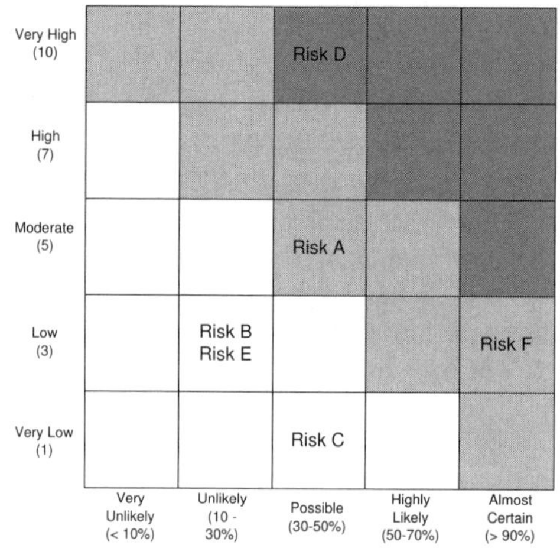

FIGURE 6.3
Risk matrix.

- Risk response planning – the activity to develop options and determine actions to enhance opportunities and reduce threats to the project's objectives from risk. Individuals are assigned to take responsibility for each agreed-upon risk response. There are four basic responses:

    1. *Avoidance* whereby the project plan is changed to eliminate the risk or to protect the project objectives from the risk impact.
    2. *Transference* whereby the consequences of the risk are shifted to a third party with ownership of the response.
    3. *Mitigation* whereby the project team takes action to reduce the probability and/or consequences of the risk event.
    4. *Acceptance* whereby the project team does nothing in response to the risk.

- Risk monitoring and control – the activity to keep track of the identified risk and ensure execution of the risk plan.

Example Risks

- *Technical* – reliance on unproven technology.
- *Technical* – the technology will not integrate as expected with legacy systems.
- *Organizational* – the project team is missing needed skills or lacks training in some area.
- *Organizational* – turnover of key employees because working on the project makes them highly marketable.
- *Market* – a competitor is developing a product that might diminish the expected market size.
- *Financial* – the organization will be stretched to afford the project.

6.3 Project Initiation

Project initiation is the process to define and obtain formal authorization for a new enterprise engineering project. The output of the project initiation activity is the Project Charter and the Business Case. The *Project Charter* describes the purpose and scope of the enterprise engineering project. The *Project Charter* is the document that is approved by the executive steering committee, which has the power to allocate resources to the project. The *Business Case* describes why it is beneficial to do the project. Both the Project Charter and the Business Case are presented to the executive steering committee during the kick-off meeting – the milestone marking the end of the Project Initiation phase. The goal of the project initiation phase is to ensure a successful project by starting out on the right foot.

Activities in this phase are:

1. Project definition
2. Project scope definition
3. Project scheduling
4. Project budgeting
5. Create business case

Figure 6.4 shows the activities, inputs, and outputs of the Project Initiation phase. Project Initiation has three main inputs: the project stimulus, the strategic plan, and the enterprise architecture. The project stimulus is what triggers the project. The project stimulus can be:

1. Problem – One or more perceived problem situations that the enterprise wants to correct (e.g., the enterprise takes too long to respond to customer requests for quotations and wishes to dramatically decrease response time).
2. Opportunity – The enterprise perceives a market opportunity, technology opportunity, or other opportunity. The purpose of the project is to let the enterprise take advantage of the opportunity (e.g., the enterprise observes a new market opportunity but needs to develop the processes, information, and organization in order to enter the market).
3. Directive – A directive is imposed by upper management, government, or other external organization. The enterprise must do a project in order to comply with the directive (e.g., the government introduces new regulations that require the enterprise to document its quality assurance process according to a prescribed format).
4. Planned – The project can be part of the overall strategic plan of the enterprise (e.g., the enterprise has a long-term plan to incrementally implement its enterprise architecture and the project is one of these pre-planned projects).

> The *Project Charter* describes why it is important to do the project and what the project will do.

The other inputs to project initiation are the strategic plan and the enterprise architecture. Both documents are necessary so that the project team can ensure the project is aligned with the overall enterprise strategy and architecture.

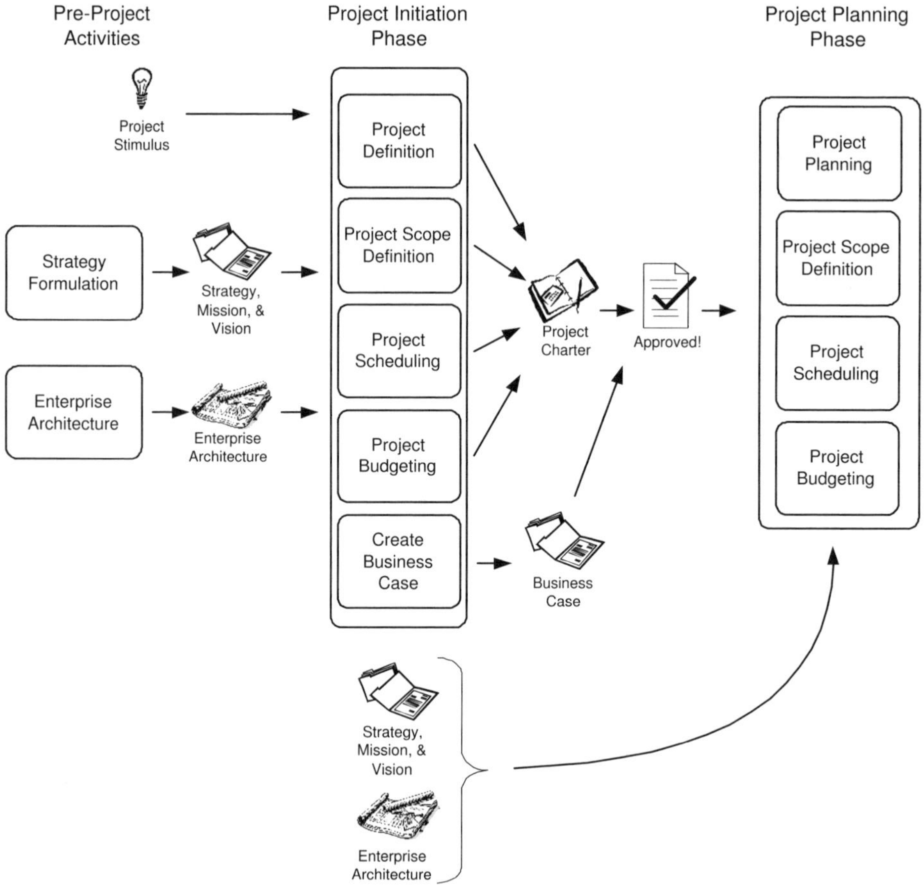

FIGURE 6.4
Project initiation.

Usually, the project stimulus is not formally defined, and it must be defined by the project team during project initiation. For example, the enterprise might have conducted a benchmarking study that showed their competitors provide superior customer-service. The executive management might then charge a project manager to define a project to improve the enterprise's customer service in order to remain competitive. "To improve customer service" is not in itself a sufficiently defined concept to start a project. Rather, the project team must use data-gathering techniques to clarify and formalize the project goals, project scope, project schedule, project budget, and identify any constraints or assumptions imposed on the project.

During project initiation the team defines the project scope, schedule, and budget, but at a high level. The reason is, at this early stage, the enterprise owners only need project details sufficient enough to determine whether it is worthwhile to fund the project. If the project is approved, then during the next phase, Project Planning, the project team will again do scoping, budgeting, and scheduling but at a very detailed level. The more detailed project plan including the budget and schedule will be reviewed again and subject to approval the project will continue. This approach of developing a high-level plan, getting it approved, and then doing a detailed plan that is also reviewed is called *creeping commitment.* The principle is that in the initial phases of the project the enterprise can abandon the project if it does not prove beneficial or cost-effective without having sunk too much money into the project. Creeping commitment is part of an overall risk management plan and is a best practice for project management. It recognizes that in the beginning phases of a project there remains significant uncertainty, and requiring detailed plans would require sinking a significant investment in a project that might not be approved.

6.3.1 Project Definition

Project Definition defines the purpose of the project, the project vision, the project objectives, and the organization that will carry out the project. The project purpose answers the question, "why should we do this project?" The project vision describes the desired end-result of the project. The project objectives are specific outcomes the project will achieve. The number of project objectives should be kept relatively small – four or five objectives are sufficient for most projects. The project team defines each objective so that the enterprise can measure whether the project is successful.

An important component of defining the project is to create a *project vision*, which is a succinct, detailed description of the project that will provide direction and align all the project stakeholders. This unifying vision, described by Goranson [1] as a high concept, is successfully used in Hollywood to align producers, actors, directors, and other participants throughout the duration of a film. This vision, if not provided by the project's sponsor, needs to be generated during the product initiation phase. A good project vision creates a picture of the project's deliverables, and provides a compelling justification for spending time, money, and resources on the project. A project vision is almost a requirement for a successful project – it is part of the leadership that the project manager needs to provide. The project vision also communicates to the stakeholders that their needs are understood and will be addressed by the project.

The project manager needs to assemble a project team. This involves identifying what knowledge, skills, and experience are needed on the project. The project manager has to decide how the team will be organized and managed. The project manager identifies reporting lines, and outlines specific roles that will be filled. The roles are defined such that everybody on the team has a clear understanding of what is expected of them. If it is a long-term project, then the project manager might even consider developing job descriptions for the team roles.

Figure 6.5 shows a typical project team structure. The Project Sponsor has the ultimate authority and control over the project and its implementation. The steering committee is a group of executives who oversee the management of the project, approve the release of resources for the project, and accept the major project deliverables. The project manager is the person who manages the project and has ultimate responsibility for the project's performance. To help the project manager there will usually be some support staff such as a secretary, assistant project manager, accountant, or others depending on the needs of the project. The QA team usually reports directly to the project manager as shown. Under the project manager are the sub-teams. These are functionally oriented teams that have responsibility for a major part of the project. Each sub-team is responsible for a major part of the project deliverables or for a major project activity. The use of sub-teams, each led by a team supervisor, reduces the number of people the project manager must directly supervise. The structure also reduces the amount of vertical communication between the project workers and the project manager. It is possible that an individual is on several sub-teams; for example, Joan a business analyst might be assigned to the accounting sub-

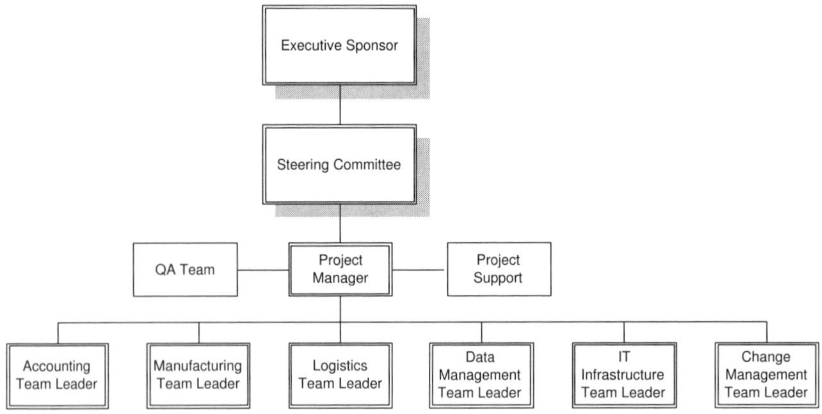

FIGURE 6.5
Project organization structure.

team and the manufacturing sub-team. Assigning people to multiple positions can enhance integration between the work being done by each sub-team.

The example project organization shown in Figure 6.5, is designed for an enterprise resource planning (ERP) implementation project. The project has a sub-team for each component of the ERP system (Accounting, Manufacturing, and Logistics). The project has separate sub-teams to manage the large data requirements, to develop the IT infrastructure, and to plan and perform the necessary change management. The type of sub-teams depends on the type of project. Moreover, the composition of the project sub-team will change during the duration of the project, especially for long-term projects.

6.3.2 Define Project Scope, Schedule and Budget

The project scope defines the boundaries for the project of what work will be included and what work will not be included in the project. The project schedule defines the time-table for completing the work. The budget defines how much the activities will cost. The three aspects of project scope, schedule, and budget are interdependent – you cannot change one without affecting one of the other two. For this reason, they *must* be defined and managed together. In project management, they call this the triple constraint.

The main output of project initiation is the Project Charter. The Project Charter describes:

1. Project Definition

   (a) Scope – What are the boundaries for this project (for example, type of work, type of client, type of problem, geographic area covered)? List any areas excluded that you believe stakeholders might assume are included, but are not. The more specific you are, the less opportunity there is for misunderstanding at a later stage in the project.

   (b) Deliverables – What will the project deliver as outputs? Where you can, describe deliverables as tangible items like reports, products, or services. Remember to include a date that each deliverable is expected. You'll use this information to monitor milestones.

   (c) Constraints – What things must you take into consideration that will influence your deliverables and schedule? These are external variables that you cannot control but need to manage.

(d) Assumptions – What assumptions are you making at the start of the project? If necessary, schedule work to confirm these assumptions.

2. Provide broad information about how the project will be implemented. Outline how the project will roll out by defining time-lines, resources, and management stages. This is a high-level overview that will, as the project proceeds, be supported by more detailed project planning documents.

   (a) Project Control – How will progress be monitored and communicated?

   (b) Quality Control – How will the quality of deliverables be evaluated and monitored?

   (c) Risks – Consider any risks that may affect the project, the likelihood of their occurrence, and their possible impact. Include mitigation strategies against the risks that the team has identified.

   (d) Project Controls – To help schedule and measure projects. Think about whether the project requires Key Performance Indicators (KPI)?

The second main output is the business case. The business case argues why the project should be approved. The business case describes the effect the project will have on the business, and supports it with a detailed account of the risks that should be considered. The business case includes:

1. Benefits – Why are you carrying out this project, and what benefits do you expect it to deliver? Include information on how these benefits will be measured.

2. Options – What other courses of action were considered as this project was designed and developed?

6.3.3 Project Scope Definition

Project scope is the definition of what the project will do by defining the project boundaries in the information, process, and organizational view. The project scope defines:

1. Functions that will be analyzed and designed.

2. Project schedule.

3. Project budget.

4. Quality.

5. It is sometimes useful to also explicitly specify what is out-of-scope. The output is a scope statement or sometimes a statement of work (SOW).

Confusion about scope

In a project to design a small manufacturing facility in Nicaragua, one of the in-scope tasks was facility layout. Throughout many discussions between the clients and the consultant team the term "facility layout" was used. Just prior to going to contract, the consultant team became aware that the clients had a very different notion than the consultants of what facility layout entailed. When the clients said facility layout, they meant the entire design of the facility, including the layout of the machines, workstations, and so forth, but also, the design of the HVAC, layout of the water, electrical supplies, and all the systems required in the building. The consultant team had a more narrow, industrial engineering definition of facility layout that pertains only to the layout of the machines, workstations, and definition of the workflow. While this misunderstanding was cleared up prior to going to contract, it illustrates the importance of a clear, thorough scope statement. The final scope statement explicitly stated that plumbing, electrical design, and structural design were not included in the project.

When defining scope the enterprise engineering team needs to identify all the system interfaces between the system of study and external systems. Generally, the system being designed is part of a larger system and interacts with other systems. How do changes to this system affect other systems? While this might be considered out of scope, it is important to understand how design decisions impact globally the super-system.

6.3.4 Project Budget

A high-level estimate for all work elements is calculated during the Project Initiation phase to support the business case. If the project is approved, then more detailed cost estimates are accumulated in the Project Planning phase.

6.3.5 Project Schedule

The time-line denoting when activities start and end, when deliverables are due, and when major milestones are reached. The project schedule includes:

1. Task definition – What major tasks (with milestones) will be completed during the project?
2. Schedule – Provides a timeline of the estimated time for the project activities. The schedule is often provided as a Gantt chart.
3. Effort estimation – Determines how many person-hours are needed for each activity. Estimates the number of support staff required. The end result is an estimate of the total person-hours for the project.

6.3.6 Project Approval

The first review and approval is after Project Initiation. The project manager submits the Project Charter and Business Case to the executive steering committee. The executive steering committee reviews the documents and decides whether the project is worthwhile to proceed. The outcome can be either proceed, make revisions and resubmit, postponement, or outright denial. It is possible the project charter and business case are good, but the executive steering committee might be reviewing multiple projects and must allocate scarce resources. One large South Florida corporation generates almost 100 project proposals each year but only has the resources to do about 20 of them. So, some projects are never done, even if they have some merit because there are more critical projects that need the same resources. The approval is the first project milestone.

6.3.7 Business Case

The business case is a document that makes the argument why the enterprise should do the project. The business case includes information on the project background, expected business benefits, expected project costs, and the expected risks of the project. For an enterprise engineering project, the business case explains how the proposed project supports the enterprise strategy and conforms to the enterprise architecture.

The business case almost always contains a cost-benefit analysis. A cost-benefit analysis calculates the net present-value of a stream of cash flows over a time horizon. A project is only economically feasible if it shows its worth in a time-discount calculation (the net present-value being the most common) against the alternative uses of the investment capital. To calculate the net present value, the project team must determine four parameters: costs, benefits, time horizon, and discount rate also called the minimum attractive rate of return (MARR). The time horizon is set to be the expected operational life of the proposed system. The discount rate is the return below which the enterprise will not consider investing in the project. The discount rate takes into account other uses of the capital and the project's risk. The discount rate is established by the enterprise, usually the Chief Financial Officer. If the project was considered to be low risk, then the discount rate used might be 9%, whereas for a higher risk project, the discount rate used might be 15%.

The project costs are composed of non-recurring costs and recurring costs. The project costs are non-recurring, they are done once during the project and not repeated. The estimate of non-recurring project costs are derived from the project budget. Recurring costs are those costs that will occur each year the system is in operation. These costs include maintenance costs, operational costs, labor costs, and material costs expected to be incurred over the time horizon.

6.4 Project Planning

Project planning defines all the work activities, the schedule, and the budget needed to complete the project. Project planning ensures all the prerequisites are in place for successful project execution. Project Planning builds upon the work completed during Project Initiation.

Figure 6.6 shows the activities, inputs, and outputs of the Project Initiation phase. Project planning is only done if the Project Charter is approved by the executive steering committee. The Project Charter is the primary input to project planning. Project planning takes the project charter and expands on it to provide sufficient detail to execute the project. The primary output is a Project Plan. The project plan is a document used to manage the project execution. There are many individual plans within the project plan. Important elements of the project plan are: the selected enterprise architecture, the selected life-cycle methodology, the work breakdown structure, the project schedule, project budget, project policies, and identification of project resources. Here we focus on the schedule and budget because the time and cost to do a project is often an important consideration in whether to even do the project.

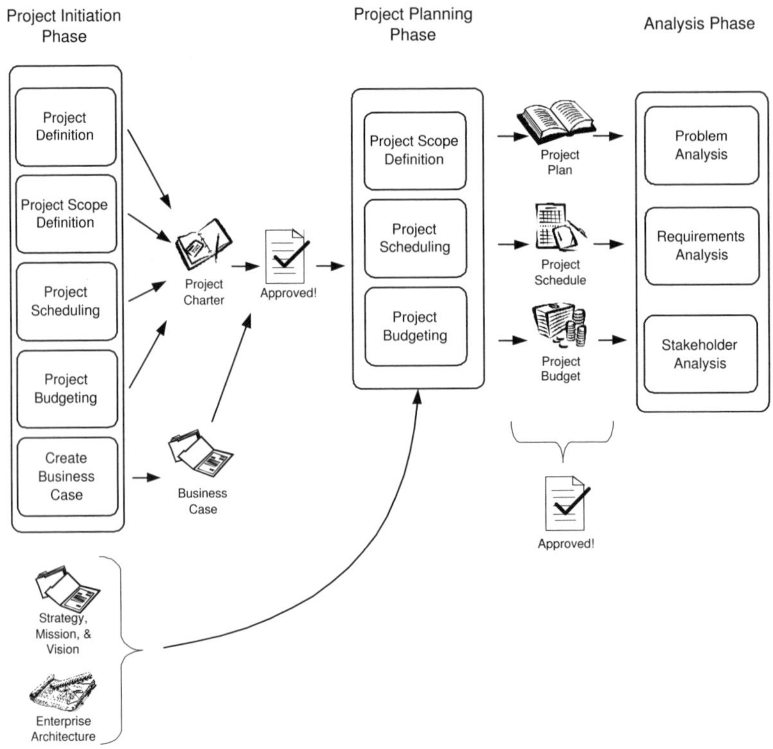

FIGURE 6.6
Project planning.

6.4.1 Work Breakdown Structure

The difficulty in project planning is how to estimate the schedule and budget prior to starting the activities. To estimate the schedule, the project team first creates a *work breakdown structure* (WBS). The WBS describes all the work tasks that will be performed during the project. The WBS takes the project definition of "design an enterprise system" and decomposes it into smaller and smaller subtasks. It is yet another application of the principle of divide and conquer. The WBS is a foundational document since schedules, budgets, and team responsibilities are all derived from it.

The enterprise design methodology is a starting point to identify the work activities to include in the WBS (e.g., project initiation, project planning, problem and requirements analysis, etc.). Using the enterprise design methodology, each activity is further broken down until each task defines a unit of work that will be the responsibility of a single individual. Each WBS task is a discrete unit of work that can be separately scheduled and budgeted. Figure 6.7 shows a partial WBS for a project. Some guidance to define a WBS includes:

1. Each task should have a well-defined start and end.
2. Each task should be easy to track in terms of the schedule and budget.
3. Each task should have a single individual who is responsible for that task. However, multiple team members might work on the task.
4. Each task should have a budget allocated to it.

Tasks are defined at a level of granularity that is neither too small nor too large. If too

1. PROJECT MANAGEMENT
1.1. Project Planning
1.1.1. Project Initiation (Pre-Kickoff)
1.1.2. Project Planning and Support
1.2. Project Support
1.2.1. Project Management (From Kickoff)
1.2.2. Integration/Architecture Support
1.2.3. Change Management
1.2.4. Project Closure
1.2.5. Project Team Training
1.2.6. Project Team Travel
1.2.7. End User Preparation & Delivery
1.2.8. Functional Consultant
1.3. Project Control
1.3.1. Requirements Management
1.3.2. Configuration Management
1.3.3. SQA Reviews
1.3.4. Peer Reviews
1.3.5. Informal Defect Reviews
2. PROJECT INFRASTRUCTURE
2.1. Infrastructure / PE Support
2.2. Hardware
2.3. Software Licenses
3. PROJECT EXECUTION
3.1. Project Miscellaneous

FIGURE 6.7
Partial WBS.

small, the tasks will require too frequent reporting. If too large, the task might go months before anybody realizes something is wrong. The decision is subjective and depends to some extent on the overall project duration. If the project is a 1-year project, then having a WBS task with a 1-day duration becomes tedious to manage without apparent value, whereas having a task of 6-month duration might lead to problems not being recognized in time for corrective action.

The overall WBS is complete in the sense that if a task consumes resources or time, then it is included in the WBS either explicitly as its own task or as part of another, larger task. This means that every activity such as writing reports, training personnel, or acquiring software is included because all of them either consume resources or take time (consume schedule). The WBS adheres to what the Project Management Institute calls the 100% rule. This means that once the WBS is completed, it includes 100% of the work required by the project scope and captures all deliverables. As a result, the WBS defines the total work scope of the project.

6.4.2 Estimation

For each WBS task, the project team needs to estimate the number of person-hours required to complete the task. Estimating is the process of forecasting how much time and resources are required to complete the task at an acceptable level of performance. Estimating is about predicting the future, which cannot be done with 100% accuracy. Moreover, lower-level WBS tasks will evolve as the project progresses because it is difficult to foresee every task and the duration of the task. However, the top-level activities in the WBS should be stable. Change should only really occur in the breakdown of the main activities.

To develop an estimate for a task's duration, the project manager assigns tasks to team members for estimation according to their areas of expertise and/or responsibilities. Each team member develops an estimate based on the task size, complexity, and experience on past projects. See Stellman and Greene [4] for a review on estimation techniques. Team members can review previous projects and find out how long similar tasks took to complete. Depending on the complexity, team members may adjust the time upwards or downwards. In preparing the estimates, each team member provides justification for how they developed the estimate.

The task-duration estimates are uncertain because we are forecasting into the future. A good practice is to provide bounds on the estimate in terms of the worst case and best case scenarios. This allows the expression of the uncertainty in the estimation, and this data can be used by the project manager in a Monte Carlo simulation to estimate overall project performance.

Estimation Example

To estimate the time required to convert paper medical records to electronic format for an electronic medical records project, the team member in charge reviews previous projects to see what the actual times were and the number and size of medical records converted. The data from three previous projects are:

project	# medical records	average # of pages	total duration (hrs)
#1	5000	7	238
#2	2300	12	138
#3	8600	6	444

Based on this data, the average time per page is a total of 819.833 hours / 114200 pages = 0.00718 hours/page or 0.43 minutes per page. She then reviews a sample of the medical records from the previous projects vis-a-vis the current project to see if she should adjust the time per page for differences in complexity. She estimates there are approximately 4000 pages to convert. She multiplies the average time per page (0.43 minutes/page) times the 4000 pages to arrive at an estimate of 28.67 hours.

Two common problems are over-estimation and under-estimation for reasons other than the uncertainty involved. Team members sometimes intentionally over-estimate because they think that they will be held strictly to the time regardless of how the project unfolds. The project manager can help avoid over-estimation due to this reason by reassuring team members that they can revisit and change the estimate if the scope changes or if they made an honest mistake. Under-estimation is more likely due to inexperience, or if the system owners are external and you are competing for the project, then sometimes people intentionally under-estimate to win the project. Again, this is not a good practice and needs to be discouraged. Once the project is approved, under-estimation to win the project is no longer a factor. If there are inexperienced team members, then the project manager needs to make considerations for this inexperience by bringing in experienced SME to review the estimates or adding this to the risk management plan (discussed in Section 6.2.5).

Some activities are estimated according to a parametric approach. A parametric estimate looks at the history of similar projects and relates the duration of an activity to some parameter of the project such as size, complexity, or technology. A regression is performed using each past project as a data point to arrive at an equation that estimates the project duration as a factor of other measurable project characteristics. In ERP projects, configuration management and control is often estimated parametrically as roughly 3% of the total project duration. Other activities that are estimated parametrically include training,

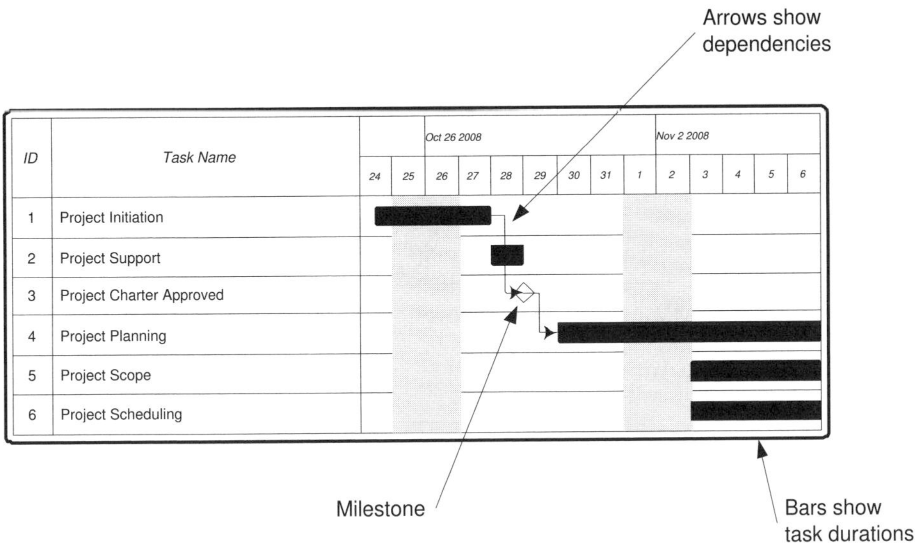

FIGURE 6.8
Gantt chart.

project management, and quality assurance. A parametric estimation method can also be used to verify the reasonableness of the bottom-up estimate.

6.4.3 Scheduling

The schedule defines the start and end for each item of the WBS. The estimates of each individual task duration are meant to account only for the actual work content of that task. The project manager needs to account for less than full efficiency, a usual factor is 80% efficiency. The task estimates are modified by this efficiency. Then the project project manager converts the time estimates for each task into durations for the project schedule. For example, if the time estimate provided by the team member is 28.67 hours to convert paper medical records into electronic format, then the project manager adjusts this by the efficiency rating of 80% to arrive at 35.8 hours (28.67 hours / 0.80). If the work day is 8 hours per day, then the conversion task should be scheduled for 5 days.

The project manager must also assign resources to each WBS task. Each WBS task requires a certain skill, knowledge, or capability. The project manager identifies what these task requirements are, and then assigns the project resources that have the job classifications meeting the task requirements. Assigning resources to tasks will help the project manager determine whether there are sufficient resources and if there are any project risks associated with too few resources. For example, suppose a project has three tasks that will be done in parallel, and each task requires a database administrator. If the project has only a single database administrator, then clearly there is a problem because the person cannot simultaneously do all the tasks. In this way, the project manager identifies problems early and can take action to avoid or rectify the problem.

A common way to present a schedule is the Gantt Chart shown in Figure 6.8. The Gantt Chart shows each WBS element (task), its start and end dates, milestones, and the dependencies between tasks. Time is shown horizontally, and a vertical line can be drawn to show the current date.

6.4.4 Budgeting

The Project Budget is an estimation of the cost to complete the project. It includes labor cost and non-labor costs (hardware, software, travel, etc.). In many projects, the labor cost is the greatest component of total project cost. It is also more difficult to estimate than the non-labor costs.

To estimate labor costs, the starting point is the WBS, each task duration, and the labor rates for each job category. For each WBS task, the project manager assigns a team member, who determines the labor rate for that WBS task. The most accurate task cost estimate is according to what is called the Level of Effort (LOE), which is the labor rate multiplied by the estimated number of hours to complete the task. This estimate is done for every element in the WBS to obtain an overall budget estimate for the labor content.

Level of Effort (LOE) Cost Estimation

Suppose a project has a WBS task "conduct time studies of order-entry process." The project manager has determined the business process analyst will do the data collection. The labor rate for a business process analyst is $45 per hour. The total estimated time for the task is 50 hours. The LOE cost estimate for this WBS task is 50 hours × $45/hour = $2250.

The parametric approach is also used to budget some tasks. There is a large body of knowledge on approximately what various activities should cost. For example, in ERP projects the training cost is usually between 5% and 10% of the total project cost. The parametric approach can be used for those work tasks where labor estimates are difficult (e.g., training, quality assurance, configuration control). The parametric approach is also useful to validate the cost estimation accuracy. If the project cost estimates do not fall within ranges deemed appropriate for the work task for that type of project, then the project team reviews its scheduling and budgeting assumptions.

The non-labor portion of the budget includes software, hardware, and external consulting. The project team estimates these costs by obtaining quotations from vendors. The estimates take into consideration ancillary costs of having the equipment. For example, what use is a printer without ink or paper?

All estimates are verified to the extent possible. This can be done by checking with vendors, consultants, competitors, etc.

Parametric Cost Estimation

Suppose the company has done several related projects in the past. For each of these projects the company knows the actual cost. So, for example to estimate the cost of training in the current project the company can measure the size, duration, scope of impact, degree of difficulty, and other parameters of previous projects. Then the current project is estimated based on how it compares with these previous projects. If the current project is deemed to be half the size and complexity of a previous project, and the previous project had a training cost of $300 per person, then a reasonable training cost estimate might be 0.5($300) = $150 per person.

6.4.5 Other Project Plans

The WBS, the project schedule, and the project budget are the main components of the project plan. However, there are many more project plans that are relevant, especially for larger projects. Some of the more common project plans are:

- Risk management plan – a plan to identify all project risks, to devise strategies for dealing with the risks, and to establish a risk monitoring system.
- Communications plan – a plan to disseminate information within the project team (team members) and to all the stakeholders (external and internal). The communication plan establishes expectations and project cultural norms. It is also a key component of the change management process required to implement the project deliverables.
- Configuration management plan – a plan to identify the configuration of the system (i.e., processes, organization, and information structures) at given points in time, to systematically control changes to the configuration, and to maintain the integrity and traceability of the configuration throughout the system life-cycle. A standard describing a configuration management plan is IEEE Std 1042 – Guide to Configuration Management Plan (SCMP).
- Quality management plan – a plan to identify which quality standards are relevant to the project and how to satisfy them. The quality management plan establishes a quality assurance (QA) process and assigns a team member as the QA manager to verify the quality and compliance of all deliverables. Part of QA is defining a test strategy for the project.
- Requirements management plan – a plan on how to gather requirements, analyze those requirements, and management the requirements. In Chapter 8 on requirements, we discuss this in greater detail.

6.5 Analysis

Analysis is a systematic approach to study a problematic situation, understand it, and derive requirements for a solution to alleviate the problems. For the project team to understand the problem situation, it needs to analyze it. Analysis occurs after the project planning is completed. Figure 6.9 shows the analysis phase, its activities, inputs, and outputs. The principle inputs to analysis are a completed Project Plan that includes the schedule and budget. These documents describe the scope of the project and provide direction for how analysis proceeds.

The activities for analysis are:

- Problem Analysis – Collect data to understand the problems, document the problems, and then analyze the causes and effects for the problems.
- Requirements Analysis – Elicit the requirements from the enterprise, document the requirements, and analyze the requirements.
- Stakeholder Analysis – Identify the key stakeholders, understand their interest in the project outcomes, and devise a strategy for dealing with each stakeholder.
- Process Analysis – Identify the processes impacted by the project, model the processes, and analyze their performance.

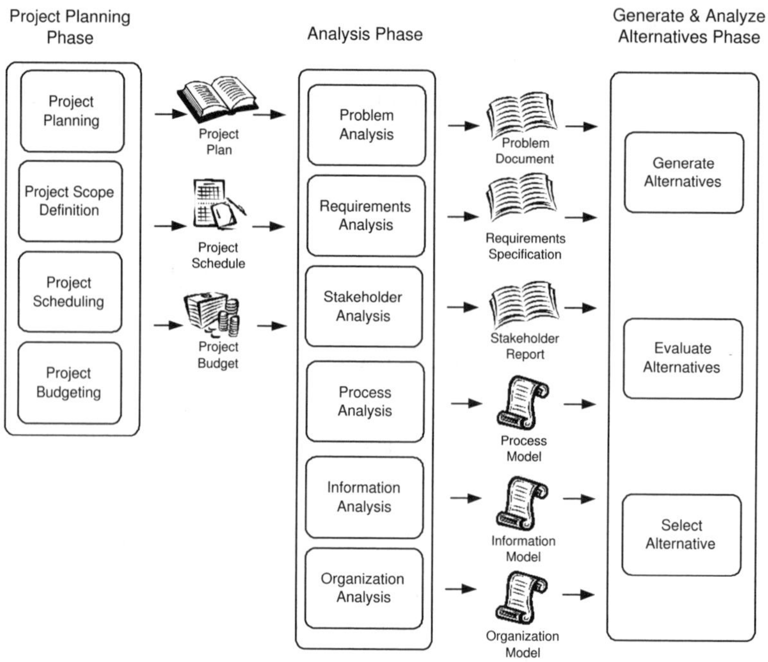

FIGURE 6.9
Analysis phase.

- Information Analysis – Identify the information objects impacted by the project, model the information, and analyze the availability and quality of the information.
- Organization Analysis – Identify the organizational units impacted by the project, model the organizational units, and analyze their structure and performance.

The outputs of the analysis phase are the documents and models for each of the analysis activities. The analysis models describe the as-is enterprise. Modeling is conducted across the three views of information, process, and organization. While analysis in each view can be conducted semi-independently of the other views, it is necessary to check whether there is consistency across all the views. For example, a process requires information input that is missing from the information model. To reveal these integration problems, the analysis sub-teams have regular meetings, use a CASE tool, and the project manager regularly ensures the requisite integration is taking place.

Analysis should be illuminating to the enterprise stakeholders. It is neither done for the sake of analysis alone, nor is it done to a rigor that does not increase understanding of the problem situation. Analysis is an art as much as a science. A good analyst can elicit the information required for analysis from the stakeholders, and then, the analysis can abstract out a process description. Some authors outright dismiss the need for analysis. Hammer and Stanton [3] say, "There are two problems with conducting analysis in the reengineering context. The first is that it is a profound waste of time ... the second is that it can inhibit change." We disagree and feel this statement misinterprets the purpose of analysis. First, analysis is not the documentation of the existing system; analysis should focus on uncovering the problems, issues, and goals of the system with the aim of developing a new system that avoids those problems. Second, the level of detail should be appropriate to understand the process. The analysis team adopts a perspective of creeping commitment, meaning that as the project progresses and it obtains more information and knowledge, then it becomes more committed to the project. A high-level analysis understands the problems, issues, and the process. Decisions are made as to whether redesign the process, reinvent a new process, or even outsource the process. Then depending on the decision, further analysis may or may not be done. This approach addresses the criticisms of Hammer and Champy [2] as well as others.

6.6 Generate and Evaluate Alternatives

The purpose of the Generate and Evaluate Alternatives phase is to identify alternative solutions, evaluate those alternative solutions, and then recommend the alternative that will become the design solution. The project team tries to generate as many alternative solutions as possible with the goal of finding the best alternative. Figure 6.10 shows the Generate and Evaluate Alternatives Phase, its activities, inputs, and outputs. The activities for this phase are:

1. Generate Alternatives – To generate alternatives requires the creative efforts of the design team to design solutions to the problems identified that satisfy the requirements. The alternatives can be to design a system completely, to purchase a Commercial Off-the-Shelf (COTS) system, to outsource activities, a combination of these alternatives, or some other creative design idea.
2. Evaluate Alternatives – The project team needs to define the evaluation criteria and then to evaluate each alternative with respect to the evaluation criteria. Common evaluation criteria are cost, schedule, risk, and performance.
3. Select Alternative – Given the alternatives and the evaluation criteria the project team needs to determine which alternative to select. Once selected, the project team then needs to present the case for the design alternative to the executive committee for approval. If approved, the alternative becomes the design solution that will be designed further in the next phases.

6.7 Design

Once an alternative is approved, the design phase needs to evolve it into a complete specification for a new enterprise system. Figure 6.11 shows the activities, inputs, and outputs for the design phase. The activities are:

- Design Process – The process goals, inputs, outputs, and activities must be defined and specified. The required resources for each activity must be specified. The process design is delivered as a collection of process models, policies, and associated process documentation.

- Design Information – The information content and structure must be defined. The information design is delivered as an information model, data dictionary, and other supporting documentation.

- Design Organization – The organizational structure, positions, roles, culture, and policies

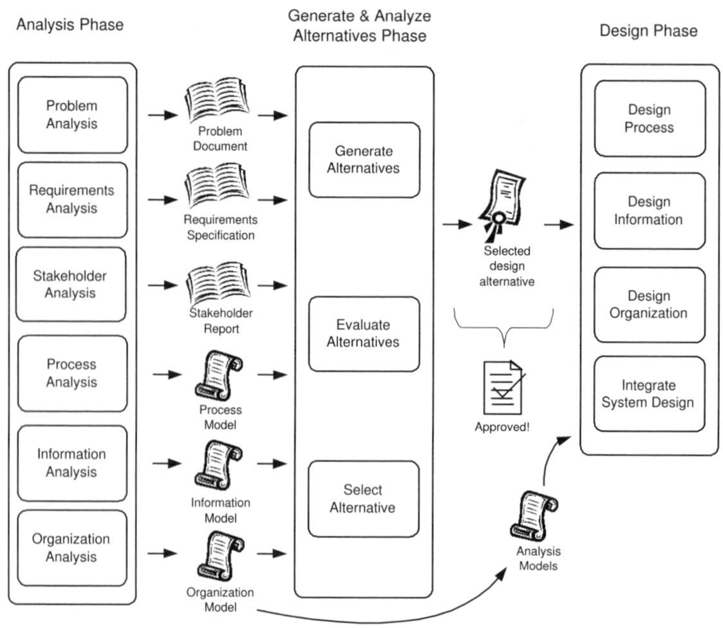

FIGURE 6.10
Generate and evaluate alternatives.

must be defined. The organizational design is delivered as an organizational chart, job descriptions, policies and procedures manual, and other supporting documentation.

- Integrate System Design – The process, information, and organizational designs must work together as a single harmonious system. Throughout the analysis and design process, the project team regularly identifies and validates interfaces between the views. The final system integration task reviews the views and ensures they are integrated.

The output of the design phase is the design specification. This is a technical document describing in precise detail all the facets of the design solution. It directly addresses the information, process, and organizational viewpoints.

6.8 Construction

The construction phase, shown in Figure 6.12, takes the design specifications, and purchases necessary systems or equipment, builds systems, writes the work policies for processes and organizations, integrates the systems, and tests the systems. The activities in the construction phase are:

- Purchase Equipment and Materials
- Write Work Policies
- Build Systems
- Acquire COTS and Software

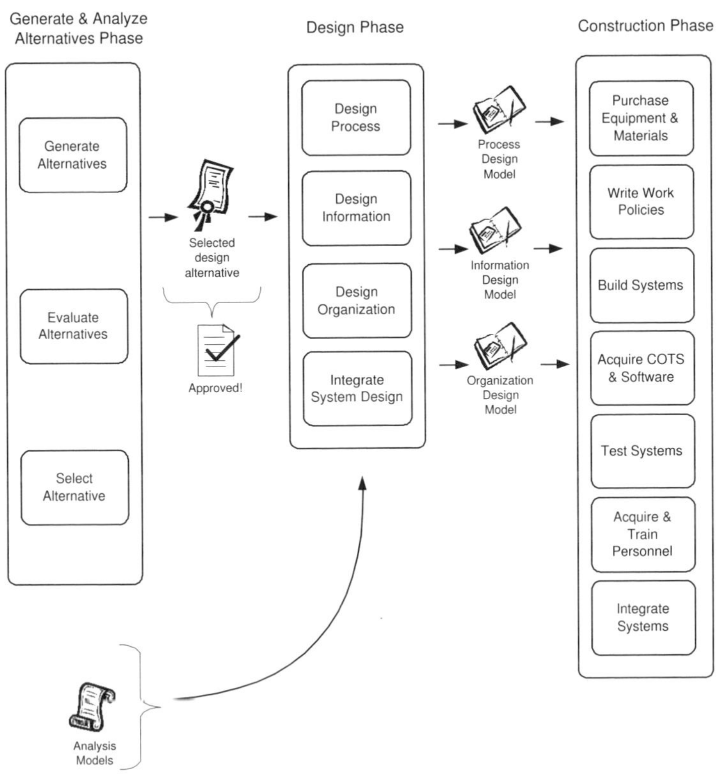

FIGURE 6.11
Design phase.

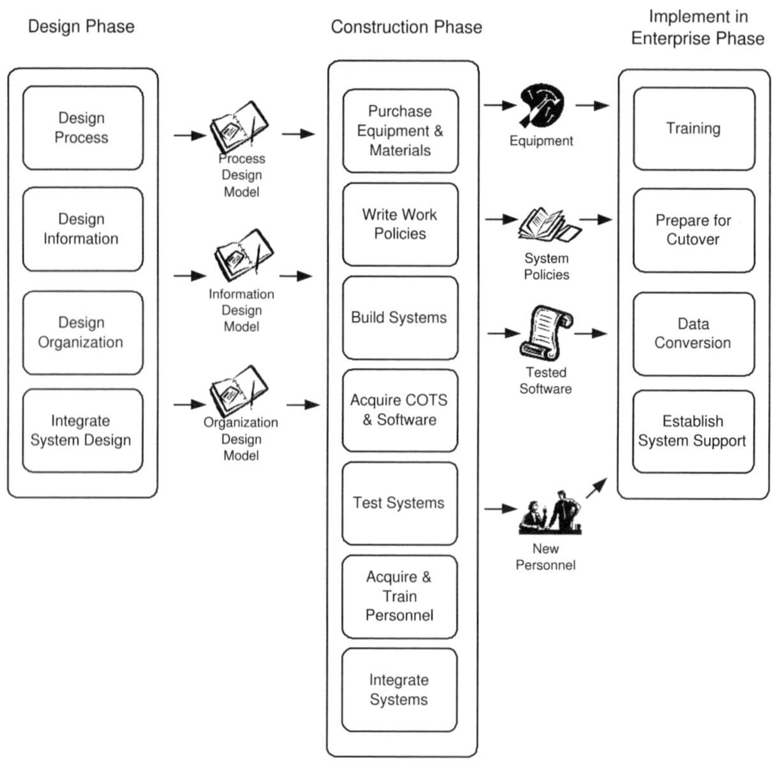

FIGURE 6.12
Construction phase.

- Test Systems
- Acquire and Train personnel
- Integrate Systems

6.9 Implementation

The implementation phase, shown in Figure 6.13, takes the new enterprise systems (processes, information, and organization) and installs them in the enterprise. Implementation of enterprise systems is a change management process. The enterprise employees need to convert from the old way of working to the new. The activities in the Implementation Phase are:

- Training. All affected employees should obtain additional training on the new system and policy changes.
- Prepare for cut-over. The cut-over or "go live" can be done in three ways:

    1. Direct cut-over is when the enterprise switches from the old system to the new system all at once.

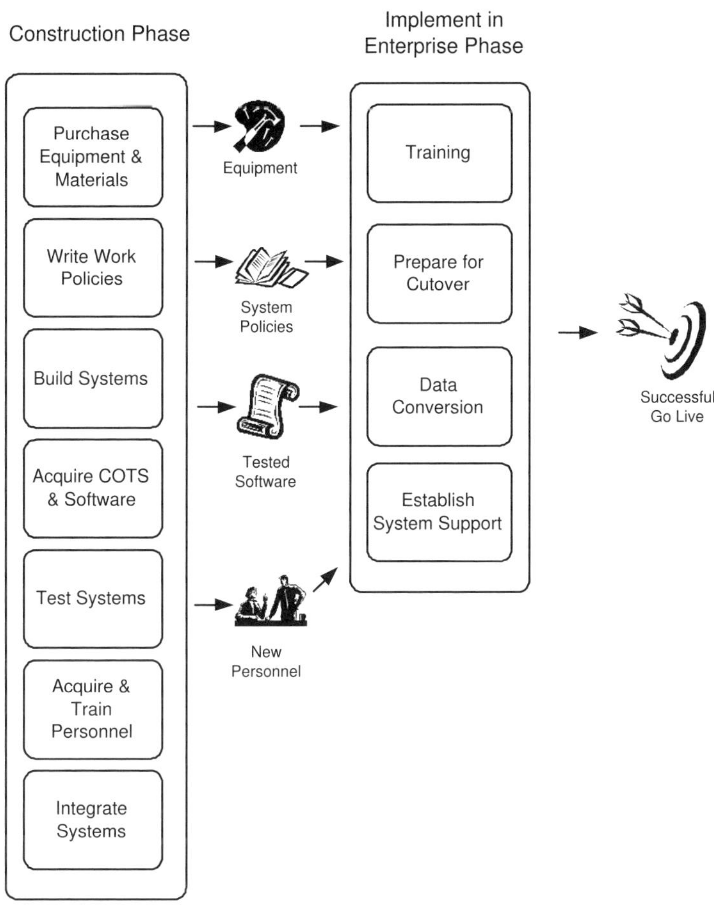

FIGURE 6.13
Implementation phase.

2. Phased cut-over is when the enterprise implements the system in phases such as the financial system first, then the manufacturing system, and so forth until the entire system is implemented.
3. Pilot cut-over is when the enterprise first runs a pilot program where it implements the entire system in a small part of the enterprise before launching it enterprise-wide.

- Data conversion. The data from the old system are extracted, transformed, and loaded into the new system.

- Establish system support. A help desk and other support mechanisms are established to provide constant user support for the new system.

A successful implementation results in the enterprise changing from the old way of operating to the new way of operating as defined by the system.

6.10 Project Tools

An enterprise will maintain project tools consisting of templates, spreadsheets, databases, checklists, and software applications to help the project team generate the deliverables. These tools are available from the accompanying Website.

6.11 Summary

This chapter presented the enterprise design methodology consisting of seven phases covering the life-cycle of an enterprise project. The chapter described the cross life-cycle activities of project management, requirements management, risk management, configuration and control, and quality assurance. The chapter described the inputs, activities, and outputs of each phase in the methodology. The methodology is based on the problem-solving approach and allows for iteration throughout the life-cycle phases. The end result of the methodology should be a successful project. A project is successful if it delivers all required functionality on-time and under-budget.

Review Questions

1. List the seven life-cycle phases of the enterprise design methodology.
2. What are the three dimensions of project success?
3. List the cross life-cycle activities.
4. Explain what makes an activity a cross life-cycle activity.
5. Explain why the problem-solving activity of design appears at some level of effort in each life-cycle phase.
6. What is a configuration item?
7. List the deliverables that the quality assurance sub-team should audit.
8. List the potential risks that could affect a large ERP project.
9. Briefly describe several different ways to deal with project risk.
10. Compare and contrast the risk-handling strategies or mitigation and transference.
11. Explain what would likely happen if the project scope is expanded and no changes are made to the project schedule or budget.
12. Why is the quality assurance sub-team linked to the project manager and not associated with the functional sub-teams of accounting, manufacturing, etc.?
13. Compare and contrast two approaches to estimating activity duration, and explain when each approach would be used.
14. Contrast the purpose of the Project Charter and the Business Plan.
15. List four stimuli for starting an enterprise engineering project.

16. What is the difference between a deliverable and a milestone? Can a milestone be a deliverable?
17. What is a work breakdown schedule?

Exercises

1. Which event is a greater risk to project success: Event A whose probability of occurrence is 20% for a $10,000 budget impact, or Event B whose probability of occurrence is 10% for a $25,000 budget impact? Show your reasoning.
2. Create a work breakdown schedule for the following tasks:
   (a) To analyze, design, and build a portal for students to access their grades at the university.
   (b) To analyze, design, and build a pre-paid card system for the student cafeteria.
   (c) To analyze, design, and build a contacts database that keeps track of names, addresses, telephone, and email.
3. Explain and provide formulas for how you would estimate the following task durations:
   (a) Time to conduct interviews of line-level supervisors. You know the number of supervisors and the questions to ask each supervisor.
   (b) Time to train nurses on a new time entry system to record their start and finish times of each shift. You know the number of nurses, departments they work in, the maximum number of nurses per training session, and the amount of time to do a single training session.
   (c) Time to conduct user-acceptance testing of a new course registration system. The system has twenty transaction types, each transaction takes approximately 10 minutes. A previous estimate is that to instruct the users will take 20 minutes, they test their assigned transactions, and then the post-test questionnaires will take 20 minutes. Each user will test four transactions. The test room has ten computers (maximum of ten users at a time). You want to test each transaction type eight times.
   (d) Suppose you work for a consultant company that implements electronic medical records for doctor's offices. You have data on the requirements analysis duration for previous projects as follows:

Number of Patients	**Number of Doctors**	**Time (hrs)**
1700	1	29.5
2500	2	44.2
1400	1	26.8
3100	3	58.9
2200	2	41.8
1200	1	21.5
4700	4	81.4
2900	3	57.6

   Do a regression analysis to develop an equation that can be used to estimate the requirements analysis duration for a project that has 1900 patients and 1 doctor.

4. Search on the Web for project management software tools. Include in your search such sites as the Project Management Institute (www.pmi.org). List and provide a brief description of some software tools that are available to help with various project management activities.

Bibliography

[1] H.T. Goranson. *The Agile Virtual Enterprise: Cases, Metrics, Tools.* Quorum Press, Westport, CT, 1999.

[2] M. Hammer and J. Champy. *Reengineering the corporation: A Manifesto for Business Revolution.* Harper Business, New York, NY, 1993.

[3] M. Hammer and S.A. Stanton. *The Reengineering Revolution: A Handbook.* Harper-Collins, New York, NY, 1995.

[4] A. Stellman and J. Greene. *Applied Software Project Management.* O'Reilly, Sebastopol, CA, 2005.

Part II

Enterprise Project

7

Strategy

> *"However beautiful the strategy, you should occasionally look at the results."* – Winston Churchill (1874-1965), Prime Minister of the United Kingdom 1940-1945 and 1951-1955.

Enterprise design is a strategic endeavor because it has a long-term, fundamental effect on the enterprise. This chapter discusses the types of goals enterprises pursue. Enterprises determine their purpose and vision, frequently to seek long-term profits and growth. Two main schools of thought on strategy are the resource-based view and the market-based view. The first says a company's competitive advantage is derived from unique core capabilities the company possesses. The second says competitive advantage is derived from the company's position in the marketplace. The concepts are complementary, and both are considered in the strategy formulation method presented. The chapter also shows how to identify the core capabilities and how to analyze the company's market position via the strengths, weaknesses, opportunities, and threats matrix. After completing this chapter, you should be able to:

- Describe how mission, vision, and strategy fit together.
- List the types of strategies an enterprise might have.
- Differentiate between the resource-based view and the market-based view of strategy.
- Identify core capabilities in an enterprise.
- Describe the steps and action necessary to formulate strategy.
- Be able to conduct a SWOT analysis.

7.1 Strategy Definition

In an enterprise there are three decision levels: strategic, tactical, and operational. Strategic-level decisions are long-term decisions that have broad impact on the enterprise. Tactical-level decisions are medium-term decisions that impact individual business units. Operational-level decisions are the day-to-day decisions made at low levels. The strategic decisions set guidance and constraints for lower-level decisions. For example, a strategy to pursue market differentiation by providing superior quality and customer service becomes the guiding criteria for tactical and operational decisions in the enterprise. The strategy is long-term, which is usually defined as five or more years out, but will differ by company, industry, and country.

As described above, *Strategy* is a plan to achieve the long-term goals of an enterprise. It is expressed in the pattern of decisions and actions taken about the operations of the enterprise. When formulating the strategic plan, the enterprise needs to evaluate the dynamic environment it operates in, consider the actions of its competitors, and determine how to meet its goals by effectively deploying its resources, competencies, and knowledge. The result hoped for in any strategy is for the enterprise to gain a sustainable competitive advantage over its competition. *Competitive advantage* is the ability of an enterprise to achieve higher than industry-average profits [11]. Sustainable means that it is not temporary, but maintained over a long period of time.

The strategic plans an enterprise makes are developed to help the enterprise attain its goals. All enterprises have goals, and there are only a handful of goals that make sense for most enterprises. For-profit enterprises must have as a basic goal to make money, or more precisely to maximize shareholder value.[1] Shareholder value is derived from the profits the enterprise generates compared to the costs to generate those profits, which is the *profitability* of the enterprise. To ensure long-term profitability, an enterprise needs to satisfy its customers and grow its business. This explains why many startups forgo profits in their early years in order to build market share, which is intended to maximize shareholder value over the long term. Non-profit enterprises might have a goal to help as many people as possible, or to have the greatest measurable impact possible. Government and quasi-government enterprises might have a goal to provide a particular service.

Achievement of long-term profitability implies market growth as an enterprise goal because in order to increase profits, an enterprise must either grow market share or must improve its products and services in terms of quality, timeliness, or cost so as to derive more profits from the same market share [1, 13]. Growth can be measured in terms of revenues, profits, return on investment, number of customers, or market share. In a growing market all companies in the market may grow. In a stable market or a declining market it is a zero-sum game; in other words, growth is at the expense of a competitor.

To communicate the purpose and goals of an enterprise, the organization develops a vision statement, a mission statement, and a statement of core values. A *vision statement* articulates the enterprise goals, describing what the enterprise aspires to be. It states audacious goals of the enterprise and reflects the values of the enterprise. The *mission statement* is a brief description of the enterprise's purpose that projects the enterprise's image to customers. Usually, the mission statement describes how the enterprise provides a product or service which satisfies the needs or desires of one or more customers. The mission must usually be accomplished in competition with other enterprises vying for the same business. A mission statement does not change frequently. For example, the mission of the U.S. Army is to protect the United States, and this has been their mission since the Army's inception more than 200 years ago. *Core values* are the deeply held values of the organization that guide the enterprise's internal conduct and how it interacts with its environment. Core values should not change as the enterprise's environmental circumstances change.

Figure 7.1 shows how the enterprise's strategy is a plan to go from the current state, to the desired state, described by the vision statement. The mission statement remains unchanged between the two states.

[1] In economics, the rent theory says the business should make more money than alternative uses of the capital tied up in the business. If the business cannot generate more revenue than alternate uses, then economically it is better to close the business and use the assets for these alternate uses. Consider the case of TWA, which was taken over by the corporate raider Carl Icahn in 1985. Once in control, Icahn proceeded to systematically sell most of TWA's assets to competitors. In 1991, Icahn sold TWA's prized London routes to American Airlines for $445 million. He also took the company private, which let him recoup almost his entire profit and saddled the ailing airline with $540 million in debt. These decisions were made not to sustain a competitive airline, but they were done to enrich the owner, Carl Icahn. In 1992, the weakened TWA declared bankruptcy and soon ceased to exist. – TWA - Death of a Legend by E. X. Grant, *St. Louis Magazine*, Oct. 2005.

FIGURE 7.1
Mission, vision, and strategy.

> **Example from Miami Children's Hospital, Miami, FL** *http://www.mch.com/*
>
> **Mission**
>
> To provide excellent family-centered healthcare to children in an academic environment that meets or exceeds the expectations of those we serve and educate. To collaborate with others in our community to improve the health status of children.
>
> **Core Values**
>
> We will always do what is best for each child. We will always value those who serve children. We will always value diversity. We will always value integrity and honesty. We will always value leadership.
>
> **Vision**
>
> Miami Children's Hospital will continue to be recognized as one of America's best children's hospitals and as Florida's academic center of pediatric clinical excellence providing a number of Centers of Specialty Pediatric Excellence nationally and internationally.

7.1.1 Strategy Hierarchy

Large enterprise will have strategies at different levels of the enterprise – one for each of their subsystems or major functions. What strategies an enterprise does formally develop depends on the particular enterprise. A common hierarchy of strategies is: corporate level, business-unit level, and function level.

Large corporations that engage in multiple, different businesses will have a corporate strategy and then each business unit will have its own strategy. Corporations do not directly make products or provide services; it is their business units that do this. The role of the corporation is to manage the portfolio of business units so that each one is competitive and contributes to the corporate mission. A corporate strategy determines what markets to compete in, what business units to have, how to allocate resources among the business units, what values to promote, how to achieve synergies through sharing and coordinating resources across business units, and how to govern the business units.

To illustrate corporate strategy, take the case of Johnson & Johnson, which has over 250 operating companies located in 57 countries throughout the world. The corporate headquarters that oversees all these companies has a corporate strategy to be a broadly based human healthcare company for the consumer, pharmaceutical, and medical-device markets. The corporate strategy provides a common set of values to unify its worldwide operations. Part of the company's corporate strategy is to broadly compete in the provision of products and services for human healthcare, to manage its businesses for the long term, to use a decentralized management approach, and to develop its people. Within the context of the overall corporate strategy, each individual business unit in Johnson & Johnson has a separate strategy developed by the senior management groups running those business units.[2]

Many large enterprises exhibit a similar corporate structure of overseeing multiple business units. Instead of the term business unit, they may be called divisions, product groups, or national operating companies. Within a single business unit, there are strategies for different functional areas. Most business units will have these strategies:

- Product and Marketing Strategy – All enterprises provide a product or service, and they need a strategy for what products to make and how to position those products in the market. The product and market strategy determines the outward face of the enterprise to the clients and the rest of the world.

- Operational Strategy – The operations unit creates the product or service. If it is a product, then it is a manufacturing strategy; if it is a service, then it is a service strategy. The operational strategy determines the overall approach of operations, defining the consistent pattern of decision making done in operations. The decisions include capacity management, manufacturing capabilities, outsourcing, supply chain operations, and inventory decisions.

- Information Strategy – The information an enterprise has is considered a valuable resource and the use of IT is required for all enterprise operations. The information strategy determines the overall approach to developing and deploying information systems to support the enterprise. It describes the relationship between technology choices and business choices.

It is important that the functional strategies are aligned with the overall enterprise strategy [6]. Alignment means the strategic decisions all support each other. When there is misalignment between functional strategies and the business strategy, then lower enterprise performance can be expected depending on other factors such as management tenure [7].

7.2 Strategy Theory

There are two opposing views on how to formulate strategy. The first view is called the resource-based theory of strategy, which emphasizes the acquisition of resources and development of capabilities as the basis of competitive advantage. The second view is called the market-based theory, and says an enterprise should identify an explicit position in the market and develop a strategy to dominate that position. The first is internal and the second is external. We examine these two theories.

7.2.1 Resource-Based Theory of Strategy

The resource-based theory of strategy is based on two premises: first, the resources and capabilities of an enterprise are the basis for generating profits, and second, the enterprise strategy should be based on exploiting these resources and capabilities. *Resources* are inputs into an enterprise's production process. The resources may be either tangible or intangible in nature. Tangible resources include real estate, production facilities, raw materials, equipment, and other readily identified objects. Although these tangible resources may be essential to the enterprise's strategy, due to their standard nature, they can rarely provide competitive advantage. It is just too easy for competitors to acquire similar resources. On

[2] 2007 Annual Report, Johnson & Johnson, http://www.investor.jnj.com/annual-reports.cfm

the other hand, intangible resources such as business culture, reputation, brand name, organizational knowledge, patents, or trademarks are difficult for competitors to imitate, and thus, strategically more valuable.

Prahalad and Hamel [12] say individual resources may not in themselves create a competitive advantage. It is through the synergistic combination and integration of sets of resources that an enterprise creates competitive advantages. The authors describe a *capability* as the capacity for a set of resources, people, and processes to integratively perform an activity. Unlike physical assets, which do deteriorate over time, core competence does not diminish with use, but grows in value. Through continued use, capabilities become stronger and more difficult for competitors to understand and imitate. An enterprise can have the capability for low cost production, high quality production, superior customer service, or fast product development as a few examples. A capability differs from a resource because it describes an end-to-end process or a combination of enterprise resources rather than a single skill or knowledge.

Peteraf [10] uses economics to describe four conditions that must be met for a resource or capability to provide sustained competitive advantage. These conditions are:

1. Valuable and rare resource – Superior resources are only possible if the resources in the industry are heterogeneous. A superior resource must be valuable in that it improves enterprise efficiency or effectiveness. Additionally, the resource must be rare such that by exercising control over it, the enterprise can exploit it to its advantage (and simultaneous disadvantage to its competitors).

2. *Ex ante* limits to competition – The resource must be imperfectly imitable to prevent competitors from being able to easily develop the resource for themselves.

3. Imperfect resource mobility – The resource must be imperfectly mobile to discourage the ex-post competition for the resource that would offset the advantages of maintaining control of the resource.

4. *Ex post* limits to competition – The resource must not be substitutable; otherwise, competitors would be able to identify different, but strategically equivalent resources to be used for the same purpose.

If a resource or capability meets these four criteria then it is a core competency. *Core competencies* are those resources or capabilities that are a source of the enterprise's competitive advantage. The core competence idea is useful to managers not only for focusing them on the essentials, but also for identifying those things that were not "at the core." The enterprise should not expend effort, money, and time on non-essential things that are not at the core. This idea has led to many companies outsourcing non-critical activities such as payroll, IT support, or food services for the employee cafeteria. This allows the enterprise to focus on core capabilities that create value.

To illustrate a core competency, consider Dell Computers' order fulfillment process that allows them to quickly assemble computers to order and deliver directly to the customer. Dell's order fulfillment process is a capability because it involves the integration of many resources, knowledge, processes, and culture. It is not any single resource, tangible or intangible, that gives Dell its competitive advantage. It is its overall business model that Dell developed and improved over many years. Dell's assemble-to-order process, tight integration with suppliers, and IT prowess helps Dell keep inventory very low. This frees up capital, which Dell used to fuel its growth. Because Dell's core competency is the integration of many capabilities, it is not easy for its competitors to replicate. Compaq created an assemble-to-order process and made other changes, but it only got inventory down to 25 days compared to Dell's 13 days. Collectively, Dell's core competency gives it a 10-15% cost advantage over its rivals, which is a good profit in a low-margin business.[3,4]

The key observation of the resource-based view of strategy is that a strategically valuable resource or capability cannot be easily duplicated by rivals. Otherwise, the competitive advantage gained is only temporary. To illustrate, consider the technology that enabled FedEx to let customers track their packages in real-time via the Internet. The technology is widely available and easy to copy. Soon after FedEx introduced the service, its chief rival UPS also introduced the service. Nowadays, tracking packages is expected by customers! This example illustrates two ideas: first, tangible resources such as technology are often easy to duplicate and consequently offer only fleeting competitive advantage; second, what is at first considered a value-added service can soon become an expected service that must be provided just in order to stay competitive.

7.2.2 Market-Based Theory of Strategy

The market-based theory of strategy says that competitive advantage comes from the optimized position of the enterprise in a market vis-a-vis its competitors. In the market-based theory, it is the enterprise's position in the external competitive environment that is examined, not the internal resources or capabilities of the enterprise. The foremost author promoting the market-based theory is Porter [11], who argues compellingly that there are two basic strategies an enterprise can pursue. A company can either provide value more efficiently than competitors by being the low-cost provider, or provide greater value than competitors by differentiating how it provides that value.

Low-cost leadership – The enterprise emphasizes its cost leadership to its customers. In a market, there can usually be only a single low-cost leader.

Differentiation – The enterprise attempts to differentiate its products or services from other enterprises in its market. Differentiation may be in terms of product features, product quality, or any other characteristic valued by customers.

The *low-cost leadership strategy* seeks to be the low-cost producer by achieving economies of scale. The company works to realize the cost advantage by: using process technology to improve efficiency; negotiate for lower raw material and other input costs due to volume discounts; and maintain strict control on all costs, especially overhead costs. Examples of a low-cost strategy can be found in the airline industry with Southwest Airlines, Ryan Air, and Spirit Airlines. These low-cost carriers keep costs down by flying only one type of plane to limit maintenance and other operating costs, fly between smaller airports with lower fees and shorter waiting times, have non-union employees, and provide little services beyond the basic provision of transportation. To illustrate the low-cost mentality, a visit to Spirit Airlines corporate offices will reveal they have no receptionist and there is almost no janitorial staff.

Within a single market only one company can be the low-cost leader. If several enterprises simultaneously pursue this strategy, then you have a price war, and eventually, one of the

[3]Michael Dell: Whirlwind on the Web, by Gary McWilliams, *BusinessWeek*, April 7, 1997.

[4]Kraemer, K.L., Dedrick, J., and Yamashiro, S., Refining and Extending the Business Model with Information Technology: Dell Computer Corporation, *The Information Society*, 16(1), March 2000, pp. 5-21.

enterprises will be driven out of business. Once a company establishes itself as the low-cost leader, this creates formidable barriers to entry against potential competitors.

The alternative to a low-cost strategy must be a differentiation strategy because the enterprise must offer some unique product features or service that the customer values and is willing to pay a premium for. The *differentiation strategy* seeks to achieve a unique position in the market. A company can differentiate itself in one of several ways. It can pursue a variety-based position in which it produces only a subset of an industry's products or services. The company can then develop distinctive capabilities to excel in that one area. For example, Jiffy Lube specializes in automotive lubricants and does not provide any other car repair or maintenance services. They design their shops, processes, and technology so that cars can enter quickly, get their oil changed, and leave. Their competition, the traditional automotive repair shop, cannot compete in both cost and speed for this market. A second way to differentiate is through needs-based position. In this approach the company identifies a particular group of customers. The company then tailors its capabilities and processes to achieve excellence for that single customer segment. Ikea follows this strategy by trying to meet all the home furnishing needs of young customers. Access-based position seeks to differentiate by segmenting customers based on their access to a product or service. For example, Carmike Theatres focuses on rural movie theaters in cities and towns with less than 200,000 people.

7.3 Strategy Formulation

Strategy formulation is the process by which an organization creates a strategy. Interestingly, many academic and business sources on strategy discuss different types of strategies, case studies of successful and unsuccessful strategies, but few provide actual guidance on how a company should go about formulating a strategy. Hamel [5] says, "the dirty little secret of the strategy industry is that it does not have any theory of strategy creation." Mintzberg [9] examined the strategic formulation process and concluded it was much more fluid and unpredictable than people had thought. More recently, Grant [4] assessed this research and concludes that strategy is both due to deliberate design via a strategy formulation process and emergence as the result of multiple decisions at many levels in the enterprise.

Ulwick [14] studied how companies formulate strategy and found that successful companies searched the universe of possible strategies to find the strategy that best satisfies the largest number of desired outcomes while adhering to any constraints imposed on the solution and the competitive position of the enterprise. The approach called the Universal Strategy Formulation Model (USFM) is conceptually a straightforward application of the operations research idea of optimization: search for an optimal solution subject to constraints. The difficulty involved in strategy formulation is that the solution space is ill-defined and ambiguous; there are no easy means to evaluate candidate solutions to see if they are the best; and it is difficult to identify all of the constraints on potential solutions.

Figure 7.2 presents a strategy formulation process that gives equal weight to internal strategy development and external strategy development. The starting point is the enterprise vision, goals, and mission. Using this as the context for the analysis, the enterprise needs to examine its internal resources and capabilities and its external position in the market. To do this a frequently used technique is the SWOT analysis. The SWOT analysis is a technique to examine a company's internal strengths and weaknesses and the external opportunities and threats facing the company in its environment. SWOT stands for Strengths, Weaknesses, Opportunities, and Threats. In a SWOT analysis a team uses vari-

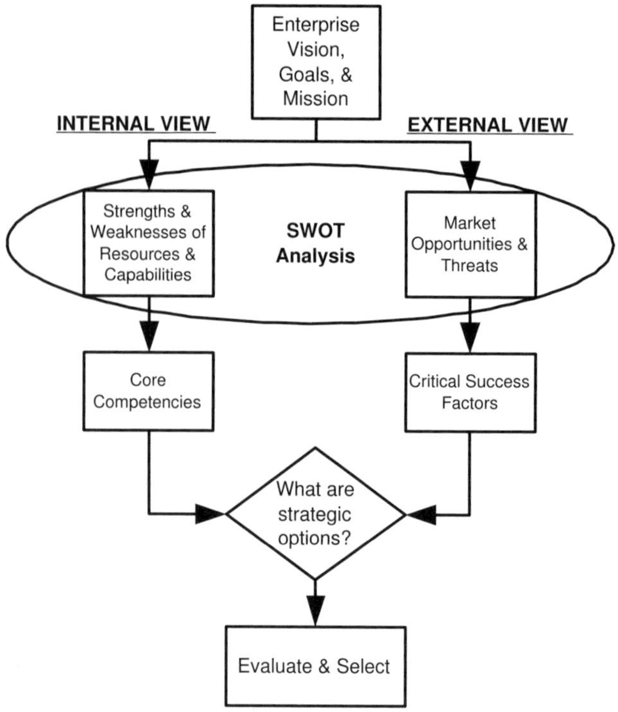

FIGURE 7.2
Strategy formulation process.

ous techniques and tools to identify their strengths, weaknesses, opportunities, and threats. Most teams use a checklist (see the SWOT checklist in Figure 7.3) to help the team identify strengths, weaknesses, opportunities, and threats.

In conducting a SWOT analysis, it is important to consider that something can be both a strength and a weakness, or an opportunity and a threat at the same time. To illustrate, in the year 2008, the world experienced a rapid rise in oil prices. This is an external trend that affects many industries, especially the airline industry. The market trend of higher fuel prices is both a threat because it is a major input to providing air transportation, and an opportunity if you can somehow twist it to your advantage. Southwest Airlines did just that. Southwest pursued an aggressive hedging strategy that locked in more than 70% of the company's expected fuel consumption at about $51 a barrel, far below the going market price of $126 a barrel.[5] All the airlines engage in some hedging but none as much as Southwest. When a company enters into a hedge contract they must pay a deposit, which can be a significant outlay of cash. Some airlines believe there are alternate, better uses of that money. Additionally, hedging involves significant risk because if oil prices drop instead of rise, then the company could find itself paying more for fuel than the current prices available on the open market.

Figure 7.4 shows the possible results of a SWOT analysis for Amazon.com. It identifies Amazon's strengths as its reputation, established customer base, and IT capabilities especially for CRM. Weaknesses are that online retail has low barriers to entry and the potential of muddling its brand name by branching out beyond books. Notice most of the strengths and weaknesses are intangibles. The opportunities and threats deal with external business

[5]Pae, P., Southwest Airlines reaps benefits of fuel hedging strategy, *Los Angeles Times*, May 30, 2008.

SWOT

Internal Factors (Strengths and Weaknesses)

1. What resources, people, and knowledge give us an advantage in the marketplace?
   (a) Tangible resources (e.g., natural resources, location)?
   (b) Intangible resources (e.g., reputation, intellectual property, brand names, cost advantage due to proprietary know-how)?
   (c) Human resources (e.g., training, culture)?
2. What capabilities give us an advantage in the marketplace?
   (a) Processes (e.g., order fulfillment, product development)?
   (b) Technology expertise (e.g., small part manufacture, chemical knowledge)?
3. What is the strongest resource?
4. Are any of the resources unique?
5. What resources, people, or knowledge can be improved?
6. In what areas are we deficient vis-a-vis our competitors?
7. What resources, people, or knowledge is lacking?

External Factors (Opportunities and Threats)

1. What trends (political, economic, social, or technological sometimes abbreviated PEST) might impact your industry?
2. What obstacles in the marketplace might affect you? Regulatory environment?
3. What is the competition doing that you are not?
4. What are the advantages/disadvantages due to location (e.g., access to markets, workforce, capital)?
5. What are your competitors doing?
   (a) Market segments? Products? Prices?
6. Are there new entrants?
7. Is there product substitution?
8. Who are the suppliers?
9. How are the relationships with partners, suppliers, distributors?

FIGURE 7.3
SWOT analysis.

FIGURE 7.4
SWOT analysis for Amazon.com.

trends. One opportunity for Amazon is to enter foreign markets. A threat is international competitors who could enter the U.S. market. Another threat is increases in distribution costs due to increasing fuel prices.

After completing the SWOT analysis, the team needs to appraise the resources' and capabilities' potential for sustainable competitive advantage. They need to determine what are the critical success factors or things that must be done to solidify a competitive position in the market.

To determine the strategy, the enterprise should generate as many alternatives as possible and then evaluate each alternative.[6] Usually there are constraints on what the enterprise can do. The basic strategy approaches identified by Porter [11] are a good starting point. The output of strategy formulation is the strategic plan to attain the enterprise's vision.

7.3.1 Strategy in Highly Dynamic Environments

Whether the aforementioned strategy formulation process will work well in dynamic, quick-changing environments is a question being investigated by some researchers. The traditional conceptualization of strategy that we discussed sees an enterprise focusing its efforts on a single strategy. Beinhocker [2] makes a compelling argument that in very complex market environments managers cannot predict with any confidence what strategies will be effective. He says an enterprise is better viewed as a complex adaptive system that must seek out the best strategy to survive. In this paradigm, an enterprise simultaneously pursues multiple strategies because it cannot predict which strategy will be successful. As an example, the author cites Microsoft's apparent "lack of strategy" in 1988 with respect to operating systems.

[6]Chapter 9 describes techniques for generating alternatives.

Microsoft was simultaneously extending DOS, the current market-leader, while promoting Windows, working with IBM on OS/2, developing applications such as Word and Excel for the Apple-MacIntosh Operating System, and building a Unix operating system. Microsoft's management could not know that DOS would eventually dominate the market, so to make sure that Microsoft remained a major provider of operating systems, they were essentially betting on all possible outcomes.

Contrast Microsoft's pursuit of multiple products simultaneously with GM's over-reliance on the sales of large trucks and SUVs for profits. In 2008, gas prices soared from an average of $2/gallon to $4/gallon; moreover, the worldwide economy suffered a financial crisis that made it much more difficult for customers to obtain credit. Suddenly, GM's profit-makers (SUVs and trucks) no longer seemed attractive to many customers. GM did not have fuel-efficient automobiles available for the market. Meanwhile, competitors such as Toyota and Honda had hybrid product lines and highly fuel-efficient automobiles. Not surprising given that GM budgeted $13.7 billion for capital spending and research and development in 2005, vs. $15.3 billion for Toyota, a smaller company.[7] In 2008, GM announced it was developing a hybrid car, but it would not be available until 2010. It is possible that if GM pursued multiple technologies simultaneously, then it would not have been caught off-guard when the market suddenly changed.[8]

The above discussion is about product strategies, not necessarily enterprise-level strategies. It may not be possible to simultaneously pursue multiple enterprise strategies because many of them are incompatible. In other words, you cannot simultaneously be both the low-cost producer and differentiate yourself by serving customers who demand high quality, luxury products. Instead of pursuing multiple strategies, a more appropriate approach is to design the enterprise so that it is agile. An agile enterprise can react quickly to environmental changes and even anticipate some changes and use the changes to its advantage [3]. Some strategy authors are studying real options theory as a way to inform management on how to invest resources for strategic advantage. A real option is an investment in physical assets, human competence, and organizational capabilities that provides the opportunity to respond to future contingent events [8]. Using real options theory, the enterprise knowingly considers all possible futures and invests its resources so that it will have the ability to quickly respond to environmental changes. In this way, the enterprise is constantly adapting to its environment and searching for ways to improve its position vis-a-vis its rivals.

7.4 Summary

This chapter described enterprise strategy and its components of mission, vision, and core values. This chapter also described both the resource-based theory of the firm and the market-based theory of the firm. The resource-based theory instructs firms to identity their core competencies that lead to competitive advantage. The market-based theory provides the SWOT tool to help firms develop a strategic position in the market. The chapter presented a method to formulate strategy.

The strategic decisions set the direction for the entire enterprise. An appropriate strategy can lead to enterprise growth and above-average profits, and a bad strategy can lead to

[7] Why GM's plan won't work, *Business Week*, May 9, 2005.

[8] GM's problems did not just occur because of changes in fuel prices – GM has been declining from its dominating position in the automotive industry for several decades suggesting a more deeply ingrained deficiency.

diminishing market position and financial losses. This can happen even if the enterprise has well-run operations.

Strategy establishes the context for enterprise design. Different strategies require different enterprise designs. This is the notion of fit. A company that wants to employ a strategy of being first to market with new products to obtain above-average profits (e.g., Intel) requires a different organizational structure, different processes, and different technology support than a company in the same industry that seeks economies of scale for commodity products (e.g., memory chip manufacturers). The strategy adopted by an enterprise sets priorities on what types of enterprise projects should be done. For these reasons, it is important to understand the enterprise strategy and how the enterprise engineering project aligns with the strategy.

Review Questions

1. Describe the three decision-making levels.
2. Contrast a vision statement from a mission statement.
3. Describe the different types of strategies that an enterprise might have.
4. Describe the relationship between resources and capabilities in the resource-based view of strategy.
5. Summarize the four conditions necessary for a resource to be considered a core competency.
6. Compare and contrast the low-cost leadership strategy and the differentiation strategy.
7. Compare and contrast the resource-based view and the market-based view of strategy.
8. Describe what a SWOT analysis accomplishes.

Exercises

1. Select a company and do a SWOT analysis of the company. Support your analysis with citations from the business literature and elsewhere as appropriate.
2. Use the Internet to find the mission, vision, and core value statements for a large corporation. What do the statements say about the company?
3. Select a company and identify what core competencies the company has.
4. Select an industry or market that you are familiar with. Identify the companies in that industry and categorize their strategies using the strategy types discussed in this chapter.

Bibliography

[1] R.L. Ackoff. *Re-creating the Corporation.* Oxford University Press, New York, NY, 1999.

[2] E.D. Beinhocker. Robust adaptive strategies. *Sloan Management Review*, pages 95–106, 1999.

[3] H.T. Goranson. *The Agile Virtual Enterprise: Cases, Metrics, Tools.* Quorum Press, Westport, CT, 1999.

[4] R.M. Grant. *Contemporary Strategy Analysis: Concepts, Techniques, and Applications.* Blackwell Publishing, Malden, MA, 2005.

[5] G. Hamel. Killer strategies that make shareholders rich. *Fortune*, June 23:30–44, 1997.

[6] J.C. Henderson and N. Venkatraman. Strategic alignment: leveraging information technology for transforming organizations. *IBM Systems Journal*, 38(2-3):472–484, 1999.

[7] M.P. Joshi, R. Kathuria, and S.J. Porth. Alignment of strategic priorities and performance: an integration of operations and strategic management perspectives. *Journal of Operations Management*, 21:353–369, 2003.

[8] B. Kohurt and N. Kulatilaka. Capabilities as real options. *Organization Sciences*, 12(6):744–758, 2001.

[9] H. Mintzberg. Crafting strategy. *Harvard Business Review*, 66(3):71–90, 1988.

[10] M.A. Peteraf. The cornerstone of competitive advantage: a resource-based view. *Strategic Management Journal*, 14(3):179–191, 1993.

[11] M. E. Porter. *Competitive Advantage: Creating and Sustaining Superior Performance.* The Free Press, New York, NY, 1985.

[12] C.K. Prahalad and G. Hamel. The core competency of the corporation. *Harvard Business Review*, 68:79–91, 1990.

[13] W.B. Rouse. A theory of enterprise transformation. *Systems Engineering*, 8(4):279–295, 2005.

[14] A.W. Ulwick. *Business Strategy Formulation: Theory, Process, and the Intellectual Revolution.* Quorum Press, Westport, CT, 1999.

8

Problem Formulation and Requirements

> *"Engineers like to solve problems. If there are no problems handily available, they will create their own problems."* – Scott Adams (1957-), American cartoonist of *Dilbert* cartoon strip.

This chapter explains how to conduct problem analysis and requirements analysis, part of the analysis phase of the enterprise engineering project. The chapter starts by describing data-gathering techniques that support all the analysis activities. Problem analysis is a matter of identifying the problems, assessing their severity so they can be prioritized, and then analyzing their causes and effects. Requirements analysis is the identification of the system requirements, their formal documentation, and then their management throughout the duration of the project. When analyzing requirements, a distinction is made between project goals, objectives, requirements, and constraints. Included is a discussion of stakeholder analysis, to understand the stakeholders' concerns, motivations, and issues. The chapter ends by explaining how the problems and requirements are related to each other. Not getting the requirements right is an oft-cited shortfall of projects that fail because if the project team does not know what the stakeholders want, need, and expect in the system, then how can it deliver it? The deliverables of problem analysis and requirements analysis are the problem analysis report, the stakeholder report, and the requirements document. After completing this chapter, you should be able to:

- Develop a data-gathering strategy for a project and conduct the data gathering.
- Identify and document the problems, issues, and opportunities facing an enterprise.
- Analyze the causes and effects of a problem using an Ishikawa diagram.
- Specify SMART requirements that are complete and consistent for an enterprise project.
- Differentiate between functional and non-functional requirements.
- Describe the requirements management process.
- Perform a stakeholder analysis.
- Differentiate between goals, objectives, constraints, problems, and requirements.
- Establish a policy to enforce requirements traceability.

8.1 Data Gathering

Data gathering is a cross life-cycle activity done throughout multiple phases but predominantly during the analysis phase. Data gathering is necessary to define strategy and project scope; to identify and analyze problems; to elicit requirements; to model processes, information, and organizational structures; and to generate alternatives. In many enterprises, the systems, processes, information sources, and organizational structures are only partially and incompletely documented, often in formats not readily useful to the project team. Moreover, much of the knowledge of how the enterprise operates and what the enterprise needs is known by the people in the enterprise. Consequently, to understand the enterprise's problems and requirements, it is necessary to identify and gather data in the early phases of an enterprise design project. This is accomplished by developing a data-gathering strategy that involves the data-gathering techniques (see Table 8.1) including:

1. Document collection and analysis.
2. Observation.
3. Interview stakeholders.
4. Questionnaires.
5. Requirements workshop.

A data-gathering strategy will typically involve most, if not all, of the aforementioned techniques. In a data-gathering strategy, the group of techniques employed must be designed to effectively and efficiency gather the necessary data to understand the enterprise problems and requirements as fast and accurately as possible. Decisions on the data-gathering strategy involve a trade-off between rigor, speed, cost, and accuracy. The analysts constantly revisit the data because as their understanding of the system grows, then their interpretation of the data will evolve. The following subsections describe the data-gathering techniques.

8.1.1 Document Collection and Analysis

Enterprises have tremendous stores of data and information concerning their systems. Document collection and analysis is the technique to identify, gather, and analyze the available documentation. Almost all enterprise projects include document collection as a data-gathering technique because it is low cost and quick to do, as well as often providing very detailed information. All effective data gathering involves some level of document analysis such as business plans, market studies, contracts, requests for proposals, statements of work, existing guidelines, invoices, order forms, analyses of existing systems, and procedures. Improved requirements coverage results from identifying and consulting all likely sources of requirements.

Nowadays, the data are often stored in electronic forms such as databases, webpages, etc. The analysis team prefers to collect completed forms because they better indicate how the form is actually used versus its intended usage. The analysis team creates a source data material log and saves examples of each source data item collected. One caveat of document analysis is oftentimes documents are old or out-of-date, and many times printed policies and procedures are not followed. For this reason, the analysis team cross-checks by using other data-gathering techniques.

8.1.2 Observation

Observation is the technique of simply observing a person while he or she is doing some task [2]. This technique may involve watching a subject and making notes, or it could involve more sophisticated techniques such as analysis of videotaped episodes. In most enterprise engineering projects, there is no substitute for simply walking around and observing the

TABLE 8.1
Data-Gathering Techniques, Their Advantages, and Their Disadvantages

Data-gathering technique	Advantages	Disadvantages
Document collection and analysis	Quick to collect data, and the source data might reveal information that people forget or overlook.	Source data can be messy and difficult to translate due to obscure codes, etc.
Observation (Job shadowing)	The analyst gets a first-hand account of the situation, and is able to observe system behavior that employees and others might not notice because it is part of his or her routine.	Observations will influence workers' behavior; takes significant amount of time to do observations.
Interview	Can react and deal with unexpected information.	To conduct interviews one-on-one takes a significant amount of time.
Questionnaires	Easy to administer; able to obtain responses from large group of people quickly.	Can be difficult to obtain a statistically significant response rate, and difficult to do in exploratory data gathering since you do not know what to ask.
Requirements workshop	More efficient than interviews and possible to reach consensus.	Some people are reluctant to speak when their colleagues are present.

facility, work, and organization in action. Observation lets the analyst become familiar with the process or system being studied in a way not possible by reading models or diagrams.

A variation on basic observation is *protocol analysis* in which the worker "thinks aloud" while solving a problem [9]. Verbalizing the steps involved may occur concurrently or retrospectively if the task was videotaped or documented. Verbalization adds the rationale for how a person acts, which can be of immense use to the analyst in understanding the system.

To conduct an observation study, the analyst needs to: (1) determine the purpose of the observation, (2) determine who and what to observe, (3) prepare recording devices (paper forms, video equipment, tape recorder), (3) do the observation, (4) record observations, and (5) analyze data obtained from observation.

A pilot observation session is frequently invaluable to uncover any issues with conducting the study, gathering data from the study, or analyzing the data collected. This is when, a walk-through makes the most sense, so the analyst can determine what tasks to observe and who to observe. When deciding who to observe, a common approach is to observe the best in class operations. For example, in a project to analyze airline operations, it was decided to observe a certain airport that had consistently outperformed all the other airports. The purpose of the observation was to note circumstances that produced superior performance. Among the findings was that the airport experienced less-crowded, spaced-out arrival of flights, and it had a higher ratio of employees per flight than at other airports. Since some of the environmental conditions cannot be replicated at other airports, then it would make sense to go and observe also other airport operations. The best operation might not be representative of the environment or situation faced by other operations, and thus, the conclusions drawn are not generalizable to the entire enterprise. In this particular case, it was useful to observe why the airport performed better.

Example of Protocol Analysis

A manufacturing company that built and installed stairs in new homes had a scheduling process to send installation crews to houses to install the stairs. The scheduling process was performed by two schedulers. It was mostly a manual process and highly dependent on the scheduler's knowledge of the crew's skills and abilities, the nature of the jobs, and the work sites. It was also an inefficient process because the number of installations has grown such that the scheduler could no longer mentally consider all the interactions simultaneously. To understand the scheduling process, the analysis team used protocol analysis. The scheduler was observed at her work area while she was creating the schedule. As she did the schedule, she would explain what she was doing and why. The analyst would ask questions to better understand the reasoning for particular steps or actions. Protocol analysis is highly intrusive, but in this case, it enabled an efficient data collection strategy for the team to understand the nature of the scheduling problem. The protocol analysis was successful because the scheduler knew of the project's purpose, agreed with the general objective, and felt secure in her job so she provided full participation.

There is a strong tendency for people to react, whether favorably or unfavorably, to being observed and recorded.[1] The degree of intrusion probably affects how people react. If the observer is non-observed (a fly on the wall), then changes in behavior are less likely, or at the very least, the observed workers will become acclimatized to being observed. This raises the issue of whether people being observed need to be informed of the observation. The extent that people are informed depends on the situation and the purpose of the observation. Full and informed consent is often required in research studies, but in the type of enterprise engineering projects considered here, it is seldom called for. More likely, a manager will tell the workers that an analyst will be around to observe work practices. The notice of observation usually indicates the purpose of the observation to allay any fears or anxieties people may have. It is essential that neither the observation nor the resulting analysis harm the people being observed.

To be useful, the observations need to be recorded. Sometimes recording is done on the spot; at other times, the analysts might record their observations after the observation. In all reports, the analyst records the date and time of the observation and a short description of the context in which the observation took place. The analyst might use prepared observation forms to record what was observed. This is useful to obtain comparable documentation from multiple observations conducted by different analysts at different locations. A structured guide makes it easier to analyze the results. If the observation is of quantifiable behavior, such as arrival rates or service times, then a recording sheet is designed to obtain the data. The least formal, structured approach to record the data is simply when the analyst records his or her field notes. The quality of the field notes will depend largely on the skills of the analyst to provide a full, rich documentation of the observed system.

Observation may provide better information and understanding of management oper-

[1] In the well-known Hawthorne studies, the factory workers' productivity improved regardless of the interventions. It was later concluded the factory workers were not reacting to the interventions as much as they were reacting to the fact they were being observed.

ations and procedures than what can be obtained from printed documentation or even interviews of participants. The reason is during interviews, some participants will recreate how the process should work ideally and leave out details or exceptions. The omission is not necessarily intentional, but it occurs at a subconscious level.

It may be important to validate your observations. Validation could be going back and repeating the observation to determine if the observed behavior is normal. It may mean observing different people to determine whether it is unique to an individual or a group-wide behavior. Videotaping a process or taking photos can help record the process so that it can later be analyzed.

Tagging Luggage

In a project to reengineer the luggage delivery process for a cruise line, it was necessary to understand the process starting with the guests' arrival to Miami until their luggage was delivered to their cabin on the cruise ship. To know the times it took for each step of the process, we needed to follow the luggage. But, this is not physically practical. Instead of the data gatherers following the luggage, we tagged the luggage. Then the data gatherers could stand at one point in the process and record when particular numbered bags passed by. In this way, we were able to estimate the durations for each stage of the process. In most process analysis projects, it is necessary to estimate task durations; this can be done by observing and timing the work, asking the workers to record the work durations, or as described here, tagging the work and then recording when it passes by critical points in the process.

8.1.3 Interview

An interview is a formal meeting with one or more stakeholders to understand some aspect of the project. Interviewing is one of the more valuable data-gathering techniques. In the beginning of a project, interviews are often the only appropriate means to gather data. Only by asking the stakeholders and speaking with them can you start to understand the system sufficiently before you can attempt other data-gathering techniques.

Interviews can be either structured or unstructured. *Structured interviews* are when you develop a set of questions that are asked of every person you interview with the aim of obtaining specific information. For a structured interview, the analyst develops a form with all the questions ahead of time. An *unstructured interview* identifies topics and maybe some specific questions, but there is more leeway to deviate from the script. The questions asked can be open-ended questions that allow the respondent to answer in any way they see fit, or closed-ended questions that restrict the type of response. The type of interview depends of course on the purpose of the interview. Good interview design identifies what information the analyst wishes to gain, and then designs the interview to best obtain that information.

To obtain initial understanding of the enterprise vision, the analysts would probably prefer an unstructured interview with mostly open-ended questions. This interview format would allow exploration of the stakeholder's visions for the enterprise without unduly restricting that vision as might happen in a structured interview. Then structured interviews could be used to gather more specific data from a wide range of stakeholders.

To interview involves five steps (see Figure 8.1):

1. Determine what information is needed

The goals of an interview are two-fold: first and foremost, the analyst needs to understand the enterprise vision, objectives, problems, requirements, and constraints affecting the en-

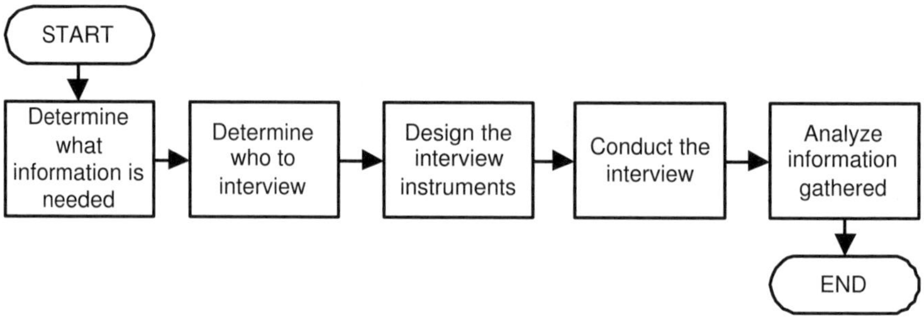

FIGURE 8.1
Interview flow chart.

terprise design. Thus, interviewing is for gathering information to analyze and design the enterprise system. The second goal of interviewing is to obtain participation of key stakeholders who can later influence the implementation of the project.

An interview requires commitment from the interviewee, and the analysis team needs to decide on an appropriate length of time for the interview. The answer depends on many contextual factors such as availability of the person, purpose of the interview, and so forth. How long you expect to have for the interview will affect what questions can be asked.

No matter who the analyst interviews, he or she schedules the interview as a matter of courtesy and to allow the interviewees to be prepared. They are told the purpose of the interview, they agree on the length of the interview, and they are provided with any materials they might need in advance. As a final courtesy, it is good practice to remind them of the scheduled interview one or two days prior to the interview.

2. Determine who to interview

The organization chart can help the analyst identify who to interview. Early in problem analysis, the analysis team will probably start interviewing system owners. Owners can provide a strategic view of how they see the problems. To understand a business process, the analysts then interview the employees directly involved in the process. To understand how customers view the system, the analysts interview customers. So, depending on the purpose of the interview, the analysis team will interview different people. Not all the people who will be interviewed need to be identified; in each interview, the analysis team can ask the interviewee, "who else should I talk to?" or some similar question.

3. Design the interview instruments

The majority, if not all, of the questions are prepared ahead of time. Asking direct questions may seem the most efficient method, but could result in unreliable answers because the stakeholders may not be accustomed to communicating in such a direct and candid manner. Questions should be clearly stated, specific, and open-ended wherever possible, requiring the interviewee to provide more than a simple "yes" or "no" answer. If necessary, several questions may be asked to obtain information on one characteristic, but doing this repeatedly runs the risk of extending the interview beyond the ideal two-hour time limit.

The questions should not be loaded or biased questions that convey the interviewer's personal opinion of the issues. (For example, never ask, "Why would the proposed ERP system improve your job?"). Instead, the use of context-free questions by the interviewer helps avoid prejudicing the response [3]. A context-free question is a question that is general enough that it could be used on many different projects, and also does not bias an interviewee towards a particular solution. For example, "Who is the client for this system?" is a context-free question because it can be used in almost any project and it makes no reference to a solution. Context-free questions are especially useful during initial interviews when the analyst knows less about the enterprise and proposed system.

One stumbling block while conducting an interview is simultaneously asking the questions and recording the responses. Huber and Power [4] suggest a technique of tandem interviewing with each interview involving two analysts to allow one to conduct the interview and the other to probe and transcribe the information.

When there is a large working group of analysts who will each be conducting interviews, they should discuss and document the protocol to be followed during the interview process. An example interview protocol could be:

- All interviews start with an introduction explaining the purpose of the interview, identifying who is collecting the information, explaining what will be done with the information, and assuring the stakeholder that all responses will remain anonymous. The definition of the project and any terms that might be ambiguous or unknown to the stakeholder should be explained during the interview.
- Two-person interview teams are used; the responsibilities are divided such that one interviewer leads the interview and the other interviewer focuses on note-taking.
- The interview should be scheduled for a set time limit.
- The interview can be terminated at the interviewee's request even if additional unanswered questions remain.
- The words and terminology used by the interviewer should be preserved for analysis.
- The interview ends by thanking the interviewee for his or her time and cooperation. The interviewers should request contact information and permission to contact the interviewee for clarification and further information if needed.

4. Conduct the interview

To begin the interview, you should introduce yourself. Next, review the purpose of the interview: explaining why you are there, how the information collected will be used, and what issues will be covered. Once the introduction is over, you start asking your questions. In order to improve the quality and usefulness of the information obtained, you should have practiced and be knowledgeable of various communication skills and strategies. Interviewees, will not answer in ways that are complete, nor will they necessarily express their answers in language that is readily understood by the analyst. As a consequence, you should be ready to ask follow-up questions to improve your understanding of the answers. One way to do this is to summarize, rephrase, and describe implications of what you understand by the interviewee's responses. The interviewee can then confirm your understanding, correct it, or elaborate. When listening to the interviewee, you need to be an active listener, which means you observe their body language and think about what they are saying. Be alert to various types of errors that can occur in communication. For example, the interviewee might be relying on an imperfect memory to answer a question, there might be interpretation errors between what the interviewee said and how you interpret it (a small discrepancy in accounts receivable and the invoice is a few cents or a few dollars?), conflicts in interest or topics for which the interviewee has a strong bias, and ambiguities because the intent of the question is broad-based while the interviewee is answering about a narrow instance. Throughout the interview you should be taking notes, and even ask to pause briefly while you finish recording the answers.

5. Analyze information gathered

Depending on how the data were recorded during the interview, it should be immediately transcribed into a computer readable format: either a word processor file, spreadsheet, or database. Interviewers should also document their impressions of the interview so that responses will not later be misinterpreted due to lack of context.

8.1.4 Questionnaire

A questionnaire is a set of predetermined questions arranged on a form and distributed to the individuals being studied who complete the questionnaire and return it. It is also possible the questionnaire can be administered verbally. An advantage of a questionnaire is that responses can be collected from a large number of people at a relatively low-cost.

The questionnaire should be designed to ensure that all respondents interpret the questions in the same way that the analyst intended. To accomplish this, the questionnaire should use familiar words in short, simple sentences, avoid technical or special jargon, avoid sensitive issues unless anonymity can be guaranteed,[2] and avoid asking leading questions. Unlike an interview, the respondent cannot request clarification of questions or get assurances of how the questionnaire will be used. So, it is important to pilot test the questionnaire, preferably with a subset of people who represent the population that will receive the questionnaire. A pilot will help route out jargon, useless questions, and other questionnaire problems.

The analysts need to fully understand what information they are seeking and then design each question with a specific purpose. Also, it helps if the analyst determines beforehand how the responses will be tabulated, analyzed, and subsequently used. This will avoid obtaining useless responses and also perhaps help speed up analysis by coding of responses for easy entry into a computer.

When designing the questions there are several choices:

- Multiple choice questions provide a small number of choices that the respondents must choose from. These questions are easy to tabulate and analyze.
- Rating questions ask the respondents to read a statement and then provide their opinion. A common rating scale is the Likert scale.[3]
- Ranking questions ask the respondents to read a statement and then rank their responses. It forces the respondent to choose between options identifying the most important or least important depending on the question.
- Open format questions leave blank lines for the respondents to write a response. It allows respondents great latitude in their response, but it can be more difficult to analyze when there is a large number of questionnaires because the analysis team then has to read each one of these questions and summarize the responses in a usable format.

Examining the Selection of Question Types

Suppose the analysis team has identified a number of problems and wants to prioritize the problems so that resources can be efficiently allocated to addressing the more significant problems. Would question #1 or #2 better elicit useful data to aid prioritization?

---

[2]Anonymity is often difficult to provide when small groups are surveyed. For example, if your questionnaire includes knowing the department the person works in, and a department has only a few people then it is easy to identify the respondent.

[3]The Likert scale is 1 = strongly agree, 2 = agree, 3 = no opinion, 4 = disagree, and 5 = strongly disagree. Sometimes, to avoid bias towards "no opinion" only four items are used.

1. For each problem, please describe your level of agreement as to whether it is a significant problem that needs immediate attention. (1 = strongly agree, 2 = agree, 3 = no opinion, 4 = disagree, and 5 = strongly disagree).

   (a) The process to create a customer order is too lengthy.

   (b) The availability of data on customer accounts.

   (c) Knowledge of promotions not widespread.

   (d) Whether a product is profitable or not is unknown.

2. Rank the problems identified below from most significant problem to least significant problem, where the biggest problem is ranked #1, second biggest problem #2, and so forth.

   (a) The process to create a customer order is too lengthy.

   (b) The availability of data on customer accounts.

   (c) Knowledge of promotions not widespread.

   (d) Whether a product is profitable or not is unknown.

The questions might seem to elicit the same data, but when the analyst team examines the results it may find the second question to be far superior. For example, out of 100 respondents, the average response to question #1 might be:

1. The process to create a customer order is too lengthy. **2.1**
2. The availability of data on customer accounts. **2.4**
3. Knowledge of promotions not widespread. **2.0**
4. Whether a product is profitable or not is unknown. **2.1**

The response from the same respondents to question #2 might be:

1. The process to create a customer order is too lengthy. **1.4**
2. The availability of data on customer accounts. **2.9**
3. Knowledge of promotions not widespread. **2.1**
4. Whether a product is profitable or not is unknown. **3.5**

Clearly, the second question provides more actionable data. In the first question, the responses are too close together making it difficult to prioritize the problems according to severity. In the second question, the respondents were forced to choose and it is evident that problem (a) is the most significant problem and problem (d) the least significant. What this case illustrates is how the questions are posed is very important, and the analysis team needs to look ahead to how the data will be used to determine whether a question will elicit appropriate responses.

In general, response rates for questionnaires can be quite low, on the order of 10% - 20%. There are strategies to increase response rate. First, the questionnaire should be short, easy to complete, and not inconvenience the respondents. Having supervisors announce the questionnaire and explain why it is needed can also help improve the response rate. At the other extreme, a U.S. Air Force organization was doing a project and the commanding officer "strongly encouraged" participation in the questionnaire. Given the military's organizational culture, this ensured a response rate approaching 90%.

8.1.5 Requirements Workshops

A requirements workshop gets a group of stakeholders together in a meeting to determine the system requirements. When the data gathering is focused on eliciting requirements, then requirements workshops are a powerful technique because they can be designed to encourage consensus concerning the requirements for a particular system capability or feature. They are best facilitated by an outside expert and are typically short (from several hours to a couple of days). Other advantages are often achieved – participant's commitment to the work products and project success, teamwork, resolution of political issues, and reaching consensus on a host of topics. Benefits of requirements workshops include the following:

- Workshop costs are often lower than are those for multiple interviews.
- They help to give structure to the requirements capture and analysis process.
- They are dynamic, interactive, and cooperative.
- They involve users and cut across organizational boundaries.
- They help to identify and prioritize needs and resolve contentious issues. When properly run, they help to manage user's expectations and attitude toward change.

Example of Requirements Workshop

A cruise company was building a new reservation system. To determine the requirements for the reservation process, they conducted requirements workshops with reservation agents, marketing, sales, and IT personnel. They would gather small groups of 6-8 people in a conference room. Each group would be tasked with a single goal; for example, reservations for a honeymoon couple, reservations for a group, reservations for frequent cruises, and so forth. The groups would discuss what needed to be done for each guest category. For example, groups like to have cabins close by, so the agent needs to be able to block out adjacent cabins, dining together, and so forth. These requirements workshops were extremely successful in identifying all the system requirements. Moreover, the requirements workshop introduced the project early to employees, and they felt they were part of the project, which builds buy-in early on and facilitates work system changes later on.

A special category of requirements workshop is a Joint Requirements Planning (JRP) workshop. JRP is a method for developing requirements through which customers, user representatives, and developers work together with a facilitator to produce a requirements specification that both sides support. This is an effective way to define user needs early. Wood and Silver [8] assert that quality systems can be built in 40% less time utilizing JRP. They explain how to perform JRP and provide diagrams, forms, and a sample JRP design document.

Some of the best practices derived from JRP experience is to hold the sessions away from the normal workplace at an off-site location. This allows the participants to focus on the JRP session. The room should be equipped with whiteboards, large paper pads, and a projector so that the participants can draw sketches of models and make lists. The layout of the room should encourage participation; this can be done with circular or U-shaped arrangement of the chairs and tables. The room should also be set up for teleconferencing so that participants can contact employees in the organization not participating in the JRP session with any questions or requests for information.

8.1.6 Putting It Together

In data gathering, you should try to triangulate the facts by obtaining data from multiple sources to verify what is really happening [10]. Verification here is to check the consistency in facts obtained from different sources and to hash out a view that everybody will agree on (or a view that the design team finds most useful).

To gather data for the cruise line reservation system project, the project team developed the following strategy:

1. *Interviewing Key Personnel.* First, the team will interview the Director of Marketing, the Manager of Revenue Management, the Manager of Reservation, the Director of Shipboard Operations, the Director of Hotel Operations, and the Vice President of Information Technology. For each person, a 45-minute structured and unstructured interview was developed. The purpose of the interviews is to specify the project scope in terms of functionality, organization, and data. The interviews were to establish the cost and desired schedule. Additionally, the interviews are designed to elicit the project goals and understand upper-management's vision of what will be accomplished. The upper-level managers will be asked who in their organization should be interviewed and who are the candidates for subject-matter experts to serve on the project. Lastly, the interviews will help the team understand the issues faced by upper-management and start the stakeholder analysis.

2. *Interviewing Managers.* The people identified by upper-management will be interviewed next. Here the interviews are designed to better define the project scope. Also, any problems will be identified and an understanding of the organization, process, and information will be started. Lastly, the managers will be asked to identify what documents and other papers can be collected and what processes can be analyzed.

3. *Source Data Collection.* The team will collect completed forms and documents as identified by the management team. For such documents such as client reservations, a suitable sample size will be collected to ensure all possible variations of reservations are covered.

4. *Follow-up Interviews.* After the source data are collected, the team will have follow-up interviews to ask questions about the source data.

5. *Observation.* The team will observe the main processes identified by the management team.

6. *Questionnaire.* Depending on what is learned from the previous data-gathering efforts, the team may use questionnaires to understand the issues faced by the cruise line guests.

7. *Requirements Workshop.* The team will hold a requirements workshop with reservation agents, hotel staff, and others to determine how to handle the reservation process.

In the single project, almost all the data-gathering techniques are employed. What is significant is the order in which they are applied, what type of information they seek, and how they are interdependent on each other. What is not shown is that the data-gathering techniques are often done in parallel, and they are iterative in that some techniques may be repeated.

Data-gathering formalities – why informal techniques fail

To illustrate the problems that can arise from not planning and formally executing a data-gathering process, we describe an experience with a project for a telecommunications company. The analyst needed to interview stakeholders (managers and employees) involved in a process for providing pre-paid telephone cards that are sold in supermarkets, grocery stores, and convenience stores. The analyst developed only a general, sparse outline of the interview. The analyst made appointments for interviews as advised by good practice. Unfortunately, due to the poor preparation, the interview would go beyond schedule. Then when analyzing the data collected from the completed interviews there were problems because important questions were not asked, the notes taken during the interviews were too brief, or they were too ambiguous to be useful. It became apparent that many interviews would have to be redone. The analyst conducted several follow-up interviews or posed questions by telephone. Eventually some of the stakeholders lost confidence that anything good was going to come from the analyst's work, and they refused to meet with the analyst anymore. It was not an outright refusal, but more of being unavailable to ever meet. A large part of the difficulty encountered by the analyst could have been avoided if a more formal data-gathering approach was followed.

8.2 Issue and Problem Analysis

If design is a problem-solving process, then the first task is to understand the problems. A *problem*[4] is a situation that needs to be corrected because it performs poorly. A problem occurs when there is a difference between what "should be" and what "is"; between the ideal and the actual situation. In addition to problems, there might be opportunities. An *opportunity* is when there is a market, technological, or organizational opportunity the enterprise wants to take advantage of. So, the current situation might not be problematic, but the enterprise recognizes an opportunity to improve the current situation – it is the difference between what "could be" and what "is."

There might be regulations, standards, certifications, or other externally imposed rules the enterprise must follow. In these cases, part of the problem analysis or requirements analysis needs to understand the regulations to ensure enterprise compliance with those regulations.

The first activity in problem analysis is to identify the problems, issues, and opportunities. The aforementioned data-gathering techniques are used, but simply asking stakeholders, "what the problems are?" is usually not sufficient in itself. A checklist can help ensure

[4]The word "problem" has a negative connotation and many consulting companies say instead that their clients face challenges or issues, not problems. In one consultant engagement with the U.S. Air Force, we had asked the commanding officer in charge of logistics what were the problems that the officer wished to solve with a new ERP system. The officer made it very clear there were no problems since the Air Force was able to complete its mission of providing logistics support. What the Air Force had was a challenge to do it more efficiently and effectively.

TABLE 8.2
Problem Types

Problem type	Description
Performance	When the system fails to provide the performance demanded by the stakeholders.
Economy	When the system does not meet either profit or cost requirements.
Security	When the system does not provide adequate security.
Service	When the system does not provide a service required by the stakeholders.
Safety	When the system lacks safeguards or is dangerous for either employees or customers.
Maintainability	When the system is difficult to maintain.
Flexibility	When the system fails to exhibit sufficient flexibility.
Reliability	When the system is unreliable because it fails too frequently or the failures are for too long a duration.

complete coverage of all possible problem types. Problem types are shown in Table 8.2. The checklist is used to generate ideas for investigating problems, to devise questions for interviews, to determine what processes to observe, to determine what documentation to gather, and so forth.

> *Problems* are situations that must be changed. *Opportunities* are chances to take advantage of new technologies or markets to improve the enterprise's operations. *Directives* are externally imposed requirements made by the government, other regulatory agencies, by parent companies, or by senior management.

Before attempting to resolve a problem or issue, the analyst needs to understand what is causing the problem situation, how severe is the problem, and what are the effects of the problem. The analyst does this so he or she can understand how the problem affects the system being examined. To help the analyst describe the problem, he has the stakeholders answer certain questions about the problem, such as:

- How frequently does the problem occur?
- When does it generally occur (for example, at a particular time of the year, on a set week, on a specific day)? Is the problem related to time?
- Where does it generally occur (for example, is it limited to a particular area, or is it everywhere)? What relation does the problem have to location?
- Who is most affected (for example, individuals or families, people of different gender, race, age, or socioeconomic status)?

TABLE 8.3
Problem List

Prob ID	Problem Description	Frequency	Affected Stakeholders	Annual Cost Impact	Quality Impact
1	Medical records are incomplete (missing papers, etc.).	1 out of 15	RN, doctors, patients	$2100	Potential for medical error
2	Medical records can be difficult to find (they are in doctor's office, RN station, lab, etc.).	1 out of 10	RN, doctors, patients	$3150	
3	RNs must wait for doctors to review charts before making telephone calls to patients. Doctors do this late in the day causing RN overtime.	Twice a week	RN, patients	$15,600	Minimal
4	Insufficient space in registration area to help more than a single patient resulting in delays.	2-3 times a day	Secretary, patients	Minimal	Poor patient service
5	Data inconsistencies between financial records and medical records.	1 out of 20	Accounting, patients, RN	$8500	Potential for billing errors
6	RNs complete documentation to meet regulatory requirements.	Every patient	RN, lab technicians	$3500	None

The output of problem identification is the Problem Statement List (see Table 8.3), which lists each problem, its description, and characteristics of the problem that can be used to prioritize the problems. To help decision-makers, the list should provide an estimate of the cost impact. Sometimes this is difficult to do, but an estimate is better than leaving this field blank (which usually represents a greater error than what a rough estimate represents). The list should also contain any possible actions that should be taken – these are usually quick fixes suggested by stakeholders. An example Problem Statement List is shown here:

The cost estimate for problem #2 is based on the amount of time an RN spends searching for the missing information. The health clinic processes approximately 30 patients per day. If the average RN salary is \$30/hour and it takes the RN about 6 minutes on average to search for the missing information, then the annual cost impact is calculated as

$$\text{cost} = \frac{1}{15}\left(\frac{30\,\text{pts}}{1\,\text{day}}\right)\left(\frac{350\,\text{days}}{1\,\text{yr}}\right)\left(\frac{\$30}{1\,\text{hr}}\right)\left(\frac{1\,\text{hr}}{60\,\text{min}}\right)(6\,\text{min/record}) = \$2100/\text{yr}$$

This estimate, while "rough" gives a better sense of the magnitude of the problem than to just put "unknown." The analysis team should strive to quantify as much as possible the problem characteristics.

8.2.1 Fishbone Diagram of Cause and Effect

Once the problems are identified, it is time to analyze their causes and effects. The fishbone or Ishikawa diagram[5] is used to identify, explore, and depict all the possible causes of a problem. It is called a fishbone diagram because it resembles the bones in a fish skeleton. To use a fishbone diagram, you first identify the problem and write it down on the right-hand side of the drawing area. Then each major "bone" of the diagram denotes a category of possible causes. Categorization is critical to deal with large amounts of information to avoid information overload. Common categories are: people, policies, processes, materials, machines and equipment, and information. For each main category, the analysis group brainstorms possible causes of the problem. These causes are drawn as smaller lines coming off the "bones" of the fish. When a cause is large or complex, then it may be best to break it down into sub-causes. Show these as lines coming off each main cause line.

After doing this, you should have a diagram showing all the possible causes of the problem that you can think of (Figure 8.2 shows an example). This is the primary objective of fishbone diagrams – to exhaustively identify problem causes. Depending on the complexity and importance of the problem, you can now investigate the more likely causes further. This may involve setting up investigations, carrying out surveys, etc. The purpose of the data-gathering is to provide more evidence of which causes contribute most to the problem and how they can be dealt with.

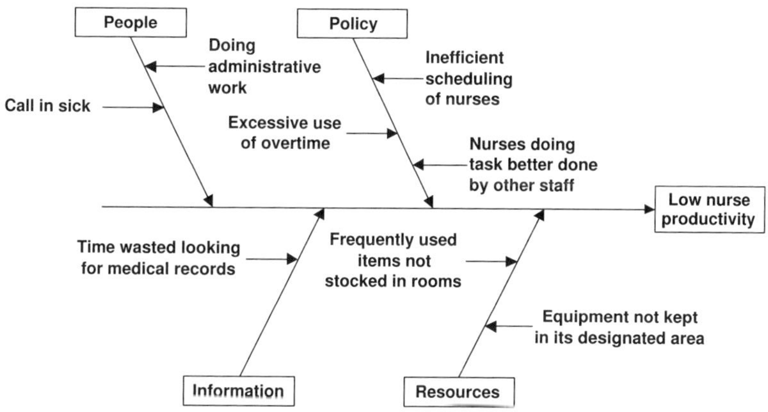

FIGURE 8.2
Fishbone diagram.

A limitation with traditional cause/effect diagramming techniques as exemplified by fishbone diagrams is the linear thinking underlying the technique. Most problems are not well described by simple explanations of *A* causes *B*. More complicated causal relationships exist between variables, which cannot be well represented in fishbone diagrams. Causal loop diagrams can represent more complex cause-effect relationships.

8.2.2 Causal Loop Analysis

Causal loop diagrams help you to thoroughly think through the root causes of a problem. Causal loop diagrams push you to consider all possible causes of the problem, rather than just the ones that are most obvious. Moreover, they encourage a holistic viewpoint and the identification of feedback between causes and their effects. The primary benefit of creating

[5]Named after Kauro Ishikawa of Tokyo University.

a causal loop diagram is the learning that occurs while trying to understand the system. This is especially evident when a small team collaborates to generate the diagram. Follow these steps to analyze a problem with a causal loop diagram:

Identify the problem

Write down the exact problem you face in detail. Where appropriate, identify who is involved, what the problem is, and when and where it occurs. Define a noun that can be used in the diagram to represent the problem. Avoid using adjectives or verbs. For example, if the problem is the process provides poor service, then a good variable name is "service quality" instead of "poor service quality." The reason is the adjective "poor" will be represented by the polarity of the arrows pointing to the variable. Write the problem variable in the middle of a sheet of paper (or better, the middle of a white board when you are leading a group to analyze the problems).

Identify causes and effects related to the problem

Start by identifying those variables that may contribute to the problem situation. Once a variable is identified, it must be named. In the diagram, the variables should be nouns as was done for the problem variable because the actions are implied by the arrows connecting the variables. Try to draw out as many variables as possible. If you are analyzing the problem as a group, then this is a good time for some brainstorming.

The variables should describe things that can be measured and that can vary over time. It is acceptable to define variables that are subjective or difficult to measure as long as in theory they can be measured. This is why in causal loop diagrams you will see variables such as "employee morale." In theory, employee morale can be measured, even if its measurement is difficult, and it can also vary over time as events in the company or environment affect morale.

A common modeling technique is to distinguish between perceived and actual states. The reason is people's perception often lags changes in the actual state of the system. Moreover, it is the perceived state that people react to. In a healthcare system, you might have both "perceived wait time" and "actual wait time."

As you add variables, you need to link them to one or more variables already in the model. To do this you should think about all the possible consequences of a change to the variable on other variables, as well as how changes on the other variables will affect the current variable. The diagram may have closed loops. Each loop should be labeled with a "+" for positive feedback or "-" for negative feedback. Additionally, good practice is to name the loop. Sometimes the effects of a change in a variable are delayed. This is noted with two short lines diagonally crossed through the relationship.

Illustrative Example

First, turn back to Section 2.2.1 to review the syntax and semantics of causal loop diagrams. Now, suppose an analysis group is working for a healthcare provider that has identified a problem that clinic capacity is low. Clinic capacity is the number of patients seen per day. An analysis group is formed to analyze the causes and effects of this problem. One cause of low clinic capacity is identified as the number of nurses available to serve patients. If the number of nurses increases, then capacity increases. If it decreases, then capacity decreases. This is shown as a positive relationship between the two variables.

Note that both variables are nouns describing quantities that can be measured and can change over time. The group might next explore what influences the number of nurses. Two potential causes are identified as the number of nurses who call in sick and the number of nurses doing other than clinical work (such as administrative work). Both of these variables have negative relationships with the nurses' availability to serve patients. Moreover, the analysis group realizes it is not the number of nurses but the number of nurses serving patients, which is called the full-time equivalents (FTEs). These two relationships are added to the diagram as shown.

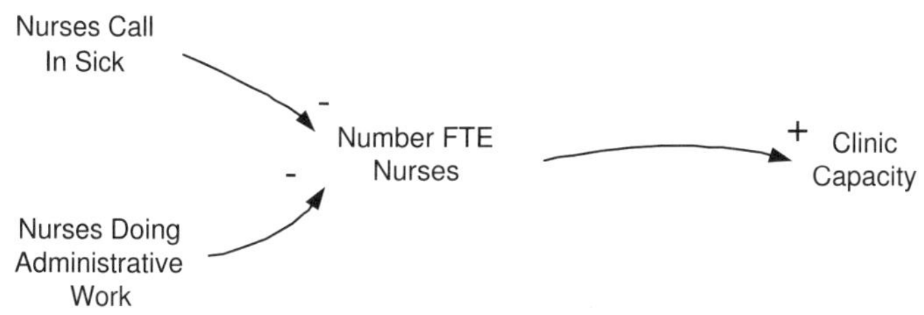

The group might now explore why nurses call in sick. Discussions reveal that nurses are asked to do overtime on top of their 12-hour shift, and they then frequently call in sick later in the week when the overtime becomes excessive (they are not really sick, just fatigued). High usage of overtime reduces nurse morale, which makes them more likely to call in sick for minor reasons. The reason for overtime is to increase clinic capacity. Understanding these relationships, the diagram is updated as shown.

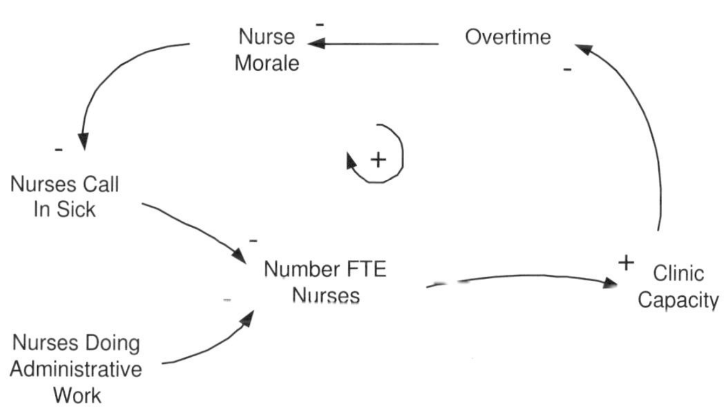

Adding these variables creates a closed loop. Tracing through the loop, we can determine that it is a positive feedback loop – meaning it reinforces itself. This can be determined by starting at the Number FTE Nurses and saying if it increases what happens to Clinic Capacity? Then continue around the loop determining how each variable reacts until you return to the Number of FTE Nurses.

The analysis group might now want to explore other causes of low clinic capacity. So far, we have only considered people causes (recall in the section on fishbone diagrams the cause categories). What about material causes? The analysis group might recognize that clinic capacity is caused by the availability of resources such as medical equipment, medicines, etc. If they are not available this will affect clinic capacity. The analysis group could continue to analyze the links to availability of resources. Then they can try to identify other variables that can affect clinic capacity. What the analysis group should find while creating the causal loop diagram is how it facilitates group learning of the problem and its potential causes.

Once a diagram is completed, it should be verified and validated to the extent possible. To verify the diagram, the group should check the syntax – are all the variables measurable? Check all the relationships and their direction (positive or negative). Have people close to the problem review it to see if they agree with its representation of the problem situation (face validity). If needed, the analysis group could also collect data to validate whether the implied relationships actually exist in the real system. Data that can confirm the relationships provides the strongest validation.

8.2.3 Document Problem Analysis

The documentation for problem analysis consists of the Problem Statement List, the causal loop diagrams, and the fishbone diagrams. The diagrams would only be done for the problems the project team has identified as the most problematic, or the problems they wish to directly resolve in the project.

8.2.4 Stakeholder Analysis

A *stakeholder* is a person or organization that has a vested interest in the enterprise system. Stakeholder analysis is a process of systematically gathering and analyzing qualitative information to determine whose interests should be taken into account when conducting an enterprise system project. Stakeholder analysis includes such stakeholder characteristics as knowledge of the policy, interests related to the policy, position for or against the policy, potential alliances with other stakeholders, and ability to affect the policy process (through power and/or leadership).

Stakeholder analysis serves two purposes in the enterprise engineering project. First, it is done to identify stakeholders and to assess their knowledge, interests, positions, alliances, and importance related to the project. Knowing who the stakeholders are and their interest is required for strategy formulation, problem analysis, and requirements analysis. Second, by understanding the stakeholders the project team can more effectively interact with key stakeholders and determine how best to implement the enterprise engineering project. Change management requires understanding of the stakeholders and overcoming any resistance to the new project. Unless the team understands who the stakeholders are and their position regarding the project, then they cannot properly develop a change management plan for system implementation. When a stakeholder analysis and other key tools are used to guide the implementation, the policy or program is more likely to succeed. This information can be used to provide input for other analysis; to develop action plans to increase support for a reform policy; and to guide a participatory, consensus-building process (by sharing the information obtained with the stakeholders and encouraging discussion about how to address the concerns of the opposition).

Identifying the key stakeholders is extremely important to the success of the analysis. Initially, the working group should identify all actors who could have an interest in the project, including actors outside the enterprise that could affect or be affected by the project. The working group should plan to interview the priority stakeholders identified to gain accurate information on their positions, interests, and ability to affect the project. The working group defines the exact stakeholder information or characteristics to be considered.

Potential stakeholders in an enterprise engineering project are shown in Table 8.4. Within each stakeholder category, there may be significant diversity and sub-categories will be warranted. For example, employees might be sub-divided into administrative employees, technical employees, and manufacturing employees.

The purpose of stakeholder analysis is to use the information gathered about the stakeholders to influence the project plan and system design. Some of the stakeholders will be

TABLE 8.4
Potential Stakeholders

Senior executives	Managers
Employees	Shareholders
Suppliers	Customers
Prospective customers	Distributors
The community	Worker's union
Alliance partners	The government

identified as key stakeholders because they wield significant influence on the project's success. It is important to maintain a constant dialog with these stakeholders.

Understanding Stakeholder Motivation

In a project to form an international alliance between the American and Korean divisions of a company, both country divisions had to identify a small group of key employees who would lead the effort. It was decided that these employees had to have certain qualifications, be in a certain position or higher, and be actively supportive of the project. The Korean division had difficulty in recruiting these employees. Some investigation revealed many of the Korean employees were reluctant not because they did not support the project's goals, but because they assumed the employees would be relocated to a new building that was far from the Seoul headquarters. These employees did not want to move from their homes or endure the impossibly long commutes. For this reason, recruitment was difficult.

Stakeholder Documentation

The Stakeholder Matrix[6] in Table 8.5 shows each stakeholder and his or her information of interest to the project. The columns are:

- **Position**. The position the stakeholder has and the organization.
- **Int/Ext**. Internal (Int) stakeholders work in the enterprise that is doing the project. External (Ext) stakeholders work outside of the enterprise.
- **Knowledge**. The level of knowledge the stakeholder has concerning the project. A useful scale is: 3 = significant knowledge, 2 = some knowledge, and 1 = little or no knowledge. Additionally, this column can contain a brief statement of the stakeholder's definition of the project goals and purpose.
- **Position**. The position the stakeholder has as either a supporter, opponent, or neither on the project. This column is sub-divided into the position per statements taken directly from the stakeholder and the position as inferred through other information.
- **Interests**. The interest the stakeholder has in the project, or the advantages and disadvantages that implementation of the project may bring to the stakeholder or his or her organization. The advantages and disadvantages should be directly listed in this column.
- **Alliances**. An alliance is when the stakeholder collaborates with other stakeholders to support or to oppose the project.

[6]This matrix draws heavily on the stakeholder analysis recommendations from the *Policy Toolkit for Strengthening Health Sector Reform*, Kammi Schmeer, Pan American Health Organization, September 2000.

TABLE 8.5
Stakeholder Matrix

Position	Int/Ext	Knowledge	Position	Interests	Alliances	Resources	Power	Leadership

- **Resources**. The resources (human, information, political, technological, knowledge, money, or material) that the stakeholder controls and can make available (or unavailable) to the project. The column is sub-divided into two parts: the quantity of resources the stakeholder has available and the ability of the stakeholder to make use of those resources for the project.
- **Power**. The ability of the stakeholder to affect the project implementation due to the strength or force he or she possesses.
- **Leadership**. The willingness and innate ability of the stakeholder to initiate and lead an action for or against the project. The stakeholder either has this ability or not (yes or no).

8.3 Goals and Objectives

Sometimes the words "goal" and "objective" are used interchangeably. In practice, goals are the high-level, broad intentions of the organization. Objectives are specific, narrow achievable states of the organization. Goals capture the reasons why an enterprise engineering project is needed and guide decisions at various levels within the enterprise. Goals should be mapped to stakeholders. Figure 8.3 shows how the vision statement, goals, and objectives form a hierarchy. The figure shows that the objectives are derived from the goals. The objectives are measurable, specific outcomes that management can verify whether or not they are achieved.

Sage and Armstrong ([7] p. 106) makes the following advice to state an objective. The sentence structure should be

$$To\ (action\,word) + (object) + (qualifying\,phrase)$$

For example, good objective statements are:

- "To improve system output by 20% without increasing the number of employees."
- "To be ranked the number one cruise line by the Cruise Line Traveller magazine's annual review of cruise lines."

An oft-quoted phrase is that if you cannot measure it, then you cannot manage it. Not only is measurement important for management of existing enterprise systems, measurement is needed to specify and then determine attainment of an objective or goal. The above objective is stated such that progress against the objective can be measured.

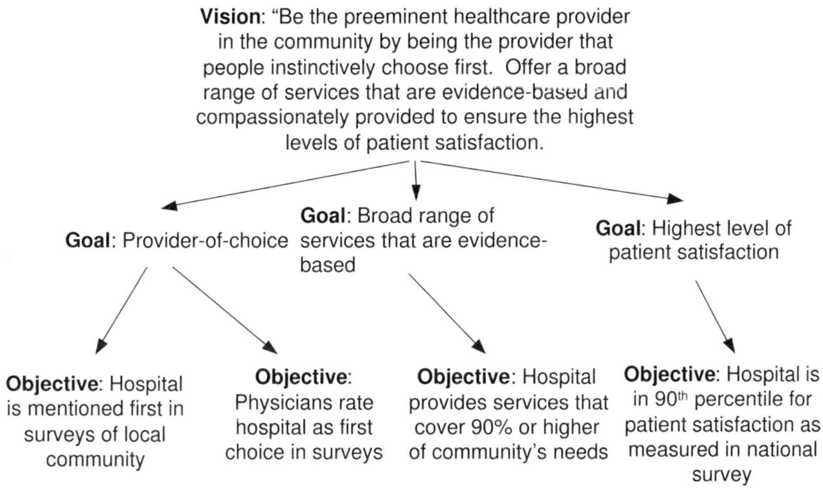

FIGURE 8.3
Goals and objectives.

8.4 Requirements Engineering

Requirements engineering is a systematic approach to specify the requirements and constraints that a new system must satisfy. The goal of requirements engineering is to obtain a complete but minimum set of requirements the system must satisfy. Requirements are statements of "what" functions, capabilities, and features the system shall possess. The idea is to create an implementation-neutral specification so as to not overly constrain the creative design process during which the "how" will be specified.

Requirements drive the design process; consequently, each requirement statement has a schedule and budget impact on the project. A complete but minimum set of requirements is developed early in the project so that project planning, scheduling, and budgeting can be performed. Throughout the project, more requirements might be discovered, so a formal requirements engineering activity is needed to evaluate whether to include each new requirement, and if it is incorporated into the project, then the project team must assess its impact on scope, schedule, and budget.

Many studies have determined that developing a complete, accurate requirements specification early in the project is one of the most important project success factors [1]. The benefits of good requirements are:

- A detailed specification to enable more accurate estimation of project costs, personnel, and activity durations.
- A detailed specification provides for clear understanding of what will be done and thus agreement between the project team and system stakeholders.
- A detailed specification is more likely to lead to a successful project that is on time, under budget, and provides all desired functionality.
- A detailed specification is more likely to lead to better system reliability, usability, and maintainability.

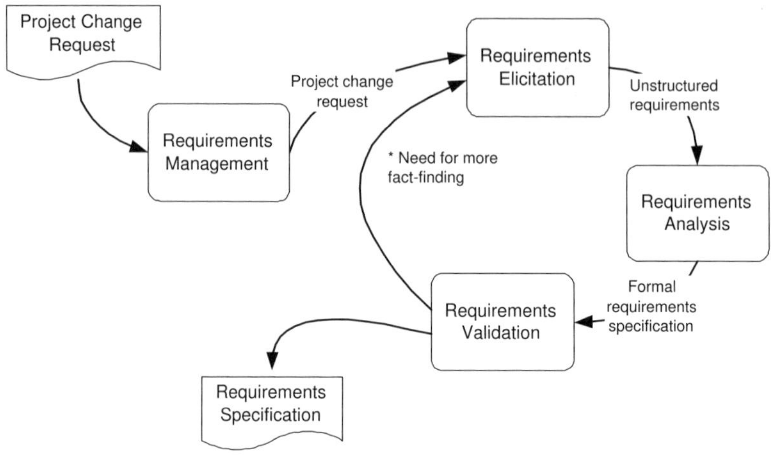

FIGURE 8.4
Requirements engineering process.

Requirements engineering manages the requirements process and the changes to the requirements. Requirements engineering starts with understanding the goals, strategies, and problems facing the enterprise. The goals and strategies are too vague to design an enterprise system to. In fact, the greatest struggle in requirements engineering is converting vague goals (e.g., "user friendliness," "security," "reliability") into detailed, precise specifications of system properties or behavior that an enterprise system can be designed to. The requirements engineering process is intended to bridge the gap from informal, fuzzy impressions of need to a rigorous, formal specification.

Over the lifetime of the project it is very common for new requirements to emerge and existing requirements to change. Studies have shown that over the life of a project as much as 50% or more of the requirements will change before the system is operational. Changes come about due to more accurate, subsequent requirements elicitation, changing user needs, changes in the relative importance of various user requirements, and technology changes. Although it is true that requirements are volatile and constantly changing, much iteration in the requirements refinement process is simply due to misunderstood or misinterpreted requirements. Good requirements engineering will ensure the project starts with a complete and accurate requirements specification, it will minimize the volatility of the requirements specification, and also ensure that changes are incorporated correctly into the requirements specification as needed.

The requirements engineering activities are (see Figure 8.4):

1. Requirements Elicitation

   The first step is to elicit the system requirements. The term "elicit" is used because it conveys the idea that requirements must be actively drawn out by the project team. You cannot simply ask stakeholders what the system must do. To elicit requirements the team must use data-gathering techniques to understand the problem and the application domain and to establish their requirements. These requirements may not be complete and may be expressed in a vague and unstructured way.

   (a) Refine requirements – Requirements need refinement for a number of reasons, including the need to: merge input from multiple sources, discard irrelevant information, and distinguish between inputs relating to system implementation requirements and those relating to general domain knowledge.

(b) Understand requirements – As the project progresses, all project personnel will receive requirements. To avoid requirements creep, criteria are established to designate appropriate channels, or official sources, from which to receive requirements. The receiving activities conduct analyses of the requirements with the requirements provider to ensure that a compatible, shared understanding is reached on the meaning of the requirements. The result of this analysis and dialog is an agreed-to set of requirements.

2. Requirements Analysis

The requirements discovered during the elicitation phase are integrated and analyzed. Usually, this will result in the identification of missing requirements, inconsistencies and requirement conflicts. The analysis is facilitated by checklists to ensure "good" requirements statements. Requirements are modeled and formally specified. Part of the requirements analysis is to prioritize requirements. The final output is a formal list of system requirements.

(a) Manage requirement changes – During the project, requirements change for a variety of reasons. As needs change and as work proceeds, additional requirements are derived and changes may have to be made to the existing requirements. It is essential to manage these additions and changes efficiently and effectively. To effectively analyze the impact of the changes, it is necessary that the source of each requirement is known and the rationale for any change is documented. The project manager may, however, want to track appropriate measures of requirements volatility to judge whether new or revised controls are necessary.

(b) Allocate requirements – In large systems requirements are allocated to particular subsystems or components of the overall system. It is also likely that some requirements are fulfilled not by direct design of a single component but by the interaction of many different enterprise components.

3. Requirements Validation

The team confirms that the system requirements meet stakeholder needs and that the requirements are valid, correct, and complete. Use peer reviews and inspections to reduce defects in all your requirements representations. Peer reviews and inspections are a best practice way of eliminating defects. Validation may involve negotiation between the team and stakeholders to try to reach a consensus about the resolution of inconsistent or conflicting requirements and hence reach acceptable "win" conditions for all stakeholders.

(a) Obtain commitment to requirements – As the requirements evolve, this specific practice ensures that project stakeholders commit to the current, approved requirements and the resulting changes in project plans, activities, and work products.

(b) Maintain bidirectional traceability of requirements – When the requirements are managed well, traceability can be established from the source requirement to its lower-level requirements and from the lower-level requirements back to their source. Such bidirectional traceability helps determine that all source requirements have been completely addressed and that all lower level requirements can be traced to a valid source. Requirements traceability can also cover the relationships to other entities such as intermediate and final work products, changes in design documentation, test plans, and work tasks. The traceability should cover both the horizontal and vertical relationships, such as across interfaces. Traceability is particularly needed in conducting the impact assessment of requirements changes on the project plans, activities, and work products.

(c) Identify inconsistencies between project work and requirements – This activity finds the inconsistencies between the requirements and the project plans and work products and initiates the corrective action to fix them.

4. Requirements Management

   The requirements engineering team establishes policies to keep track of the requirements process and maintains a central, shared repository for storing all data gathered, open issues, and requirements. The requirements engineering policies include methods to keep track of open issues, reminders to consider recurring questions, illustrations of the rationale for previous decisions, etc. The database is a best practice and is needed to handle the large volumes of informal information [5]. The requirements engineering process also interacts with the configuration control process to manage any changes to the requirements specification.

The output of the requirements engineering process is the Requirements Specification document – a complete set of valid enterprise system requirements. The Requirements Specification, once approved, becomes a baseline document that is controlled by the configuration control process.

8.4.1 Requirements Definition

A *requirement* is a specification of a system function, capability, or an essential attribute of the system. Requirements are classified as either being functional requirements or non-functional requirements. A *functional requirement* specifies an action the enterprise must be capable of. A functional requirement describes enterprise behavior in terms of functions of the system that support stakeholder goals. Functional requirements are specified with verbs to describe the action.

Non-functional requirements describe properties or characteristics of the system that stakeholders deem important and will affect their degree of satisfaction with the enterprise. Non-functional requirements describe the attributes and characteristics of the enterprise (see Table 8.6).

Related to requirements are *constraints*, which are restrictions on what can be done. Constraints can be imposed due to physical limitations of resources, due to the environment such as regulatory rules, or due to any reason that justifies defining restrictions on other requirements.

8.4.2 Good Requirement Statements

A good requirement statement specifies an attribute or characteristic of the system, a relation, and the required value or units. A single statement addresses a single requirement – avoid conjunctives to tie together two or more requirements. Requirement statements are written in the active voice and use strong verbs.

The SMART requirement criteria [6] is an acronym for **S**pecific, **M**easurable, **A**ttainable, **R**ealistic, and **T**ime-bound or sometimes the "T" represents **T**raceable. SMART is used to analyze each requirement statement to verify it and determine whether it is a good requirement statement. Note, this does not prove that the requirements are true in the sense

TABLE 8.6
Non-functional Requirements

Requirement type	**Description**
Interface	Describes how the system must interact with other systems.
Quality	Describes the level of measurable quality (statistical) or perceived quality.
Performance	Describes how the system performs (productivity, profitability, efficiency, effectiveness, innovativeness).
Reliability	Describes the consistent performance of the system and its dependability.
Supportability	Describes the ease of supporting the system.
Regulatory	Describes requirements that are to meet regulations imposed by government, market, or other agencies.
Infrastructure	Describes requirements to provide for a structure to support the system.
Usability	Describes requirements for how easy it is to use the system, learn a job, etc.
Safety	Describes features of the system that provide for people's safety from harm, information safety from theft or misuse, or safety of equipment.
Flexibility	Describes requirements for the system to be able to handle a variety of products, services, resources, or information as inputs and outputs.
Sustainability	Describes requirements for the ability to keep a system going without needing a constant stream of capital or other inputs.
Maintainability	Describes requirements of the system to repair failures and to do preventive maintenance.
Scalability	Describes the ability of a system to efficiently and effectively grow larger or grow smaller as the need arises. Scalability can be achieved by adding new resources, collaborating with other organizations, or whatever strategy makes sense.

that they accurately represent what is needed. It only establishes whether the requirement is stated well and something that could potentially be achieved.

Specific

A good requirement is clear, concise, and as simple as possible. A good requirement should not be open to misinterpretation when read by others. A single requirement should specify only one thing. Having specific requirements is the most important attribute to get correct. The requirement should also be complete by specifying any necessary conditions or qualifiers. Avoid using conjunctions (and, or, but) as they can confuse or misconstrue the meaning. Avoid indeterminate amounts of time (soon, fast, later, immediately) and indeterminate quantities of things (several, a few, most) as they are open to wide misinterpretation which results in dissatisfied stakeholders. Avoid subjects that can refer to multiple things (it, they, this, these, that). Avoid subjective phrases (if feasible, when cost effective).

> WEAK REQUIREMENT: The customer service agents should answer all calls quickly.

This requirement is weak because the word "should" is weak, more like a suggestion than what must be done. Also, anytime you see the words "all, always, never" and other global adjectives, then you should beware of imprecision of thought. Moreover, exactly what is meant by quick? The problem with the statement is it is open to too much interpretation because it is not specific.

> STRONG REQUIREMENT: The customer service agents shall answer 90% of calls within the first two minutes.

This fixes the two problems cited with the weak requirement. The requirement statement defines what is called a service level. It takes into account that there is variation in the system so the service level is set for 90% of calls, allowing for unusual or rare situations in which answering a call in two minutes is not possible.

Measurable

A good requirement can be measured and thus verified that it is indeed met at the end of a project. If a requirement cannot be measured, then there is no way to verify whether the resulting design satisfies the requirement. In fact, part of the final requirement's document should specify tests for each requirement to verify at the end of the project whether it was satisfied or not. A requirement that is not specific will usually not be measurable. When a requirement is specific, then whether it is measurable or not is more subtle. Measurement might be a matter of degree. For example, what is the cost to measure the requirement? What data are needed? Or is a quantifiable measure available?

> WEAK REQUIREMENT: The system shall have an optimal response time for the end-user.

This requirement is weak because there is no way you can be successful with this requirement once the new system goes into production. It's similar to being specific, in that optimal is not defined and really cannot be defined.

> STRONG REQUIREMENT: The system shall have user response times on user click-events that are 5-seconds or less during business hours of 9AM-5PM, Monday-Friday.

This requirement is measurable because a specific time is stated for its performance.

Attainable

A good requirement is attainable if the requirement is physically able to be achieved given existing circumstances.

> WEAK REQUIREMENT: The production line will be capable of operating at 100% capacity 24 hours a day, 7 days a week.

This requirement is weak because even if possible, it is prohibitively expensive to make a system that has no down-time. Moreover, it may be inconsistent with the goals of the enterprise in terms of cycle time. An attainable requirement needs to consider the availability of technology and what is physically possible.

> STRONG REQUIREMENT: The production line will be capable of operating at 85% capacity 24 hours a day, 7 days a week.

This requirement allows for some slack time for maintenance and also observes that performance must trade off speed and capacity as is discussed later in Chapter 11 on queueing theory.

Realistic

Answers whether the requirement is realistic to deliver when considering other constraints of the project and requirements. There is arguably an overlap between attainable and realistic. The difference is that attainable considers the physical possibility, and realistic considers the project constraints of schedule, budget, and resources. To assess whether a requirement is realistic, the analyst determines whether the requirement is consistent with the project schedule and budget. The resources available must be able to deliver the requirement by either having the correct skills, equipment capability, or whatever resource is required.

> WEAK REQUIREMENT: The organization will have ISO 9000 certification when it starts operation.

This requirement is weak because while attainable it is not realistic that ISO 9000 certification can be obtained prior to starting operations. ISO 9000 certification takes time to obtain. The requirement can be corrected by simply adding a realistic time for attainment.

> STRONG REQUIREMENT: The organization will have ISO 9000 certification within its first year of operations.

Time-Bound (timely, traceable)

Where appropriate each requirement is time-bound or includes criteria specifying by when or how fast a requirement needs to be completed or executed. In enterprise engineering projects, the schedule is developed separately from the requirements document. So, the time-bound attribute is more to determine when the requirement is needed and then the team can assess whether the project schedule will meet the time requirement. Additionally, the "T" in SMART designates whether a requirement is traceable, which describes the ability of a person to trace the requirement through conceptual models, to detailed models, and to its final implementation in the system. Traceability is important for understanding the design rationale of an enterprise. If an engineer wants to understand why a particular design decision was taken, traceability helps that engineer to trace back from the system through the models to the requirements document that first specified a particular need.

WEAK REQUIREMENT: The report will be available soon after month-end close.

This requirement is weak because you cannot rely on what you consider to be reasonable expectations of the customer. You may know the time cycle of month-end close and that it takes the first 5 days of the month to complete. The customer may assume that soon means on the 1st of the month.

STRONG REQUIREMENT: The report will be available by noon on the first business day after the successful completion of the month-end accounting reports.

This clearly explains when the enterprise can expect the report (first business day, not on a weekend) and only after a successful run of the closing monthly reports from accounting.

Evaluation of the Requirements Document

The SMART criteria applies to each individual requirement. In addition, the entire set of requirements should be evaluated as a whole completeness and consistency. Completeness is that the requirements document includes the minimum set of requirements. Consistency is that none of the requirements conflict with each other, they align with the project goals and objectives, and they do not violate any constraints. The requirements engineering sub-team holds a requirements review to evaluate the requirements as a whole to answer questions such as:

- Are the requirements complete?
- Are the requirements consistent?
- Do the descriptions of different requirements include contradictions?
- Do the requirements document and individual requirements conform to defined standards?

Completeness of the requirements document is evaluated with respect to the project goals. Goals provide a criterion for the sufficient completeness of a requirements specification. If all the goals can be proved to be achieved by meeting the requirements specification, then the requirements are complete [11]. Simultaneously, goals provide a precise criteria for determining whether a requirement is really a requirement – a requirement is not needed if it does not support attainment of a goal. Goals provide a rationale for the requirements, and thus, can help explain those requirements to the stakeholders.

8.4.3 Requirements Best Practices

There are several best practices for requirements engineering. To help write good requirements templates, checklists, and standards are useful. One best practice is to have a checklist based on the SMART criteria to evaluate each requirement. Managing the requirements is best done using a database or spreadsheet. It is often useful to maintain records of rejected requirements. The requirements engineering sub-team also develops traceability policies and a tool to track traceability. Finally, it should be remembered that modeling is a means to document requirements. The process models, information models, and organizational models discussed in the next chapters all document system requirements in a way that textual descriptions cannot.

The requirements engineering sub-team needs to devise practices to prevent scope creep. Scope creep is when the requirements of the project keep changing and increasing with the result that the project falls behind on schedule or goes over budget. The practices described in this section will help minimize scope creep.

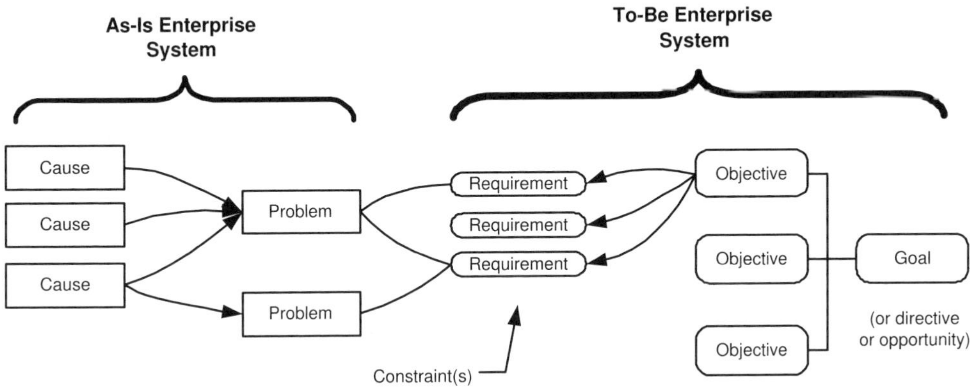

FIGURE 8.5
Relationship between analysis outputs.

8.5 Linking Problem Analysis and Requirements

The activities of problem analysis, stakeholder analysis, goal analysis, and requirements analysis were presented as separate activities. Collectively, all the analysis outputs must be consistent with each other. Figure 8.5 shows the relationships between the analysis outputs. A problem has one or more causes. A cause might be related to more than one problem. The problems and their causes are a description of the "As-Is" enterprise system.

The goals and objectives are descriptions of what the stakeholders want the "To-Be" enterprise system to possess. As previously shown, each goal is decomposed into one or more objectives. The objectives are linked to the requirements that specify functions, capabilities, or features that contribute towards achieving the objective. Constraints limit what can be done, and constraints can affect one or more requirements. The requirements may address one or more problems, or possibly no problem at all because it is a statement of the future system. One problem might be addressed by more than one requirement.

A useful activity is for the project team to review the entire analysis for consistency as presented here. The purpose is to uncover problems that might not be addressed by any requirements, inconsistency between the problem and the requirement, or objectives that would likely go unfulfilled because there are no requirements to ensure the necessary conditions for the objective.

Figure 8.5 should not be misconstrued to imply a straightforward, linear process can go from causes to problems to requirements all the way to goals. As in any complex process, there are many iterations between the activities.

Analysis in an Urgent Care Clinic

A project team was assigned the objective to reduce patient turnaround time (cycle time) to two hours or less for 90% of all patients. This project objective contributes towards an enterprise goal of achieving high-level patient satisfaction. The project team gathered data to identify and analyze the problems and their causes. Figure 8.6 shows three of the problems identified and their causes.

One interesting cause of long waiting times was identified after analyzing the waiting

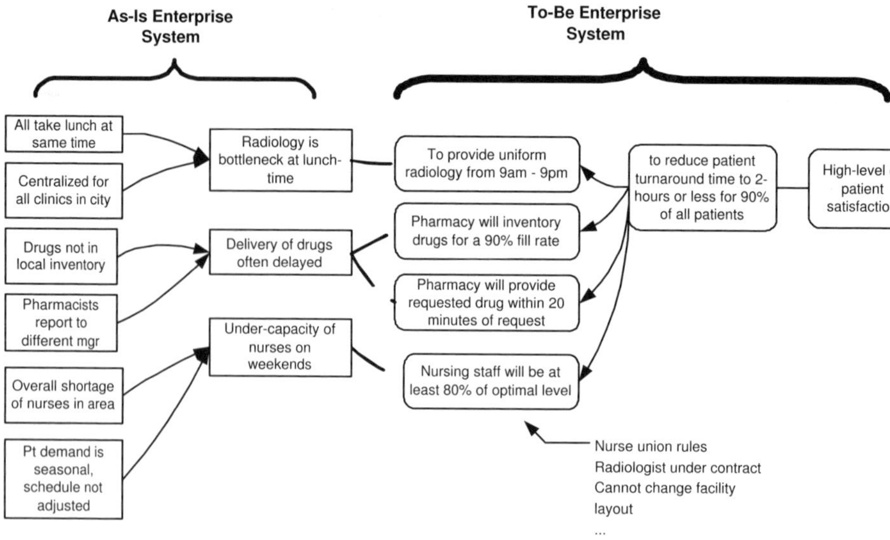

FIGURE 8.6
Example tracing problems to goals

time data by day. The clinic electronically sent the x-rays to a remote radiology department where the radiologists could view the x-rays and make their reports. The data showed that waiting times around lunch time were double the waiting times for before or after lunch. Investigation found that the radiologists all took lunch as a group together at the same time. To address the radiology problem, the new enterprise design should satisfy a requirement of providing uniform radiology during all operating hours (9am-9pm). This is what needs to be achieved; how it is to be achieved is not part of the requirement specification.

8.6 Problem Analysis and Requirements Documentation

The problem analysis, stakeholder analysis, and requirements engineering activities are recorded in reports that become part of the project's deliverables. The reports generated are:

1. Problem Analysis Report
2. Stakeholder Report
3. Requirements Document

The problem analysis report includes a summary of the data-gathering strategy used, a summation of the data, list of the problems identified in the Problem Statement List, and an analysis of the causes and effects of those problems. The stakeholder report consists mainly of the Stakeholder Matrix supported by the plans to utilize this information to aid project success. The requirements document (see box below) lists all the requirements and includes any models of process, information, or organization that diagram system requirements. This chapter did not discuss the modeling of processes, information, and the organization, but this is an important part of problem analysis and requirements analysis. Models can capture requirements for processes or information better than textual descriptions. Consequently, the models are a significant component of the requirements documentation. In the next chapters each enterprise perspective is discussed separately.

Format of Requirements Document

1. Introduction
   - (a) Purpose of Document
   - (b) Scope
   - (c) Definitions
2. Requirements
   - (a) Functional Requirements
   - (b) Non-functional Requirements
   - (c) Interface/Integration Requirements
   - (d) Constraints
   - (e) Assumptions
   - (f) Process Requirements
   - (g) Information Requirements
   - (h) Organizational Requirements
3. Appendices
   - (a) Reference Documents
   - (b) Requirements Engineering Policies and Procedures
   - (c) Traceability Matrix
   - (d) Requirements Analysis

8.7 Summary

This chapter describes the methods, management, and techniques for problem analysis and requirements analysis. The chapter first presented data-gathering techniques that are done to support many different phases of the enterprise engineering life-cycle but predominantly the problem and requirements analysis phase. The chapter described problem analysis techniques, especially has to analyze causes and effects. The requirements engineering process was described and the criteria for SMART requirements. The enterprise engineering team needs to elicit the requirements, understand them, analyze them, put them into a usable form, validate them, and then analyze them in regard to all the other requirements and the overall project. The chapter ended by describing the main deliverables of problem and requirements analysis. It should be remembered that a large part of identifying and defining system requirements results the modeling done of the process, information, and organization.

You might be put off or think the degree of formalization and detail that goes into data-gathering techniques is uncalled for. Many technical-oriented people want to start right away on designing. However, what must be remembered is that most enterprise engineering projects are large, complex, and must satisfy many stakeholders. Moreover, these projects are conducted by a project team not an individual. The person conducting an interview will probably not be the same person analyzing requirements, or modeling the process, or designing the system. Formalization improves the process, limits the risk of missing important information, and helps avoid the problems that would arise from ambiguity.

Review Questions

1. Categorize the data-gathering methods in terms of effort required versus richness of data obtained.
2. Describe the steps involved in conducting an interview.
3. Explain the difference between an unstructured interview and a structured interview. Describe a scenario where you would use each one.
4. Contrast objectives from goals.
5. Explain the relationship between a problem and requirement.
6. List the criteria that are used to evaluate the whole requirements documentation.
7. Explain why traceability of a requirement is important.
8. What is the difference between a functional requirement and a non-functional requirement?
9. Are functional requirements more important than non-functional requirements? Argue for or against this statement.
10. List the benefits of a requirements workshop.
11. Describe a scenario where it is appropriate to use questionnaires to gather data.
12. What is the reason that requirements should avoid stating what technologies to use?

Exercises

1. For each of the requirement statements below: (i) state whether the requirement is well written according to the SMART criteria; (ii) if not, explain what is wrong with the requirement and restate an improved requirement.

   (a) The system shall be user-friendly.

   (b) The order-fulfillment process should be capable of handling many different customer order types.

   (c) The system shall be able to provide real-time status of on-hand inventory.

   (d) The system must present information resulting from system queries in a format that facilitates the understanding of the information by non-technical users; and unformatted for more technically proficient users.

   (e) The system must be reliable, easy to maintain, and have automatic maintenance reminders to alert the users when maintenance is due.

(f) Patients shall not wait a long time before being taken to an examination room.

(g) The system must enforce three levels of security for administrator, manager, and employee.

(h) The system should support international operations involving different currencies.

2. The Liberal Christian's Charity in Memphis, TN has started a project to improve and expand its operations. The charity staffs collection trailers in key points throughout the city where people can drop off items they wish to donate. The goal is to place collection trailers in locations that are most convenient for people to donate items so they increase their donations. They also need to develop an efficient means to move the items from the collection trailers to a centralized sorting and distribution center in downtown. They want to determine the best places to locate the trailers, the best staffing policies, and the best way to transfer the items from collection to distribution. They also feel that better information on their collections would help them to make better decisions in the future.

   Describe a data-gathering strategy to determine the problems, constraints, and requirements for the proposed expansion of the Liberal Christian's operations.

3. For each of the following requirements, state whether they are functional or non-functional.

   (a) The system shall have an average throughput of 1.5 Mb/s.

   (b) Customers will be able to update their own information.

   (c) The system shall be available 24 hours each day and 7 days each week.

   (d) Patients must acknowledge receipt of the privacy policy prior to service.

   (e) The system shall automatically generate a departure for 8 months after a crew member is scheduled to join a ship.

4. State University is building a Web-based registration system where 90% of all registration takes place during a 1-week period each semester. State U. has 35,000 students who must register. As a Web-based system, registration can take place 24/7 (24 hours a day, 7 days a week). However, a more reasonable profile is shown below (usage as number of registrations and hours on 24-hour clock). Each registration is approximately 400 kb. Write out a reasonable throughput requirement for this system. State all assumptions and provide rationale for your requirement statement.

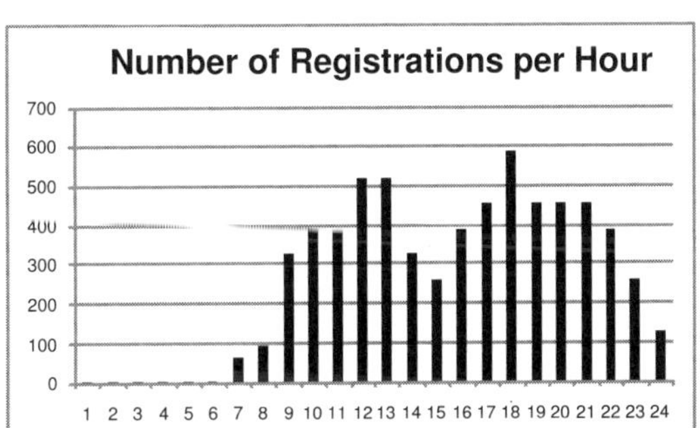

5. The Grilli Shoe Company in Sirolo, Italy makes great shoes but has a terrible method for taking inventory. Their current inventory system is a spreadsheet

where the production manager enters in the shoes made for the day and the sales representative records what was sold to retail stores/distributors for the day. Needless to say, they need to physically count the inventory each quarter because what they have on-hand never matches what the spreadsheet says. You are hired to build an inventory management system for them. First, you need to determine the system requirements.

(a) Who would you interview?

(b) What questions would you have?

Remember, the company president, production manager, and sales representative are very busy people so the interview must be efficient.

Bibliography

[1] R.N. Charette. Why software fails. *IEEE Spectrum*, September:1–2, 2005.

[2] C. G. Drury. *Evaluation of human work*, chapter Methods for direct observation of performance. Taylor and Francis, London, England, 1992.

[3] D. C. Gause and G. M. Weinberg. *Exploring Requirements: Quality Before Design.* Dorset House Publishing, New York, NY, 1989.

[4] G. P Huber and D. J. Power. Retrospective reports of strategic level managers: Guidelines for increasing their accuracy. *Strategic Management Journal*, 6:171–180, 1985.

[5] M. Lubars, Potts. C., and C. Richter. A review of the state of practice in requirements modeling. In *Intl. Symposium on Requirements Engineering,*, San Diego, CA, 1993.

[6] M. Mannion and B. Keepence. SMART requirements. *ACM SIGSOFT Software Engineering Notes*, 20(2):42–68, 1995.

[7] A.P. Sage and J.E. Armstrong. *Introduction to Systems Engineering.* John Wiley & Sons, New York, NY, 2000.

[8] J. Wood and D. Silver. *Joint Application Development.* John Wiley & Sons, New York, NY, 1995.

[9] G. Wright and P. Ayton. Eliciting and modeling expert knowledge. *Decision Support Systems*, 3(4):13–26, 1987.

[10] R. Yin. *Case Study Research.* Sage Publication, Beverly Hills, CA, 1984.

[11] K. Yue. What does it mean to say that a specification is complete? In *IWSSD-4, Fourth International Workshop on Software Specification and Design, Monterey, CA*, 1987.

9

Generate and Evaluate Alternatives

"Get your facts first, and then you can distort them as much as you please." – Mark Twain (1835-1910), American author.

The project goal is to find the best design to satisfy the system requirements. The more alternatives considered, then the more likely the best design will be revealed and selected. This chapter describes approaches to generate and then evaluate alternative designs. Once the project team is satisfied they have generated sufficient design alternatives, they need to evaluate those alternatives to decide which one will form the basis for future design work. Evaluation of the design alternatives is done through feasibility analysis that considers various characteristics of the design solution and how well they meet the system requirements. The input to generate and evaluate alternatives is the system requirements and the output is the selected design alternative that will be used in the detailed design phases that follow. After reading this chapter, you should be able to:

- Compare and contrast COTS and proprietary systems.
- Describe how the morphological box works to generate ideas.
- Describe how to conduct a brainstorming session.
- Identify evaluation criteria and create an alternative evaluation matrix.
- Describe methods to select the best alternative.

9.1 Generate Alternatives

The design team needs to generate alternative solutions to the problems and system requirements. Generating alternatives requires divergent thinking. What are all the ways we can solve the problems and satisfy the system requirements? To generate many alternative designs the team uses techniques that assist creative problem solving. Failure to generate alternatives leads to lack of system optimization and the possibility of pursuing an inferior design solution when other better design alternatives existed.

Alternatives

The contents of an alternative solution depend on the project. A design alternative can include one or more of the following:

- Process design
  - List of processes
  - Process hierarchy

  - Process sequence of activities
  - Business policies and rules

- Information model design
  - Entity-relationship model
  - List of information objects
  - Data dictionary

- Organization design
  - Organization structure
  - Job design
  - Design of decision structures
  - Design of culture
  - Design of reward and compensation system
  - Training of employees

- Information system design
  - Functions of information system
  - User interface design

- Network or IT infrastructure design
  - Network structure
  - Technologies to employ
  - Protocols and standards to adopt

The design alternative at this point is not a finished design, but a description of the design solution approach. So, for example, the project might decide to develop a Web-based reservation process and a telephone reservation process with a call center of reservation agents. The processes will be defined at a high level. The organizational design will specify the approximate number of call center agents, high-level job description, reporting structure, and reward system. The information model will identify the main information objects and their relationships. The technology specification is Web-based; included would be the hardware, software, and infrastructure necessary to support this approach. The details provided at this stage are only sufficient to evaluate the feasibility of the design alternatives. The detailed design and construction begins after a design alternative is chosen.

Enterprise design alternatives can be classified into one of several broad categories:

COTS

A Commercial Off-the-Shelf (COTS) system describes any software system that can be purchased with most of the functionality a client desires. The need for customizing the COTS system ranges from very little to very extensive customization – project risk increases with more customization. The popularity of purchasing COTS has increased dramatically over the years because it seems to be a lower-risk, less-expensive means to quickly acquire a new system compared to developing proprietary software. In the COTS category are many enterprise-level applications such as Enterprise Resource Planning (ERP) systems, Customer Relationship Management (CRM) systems, Supply Chain Management (SCM) systems, Workflow Management (WM) systems, Product Data Management (PDM) systems, Data Warehouse systems.

Of COTS systems, ERP is most widely adopted. ERP systems are large, integrated systems that serve all the departments within an enterprise. ERP applications are defined around business processes, so they have a horizontal view of the organization. The attractiveness of ERP is that it can solve many integration problems because all the functions are met by a single application, so there is no system heterogeneity to cause integration challenges.

When selecting the ERP or any COTS system as a design alternative, the enterprise is making the choice that it will change its organization, processes, and information to comply with the way the ERP system works. This implies the ERP implementation project has a major focus on change management to prepare the organization for the new system. An ERP project involves vendors for the following tasks: the ERP application vendor (e.g., SAP, Oracle, Peoplesoft). A systems integrator consultant that has expertise in ERP implementation. The system integrator may be the same as the ERP vendor, but commonly it is a different company. Finally, the ERP application is software; the enterprise will need to acquire separately the hardware and maybe even a database for the ERP system.

ERP projects change the approach to some of the life-cycle phases. Requirements analysis changes because with COTS you are not going to specify the desired "to-be" system with a set of requirements that a system is then designed to meet. The ERP system defines the "to-be" system. So requirements analysis becomes what is called gap analysis, where gap analysis looks at what the software offers and what the enterprise wants, and then determines how the gap between them can be closed. A gap can be closed by: (1) changing the organization to match the ERP system; (2) configuring the ERP system to meet the organization's needs; (3) customizing the ERP by changing the code, adding bolt-ons, or other application; or (4) live the gap. Many projects that try too much customization fail, so the trend has been towards the organization changing to comply with the way the ERP system operates. The design phase also changes to include configuration of the ERP system. ERP vendors provide built-in flexibility to the application software that their clients can configure to local needs. This might include what type of payments to accept (check, cash, credit card, debit card, bank wire, or loan). The clients by configuring the application can turn these options on/off as needed.

Proprietary Solution

The traditional approach is a proprietary solution that the enterprise develops to meet its requirements. The system can be built in-house if suitable expertise is available, or a consultant company can be hired to build the system. The greatest benefit of a proprietary solution is that it will exactly meet the project requirements. The drawbacks of a proprietary solution is these projects often take a long amount of time, there is significant risk if the technologies are new, and a proprietary system may later have maintenance and integration problems because it is a one-of-a-kind solution.

Outsourcing

Outsourcing is when an organization takes a function that is normally done by the organization and hires an outside organization to do that function for it. Since the outsourcing companies do only that function, they can often do it more effectively and efficiently than a large organization. If the company the work is outsourced to is in another country, then it is called off-shoring.

The best candidate functions for outsourcing are those functions that are not part of the core capabilities or core competencies of the organization. Commonly outsourced functions include low-technology functions such as janitorial services, business functions such as payroll, customer call centers, and technology functions such as IT support. For many companies, these ancillary, but necessary functions, are better done by an outside company so that the organization can focus on those things that make it successful.

9.1.1 Brainstorming

Brainstorming was mentioned earlier when discussing design thinking. Brainstorming is one technique for generating ideas by removing traditional barriers to creative thinking. Brainstorming is a group process to generate new solutions to a problem. In our case, the problem is a design problem and the new solutions are designs that satisfy the design requirements. A successful brainstorming session generates many ideas, which are later pared down to a few alternatives that will be explored in greater detail. To do brainstorming, you need to prepare for the brainstorming session, conduct the brainstorming session, and document the brainstorming results. A good brainstorming session identifies the right people for the session and encourages their full participation.

To prepare a brainstorming session, you identify a facilitator and the participants. The facilitator is key to the success of the brainstorming session. Usually the facilitator will be a consultant, but whoever it is should be trained and experienced as a facilitator. It is not important that the facilitator has any knowledge of the problem or how to solve it – the facilitator is there only to run the brainstorming session. The participants are chosen according to who can best contribute. Avoid having a person's supervision present because it will inhibit their full participation. A brainstorming group will be from three to eight people. The purpose of the brainstorming session must be clearly defined, time allocated for conducting the brainstorming session, and a location selected. The brainstorming session should be conducted away from the normal workplace so as to reduce interruptions and other distractions. You want to have 100% of the participants' attention during the session – no cellphones or other devices permitted.

During the brainstorming session, the group is presented with a task by the facilitator. A good facilitator encourages participation from all participants. In a brainstorming session, participants call out, in no particular order, solutions to the problem. It is important that the solutions are not evaluated or criticized at this time, no matter how ridiculous or absurd they may seem. Because if somebody criticizes a solution at this stage, then it will likely discourage others from suggesting other-than-expected solutions. In fact, the facilitator encourages unusual and strange ideas. Part of brainstorming is building on the solutions suggested by others. The facilitator keeps the free flow of ideas going and visibly records all the solutions as they are made.

Sometimes there are individuals who due to the nature of their personality dominate the session by doing most of the suggesting without letting others have a chance to speak. This is when a trained facilitator needs to intervene. The facilitator can speak to the person individually and say something to the effect that they need to let others speak too. Frequently, this little suggestion will be sufficient. Other participants will not make any suggestions; again a good facilitator will try to draw these people into the discussion.

A brainstorming session might last one hour. After solution generation, then the group can do categorization and analysis of the ideas, or this can be done by a separate group. The final results are documented and distributed to the appropriate people.

A strength of brainstorming is the ability to generate many ideas in a short period of time. The non-judgmental approach encourages free-flow thinking by the participants, which will hopefully lead to novel and creative ideas. One potential weakness of the technique is that many participants are conditioned from years of work to be reticent, not share ideas, and to stick with conservative, accepted ideas. This is why the selection of participants as well as a facilitator that can help them break away from this thinking is critical to the success of brainstorming.

9.1.2 Delphi Method

A group process to explore design spaces like brainstorming but with three differences: anonymity of the participants, iterated controlled feedback to the participants, and statistical group response. The Delphi method is usually done by mail or nowadays by email. It attempts to reach a group consensus, in this case a group consensus about the best design alternatives. The basic approach is the participants are given the problem, they generate responses, the facilitator collates the responses, and then sends the collation back to the participants for another iteration. This is done for a few iterations until consensus is reached. Since the participants are not in contact with each other, much of the success of the method depends on an experienced facilitator who can generate questions that elicit useful responses and is able to put together the response in a useful way. Further information on the Delphi Method can be found in Linstone and Turoff [1].

9.1.3 Literature Search

One way to generate design ideas is to search the available literature. New ideas are constantly being presented in books, journals, and trade publications. If the enterprise project involves designing a new order fulfillment system, then identify the key terms, identify data sources (e.g., Google) and conduct the search. Typically as you find ideas for order fulfillment, your search will take you on to other related topics. This can be a fruitful approach to generate ideas, and made easily accessible by the Internet.

9.1.4 Request for Proposals

If the problem involves technologies that can be delivered by vendors, then the problem and requirements documentation can be formatted as a Request for Proposal (RFP) that is sent to vendors asking them to bid on it. The degree of details in an RFP can vary from a statement of project goals, objectives, and constraints to very detailed requirements of what is needed. RFPs can frequently be very large documents and are usually only justified on larger projects. Vendors respond to the RFP with a bid describing the product and services they would provide with the cost and schedule. The enterprise receiving the bids evaluates the bids, contacts references of the vendors – usually previous clients of the vendors – reviews the history of the vendor, and invites a short-list of vendors to do demonstrations or give oral presentations of what they would do. The enterprise then decides whether to select a vendor and then enters into negotiations with that vendor.

In smaller projects or in the early stages of a large project, a more informal approach of calling vendors could be used. One caveat is that vendors generally see problems through the solutions they provide. For example, a database vendor such as Oracle will see all problems as variations of data storage and processing problems.

9.1.5 Benchmarking

One means to generate alternatives is to study other enterprises and see how they designed their systems. Two benefits are derived from doing this: first, the enterprise avoids "reinventing the wheel" if there is a suitable design elsewhere that it can copy; and second, the enterprise stimulates the thinking of its staff by seeing other design solutions that when combined with its own knowledge may lead to new, novel designs. A good company to benchmark is not always in the same industry. Instead, the enterprise seeks out the best in class for whatever process, capability, or system they want to benchmark. For this reason, many companies might benchmark Disney's customer service, or Dell's order fulfillment process, even though they do not compete against either company. To benchmark, the enterprise needs to identify the "best-in-class" for that system or process, and then study how they operate. One caution when benchmarking, is that many best practices are contextual and not easily adopted by other companies. The best practice is often a core competency as described in Section 7.2.1.

9.1.6 Morphological Analysis

Morphological analysis is a structured technique to solve design problems developed by Zwicky [2]. In this technique, the design team decomposes the design problem into component variables, generates solutions to each component, and then forces together components to obtain an overall design. The technique is based on two simple concepts of creativity and problem solving: decomposition and forced association. In the case of enterprise design, the decomposition could be given by the enterprise architecture: information, process, and organization. For each view, you list alternative solutions. Then you associate one solution from each view to form an integrated solution. You do this for all permutations. This "forced association" may generate what seems to be ridiculous solutions, but you should try to figure out how the solution can work. The morphological analysis helps the design team to discover new relationships or configurations which might not have been evident; it encourages the identification and investigation of boundary conditions, and it encourages exploration of design alternatives that might be too easily dismissed otherwise. Further information on Zwicky's morphological analysis technique can be found in Zwicky [2].

Morphological Analysis for Onboard Cruise Payment System

The design team randomly selects one option from each column to create a design alternative to discuss and stimulate ideas that may not have otherwise been considered.

Technology	User Interface	Data Storage
Magnetic card	Cell phone	Oracle
Credit card	Touch screen	DB2
Biometric	Card reader	MS SQL
Room key	Computer	
Voice recognition		
RFID		
Smart card		

9.2 Document Alternatives

The project team needs to fully document each alternative. The alternatives should all be documented in the same format, at the same level of detail, and for the same dimensions. The presentation of the alternatives should not bias the team during evaluation and selec-

TABLE 9.1
Alternative Matrix

Process Design	List processes affected. Describe how the process is designed. What portion is automated versus manual.
Hardware Platform	Describe the hardware platform, networks, etc. that are required to support the system.
Software Application	Describe the software applications. Is it COTS, custom, or general purpose software?
Interfaces	Describe the main interfaces between the system and system users; between system and other systems both internal and external.
Data Storage	Describe how data are stored (database, data files, spreadsheet, paper files, etc.)
Organization Structure	Describe any changes to the organizational structure; new positions; deleted positions.
Job Description	Describe the job descriptions for the employees in the system; skills and training required.
Strategy Alignment	Describe how the project adheres to the enterprise strategy.
Architecture Alignment	Describe how the project alternative complies with the enterprise architecture.

tion. The alternatives are documented in Alternative Matrix. Table 9.1 shows an example matrix with common dimensions of alternatives. The objective is to list all facets of the alternative that should be considered in the evaluation and selection activities. Additionally, all alternatives should include statements of their compliance with the enterprise architecture and enterprise strategy.

9.3 Evaluate Alternatives

Evaluation of the alternatives is a systematic approach to define evaluation criteria, their importance, and then to evaluate each alternative against the criteria. Evaluation of the alternatives is done at a discrete point of time because at the end of the Generate and Evaluate Alternatives phase the project team needs to have an approved design alternative – a major project milestone.

Part of evaluation is defining the evaluation criteria, which should be linked to the objectives of the enterprise and the project objectives. In most all cases, cost, schedule, and risk are part of the criteria. The other criteria will be related to performance, but the definition of performance will differ in each enterprise project.

To evaluate the cost of each alternative, a budget of the alternative is needed and its projected economic impact over the decision horizon. The cost of the alternative includes the purchase of any application software, development of software, purchase of hardware, training, hiring of employees, and any other costs to implement the alternative. For example, suppose one alternative on a project is for a company to purchase a COTS warehouse management system. The obvious costs are the purchase of the COTS system, hardware for the system, consultant services to install and configure it, and training costs. Other less

TABLE 9.2
Example Benefits

Benefit Category	Benefit
Improved efficiency and productivity	Reduced personnel costs Increased efficiency of resources Reduced process cycle-time Increased process throughput-rate Reduced hardware, equipment, material costs Reduced facility costs Reduced IT costs Reduced outsourcing cost Increased productivity of organization
Cost avoidance	Avoid penalities Asset protection Improved regulation compliance Avoid non-value added costs
Revenue generation	Additional revenue generated
Customer service	Reduced transaction costs Increased availability or accessibility Additional services provided Improved quality of service
Quality	Reduced error rate Increased accuracy of transactions Increased accuracy of information Reduced process waste Reduced absenteeism of employees

obvious costs may include: the cost of new paper forms for the system, the cost to do a data conversion from the legacy system to the new COTS, the cost to create any new templates needed, or the costs to modify business policies and rules. The project team strives for completeness to include every conceivable cost of the alternative.

To determine the benefits of the alternative, the team needs to identify the tangible and intangible benefits, and of the latter, make as many tangible as possible. Tangible benefits are easy to quantify and often drive the evaluation, but the intangible benefits are no less important. Table 9.2 categorizes some benefits.

The time horizon of the project is set to start when the alternative solution is implemented in the enterprise. Consequently, all project costs, or at least the majority of costs when the project is implemented over a long duration, occur in year 0. Project maintenance and other recurring costs will occur throughout the time horizon. The benefits accrue during the life of the system. Oftentimes, it is a good assumption that the benefits will not be fully realized in year 1 due to the learning curve of a new system. These benefits are brought forward to year 0 using the principles of the time-value of money and the net present value (NPV) calculation. The project sponsors and the project management select a discount rate to use for the project and a time horizon. The NPV calculation takes the two cash flows and generates a single number representing the alternative's worth. To illustrate the cost benefit analysis, suppose the time horizon is five years and the cash flows are as shown in Table 9.3. The chosen discount rate is 12% and the net present value of the cash flow is \$605,765.

TABLE 9.3
Cost-Benefit Analysis

	Yr 0	Yr 1	Yr 2	Yr 3	Yr 4	Yr 5
Project Costs	$900,000	$125,000	-	-	-	-
Operational Costs	-	$50,000	$52,000	$54,000	$56,000	$58,000
Revenue	-	$200,000	$250,000	$275,000	$280,000	$285,000
Cost Savings	-	$175,000	$300,000	$300,000	$300,000	$300,000
Net Income (Loss)	$(900,000)	$200,000	$498,000	$521,000	$524,000	$527,000

One shortfall of the NPV is it ignores uncertainty in the cash flows, yet the project team will realize how speculative and uncertain their point estimates of benefits and savings are. To incorporate uncertainty into the cost-benefit analysis, the team can use a tornado analysis and Monte Carlo simulation. A tornado diagram allows the user to analyze the relative importance of the input parameters in terms of their influence on the outcome of the cost-benefit model. This sensitivity analysis is important to understand which estimates are most important. Instead of the point estimates above, for each cost and benefit, the team can define a probability distribution and then perform a Monte Carlo simulation. Tools are available for performing this analysis in spreadsheets. The Monte Carlo output will provide an expected NPV and also a probability distribution surround the expected NPV. This allows the project manager to understand how the uncertainty in the cost and benefit estimates propagate to the uncertainty in the NPV.

To evaluate the schedule of each alternative, a schedule of the design, construction, and implementation is needed. Other information needed for evaluation depends on the evaluation criteria. A best practice is to include a measure of the uncertainty or risk associated with each measure. For example, instead of providing a single schedule duration, provide an "optimal," "most likely," and "pessimistic" duration. These data could be used in a Monte Carlo simulation as was done for the cost-benefit analysis. In all cases, the team states all the assumptions that went into developing the estimations.

In addition to the cost-benefit evaluation and the schedule evaluation, the project team will identify other evaluation criteria and evaluate each alternative with respect to the criteria. The other criteria fall into the categories of some type of risk, other system performance measures, and alignment with enterprise architecture, strategy, and goals. In the risk category are technological risk, operational risk, schedule risk, and others depending on the particular project. Technological risk is the risk the technology will not function properly, risk of losing technical support, risk of not finding qualified people to support the technology, or risk its operational cost will increase. For example, a new technology might be sound, but if there is a small user-base then there is a risk the vendor will not support it well or even drop support. Operational risk is a measure of how well the alternative will work in the enterprise. Causes of high operational risk can be cultural, skills and training of workforce, or environmental. For example, a system based on using touch-screen computers might not work well in a dirty shop environment.

Performance criteria of the alternative includes reliability, maintainability, flexibility, quality, etc. of the system. Again, each project is different, so the criteria will be different. Some of the performance measures can be quantitatively evaluated; for example, the reliability of the system can be estimated.

The alternatives are also evaluated on how well they comply with the enterprise architecture. The enterprise architecture provides a high-level design of the system. It is important that all projects fit into the overall design structure. Otherwise, there may be integration

TABLE 9.4
Evaluation of Alternatives Matrix

Criteria	Weight	Alternative A	Alternative B
Project Cost		\$0.85 M	\$1.15 M
Benefits (over decision horizon)		\$1.8 M	\$2.0 M
Net Present Value	0.2	\$0.45 M	\$0.52 M
Project Duration	0.2	16 months	20 months
Schedule Risk	0.1	Low	Low
Technological Risk	0.1	Average	Average
Operational Risk	0.1	Average	Average
System Reliability	0.1	Medium	High
Architecture Alignment	0.2	Medium	High
Overall Rating		4.8	6.0

problems or higher costs incurred later on because the alternative did not comply with the overall enterprise design.

Once the criteria that will be used to evaluate the alternatives is identified, then each alternative needs to be evaluated against the criteria. The evaluation is documented in a matrix that would look something like Table 9.4.

In Table 9.4, the verbal measures of low, medium, and high are mapped to 1, 5, and 9, respectively. The net present value and the project duration would each be mapped to a scale between 0 and 10 as well. In this case, the net present value of Alternative A is rated 5 and Alternative B is rated 6. The product duration of Alternative A is rated 6 and Alternative B is rated 5. The overall rating is the sum of the weight multiplied by the rating for each criterion.

9.4 Selection of Best Alternative

The section of the "best" design from the list of alternatives is a decision-making problem of which there are many ways to go about it. Essentially, the problem is a group multi-attribute decision-making problem. The salient characteristics of the decision problem are:

1. The decision problem has multiple criteria, specified as attributes of the design alternatives (cost, expected system life-time, delivery time, risk, technical feasibility, flexibility, etc.).
2. The criteria might conflict or be incommensurate (e.g., we desire a low-cost system with high reliability).
3. Group rankings or ratings of the alternatives might differ (e.g., one stakeholder group believes alternative A is more reliable than alternative B, whereas another stakeholder group believes the opposite). Group weighting of the criteria might differ (e.g., one stakeholder group says cost is the most important criterion while another says flexibility is the most important).

There are a few ways to make group decisions. How the decision is made depends somewhat on the enterprise culture. A very centralized, authoritative enterprise might have the team leader make the decision with little to no input from other team members. Essentially, it is not a group decision. A more common approach, at least in the U.S., is some sort of voting whereby the majority decides. The drawback of majority rule is that it leaves unresolved any concerns held by the minority and it does not obtain full support from all team members. Another method is whereby all the team members rate the alternatives.[1] Rating the alternatives is a common method to evaluate alternatives. By itself, rating the alternatives does not address any problems or issues with any alternative and it does not achieve group consensus.

Instead of each group member rating the design alternatives individually, the group can form a consensus on the rating of each of the criteria. A group consensus is the most desirable way to arrive at a decision because all team members participate in the process and nobody feels left out. The one drawback is that achieving a group consensus can take a significant amount of time and it requires a good team facilitator to avoid problems. To achieve consensus, a majority of the group should back the selected design alternative, and the minority should agree to accept it. To obtain the minority's agreement, the group should modify the design alternative to remove objectionable features to the extent possible. Collective decision arrived at through an effective and fair communication process (all team members spoke and listened, and all were valued).

9.5 Summary

The generate and evaluate alternatives life-cycle phase generates multiple alternatives, identifies the evaluation criteria, evaluates the alternatives, and selects the single alternative that the remainder of the project will concentrate on designing, building, and implementing. The chapter discussed techniques to generate ideas for alternatives such as brainstorming, morphological box, and benchmarking. The alternatives are documented in an Alternative Matrix. To evaluate alternatives, the most important criteria are the cost-benefit analysis and schedule. Other criteria deal with various types of system risk and performance. The Alternative Evaluation Matrix documents the evaluation criteria and the evaluation of each alternative. The milestone marking the end of this phase is the selected design alternative that is approved by the project steering committee.

Review Questions

1. Describe the techniques to increase group creativity.
2. List criteria that are considered when evaluating different design alternatives.
3. Explain whether a project can be economically feasible but not technically feasible.
4. What makes a project not technically feasible?
5. List the types of benefits a system can provide.

[1]Rating the alternatives can also be corrupted by a jaded team member. It is common for some academic departments to hire new faculty using this technique. Suppose there are three job applicants and five faculty members are going to rate each applicant on a scale from 1 to 10. Likely, all candidates are at least qualified, and faculty will assign rating such as 7, 10, 7 for applicants 1, 2, and 3, respectively. A lone faculty member could rate the candidates 10, 1, 1 and consequently, dominate the final rating and outcome.

6. What is a measure of how well a proposed system design will work in an organization?

Exercises

1. Suppose you are a project manager for developing a new course registration system at a local community college. The solution will be developed internally by the IT department, which has all the required skills and knowledge. List three key cost variables you need in order to derive the budget? Give an example of how the data can be obtained (e.g., maintenance cost available from software vendors).
2. Suppose you are working on a project to develop a new electronic medical records system. The clinic has three doctors, ten nursing staff, and twenty administrative and clerical staff. The clinic has seven examination rooms and a laboratory. You are tasked with estimating the hardware costs associated with the new system. The system is a client-server system, with an Oracle database. Each examination room needs a computer, the laboratory, and approximately five computers for the administrative staff. The clinic has no existing network infrastructure, so that will need to be estimated as well. Develop your estimate and explain your rationale and assumptions to make the estimate.
3. For the electronic medical record system described above, list the types of benefits the clinic might expect from the new system.
4. Do a Web search to find COTS systems for an application of interest to you (e.g., accounting, project management, online ecommerce, etc.). Describe the COTS system and what features, functions, and limitations it has.

Bibliography

[1] H.A. Linstone and M. Turoff. *The Delphi Method: Techniques and Applications.* Addison-Wesley, New York, NY, 1975.

[2] F. Zwicky. *Discovery, Invention, Research - Through the Morphological Approach.* Mcmillian Company, Toronto, Canada, 1969.

Part III

Process View

10

Process Modeling

> *"A process cannot be understood by stopping it. Understanding must move with the flow of the process, must join it and flow with it."* – Frank Herbert (1920-1986), in the book *Dune* (First Law of Mentat).

This chapter discusses the modeling of business processes. The process view of the enterprise is one of the three enterprise system views, and it is the process view that shows how value is delivered to the customer by the enterprise. The purpose of process modeling is to produce an abstraction of the process that serves as a basis to understand the process, to describe the process to others, to analyze process performance, and to specify a process design.

A process is not a physical, tangible artifact – it is our conceptualization or mental model of how work is performed. How we define and model business processes plays a major role in our perception and understanding of business processes. For this reason, one can find a wide diversity of definitions and process modeling techniques in the literature. Two dominant views of a process focus on the activity flow and the information flow of a process. The activity flow view represents the sequence of activities in the process. The information flow view represents how information flows through the system. In this chapter, we use flow charts to model the activity flow and data flow diagrams to model the information flow. Additionally, we describe the IDEF0 methodology that has a functional viewpoint and strong hierarchical structure.

After completing this chapter, you should be able to:

- Define a business process and describe how to decompose business processes.
- Describe the elements of a business process.
- Decompose a business process.
- Classify processes.
- Create a flow chart, data flow diagram, and IDEF0 diagram of a business process.
- Explain the similarities and differences between the main process models.

10.1 Definition of Business Process

A *business process* consists of a network of inter-related activities that convert inputs into an output, where the output is a product or service, valued by a customer. A business process involves the coordinated effort of people, information, and technology. Figure 10.1 highlights the elements of a business process:

1. Transformation of inputs to outputs
2. Work unit

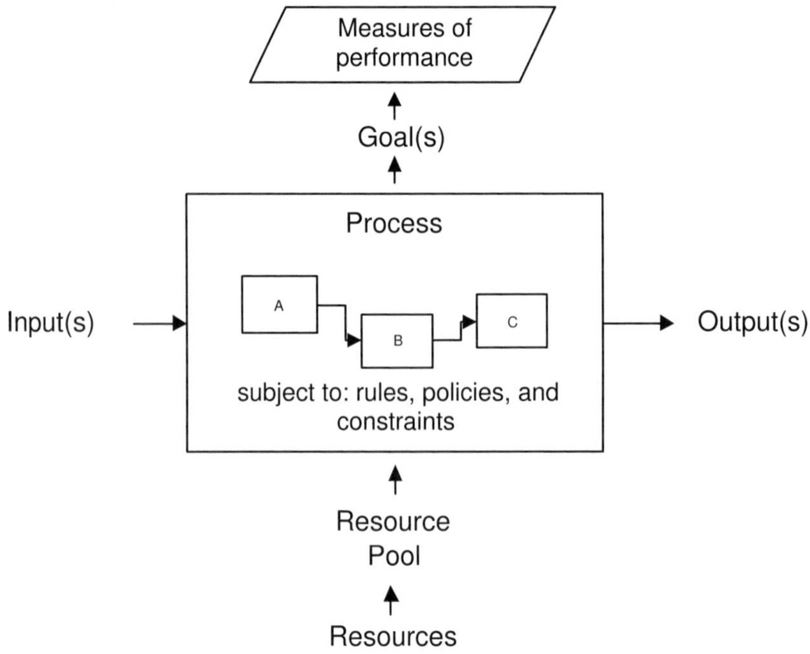

FIGURE 10.1
Business process.

3. Network of activities
4. Customer
5. Goals and measures of performance
6. Resource pool and resources

Transformation of Inputs to Outputs

The foundation of a process is the concept of the transformation of inputs into outputs. A transformation can be physical, such as occurs in manufacturing when raw material inputs are transformed by various manufacturing processes into finished products. The transformation can be of location. In logistics, the transformation changes one location into another location, such as an airline taking passengers from one city to another city. The transformation can be transactional, such as making a reservation, in which the transformation is from a reservation request to a confirmed reservation. The transformation might be informational, such as the transformation of sales data into graphs showing market trends. Whatever the transformation, the business process adds some value to the inputs which are provided as outputs to a customer.

The *inputs* are the tangible and intangible items that come from the external environment and are part of the transformation process. Inputs include: raw materials, parts, sub-assemblies, chemicals, energy, information and customers. If an item is not part of the transformation process, then it is probably a resource such as labor or machines.

The *outputs* generated by a business process can be either a tangible product, a service, or both a product and service bundled together. The design of a process depends on whether it produces a primarily tangible product or an intangible service. One primary difference

TABLE 10.1
Examples of Business Processes Listing Their Work Unit, Primary Input, and Primary Output

Process	Work Unit	Input	Output
Camera manufacturing process	Camera	Camera parts	Camera
Provide air transportation	Passenger	Passenger and fuel at one location	Passenger at destination
Cruise reservation process	Reservation	Reservation request, cruise information, payment information	Booked reservation
Generate financial statement	Statement	Financial, accounting, marketing, and other data	Financial statement

between products and services is that a product can be stored in inventory for later use while a service cannot. This means that products can be made even when there is no immediate customer for the product. It can be stored and later used. Also, products can be shipped to wherever the customer is located. So manufacturing processes can be located far from the customer. Services do not have this trait. If a dentist has no patient for a particular time slot during the day, then that time slot is lost – i.e., it is perishable because the dentist cannot save the time slot for another day. Additionally, unlike with products, the dentist cannot provide the service unless the patient is in the same location. A last difference is that the customer is often an input to a service process (the patient is an input to the dental procedure).

One repercussion of the inability to store services is the widespread use of revenue management in service industries. Revenue management includes overbooking and differential pricing to maximize the profits generated by the business. Initially, airlines started using overbooking because if a passenger did not show up for the flight, then the seat reserved for that passenger would be lost revenue. Airlines also found that they could differentiate between different customer classes of business travelers and economy travelers and price the tickets differently for each customer class for what is essentially the same service. Revenue management is now applied throughout the service industries including airline, hotel, cruise, healthcare, and other related industries that have similar characteristics.

Increasingly, in the economies of the Industrialized World, the output of a business process is a service. A service output includes outputs such as establishing a banking account, paying an insurance claim, providing legal advice, or providing knowledge and training such as provided by a university. Combined service/product outputs are delivered by restaurants, airlines, hotels, and cruise lines. For example, a restaurant provides food – the product – and it does so with a level of service. A waiter takes the order, serves the food, and removes the dinner plates and utensils. A restaurant provides ambiance and other elements of service beyond the provision of food. Thus, the output is both a service and a product. Even traditional manufacturing industries have a strong service component. Automobile manufacturers have financial divisions to provide purchase financing and they have extensive dealer networks to provide maintenance services.

In addition to the desired output, a process generates unwanted waste outputs such as smoke, chemical waste, unused materials, and intermediate data. Part of process design is determining how to eliminate or reduce the waste output, and if this is not possible, then how to handle the waste.

Work Unit

The *work unit* is the item that has done to it by the process and becomes the final process output. Consequently, the work unit changes throughout the process. For this reason, as shown in Table 10.1, we generally refer to the work unit by using the name of the final process output.

Customer

A customer is the person or group that receives the process output. Generally, we think of customers as being external to the organization, but a customer can also be internal. For example, a consumer purchasing a new television is an external customer of the television manufacturing process. Within the television manufacturing company, the production department is an internal customer of the machine maintenance process, which is performed by the maintenance department.

It is important to identify the customer(s) of the business process because it is the customer who defines the value of the process output. *Value* is the benefits a customer derives from the process output compared to the costs of the output. Here "costs" refer not only to the price of the product/service, but also to the costs in terms of time and other efforts required to obtain the product/service. For example, a restaurant might be known to provide excellent meals at a reasonable cost, but if the customer must wait longer than desired, then the value of the restaurant's output is diminished in the eyes of the customer.

The definition of value lets us distinguish between *value-adding activities* and *non-value-adding activities*. Value-adding activities do something to the work item of value to the customer. Non-value-adding activities are those activities that do something to the work item that a customer does not value. Four general types of non-value-adding activities are: storage of work items, transportation of work items, inspection and/or testing of work items, and support activities necessary to deliver work items. For example, moving an automobile between workstations in the factory does not add value in the eyes of the customer. Neither does filling out forms in the doctor's office, even when necessary for later value-adding activities.

In a business process, all activities should contribute toward providing value. Part of process analysis is to identify value-adding and non-value adding activities and seek ways to eliminate or reduce the latter.

Network of Activities

The transformation process is usually not a single-step process, but consists of a network of distinguishable activities. This leads to a natural hierarchy that a process consists of subprocesses, which consist of subprocesses, and so on. In fact, the entire enterprise could be viewed as a single process that transforms inputs into outputs. To better understand a process, it is useful to decompose the process into subprocesses to model the process at an abstraction level appropriate for whatever analysis is being conducted. In the next section, we discuss a hierarchy to create this process decomposition.

When a process is decomposed into activities, those activities are interdependent on each other. Interdependence is the degree to which one activity is controlled or contingent on another activity. Interdependence occurs due to the information flows, sharing of resources, and decision (control) flows of business processes. It is important to understand these interdependencies because they have a profound effect on the performance of a process and the enterprise. Chapter 17 discusses how to use the interdependencies in analysis and discusses technologies to manage the interdependencies.

Resource Pool and Resources

The business process is executed by one or more resources. A *resource* is a person, piece of equipment, or computer program used by the enterprise to do the business process. Resources should not be confused with process inputs. Process inputs are consumed by the process and play a direct role in the transformation, whereas resources are not consumed. A *resource pool* is a group of common, interchangeable resources. Example resource pools are a group of machines and a department of technicians. A process should make efficient use of its resources, since resources cost money to purchase or employ.

Goal and Measures of Performance

The goal of a business process is a statement of what the business process is intended to accomplish for the enterprise. A formal statement of the process goal aids process analysis by bringing the purpose of the process to the forefront so that the analysts will not lose sight of why the process is being done and what the process accomplishes. Both the process itself and the product/service it produces can be measured.

A business process is managed by measuring the output to ensure it meets customer expectations in terms of performance, price, timeliness, and quality. Customers decide among competing products/services by trading off these aspects of value. Price is what a customer is willing to pay for the product/service. Timeliness is whether the product/service is provided in a timely basis when the customer wants it. Quality is how well the product/service meets customer expectations. Additionally, the process performance should be measured directly. The process performance measures, defined in Table 10.2, are quality, efficiency, effectiveness, profitability/budgetability, productivity, and innovation [9]. These measures describe the operational performance of the process. Additionally, the product/service generated by the process must meet customer needs in terms of functionality.

10.2 Process Decomposition

Business processes are complex, and their network of activities can be very large. It is not uncommon to see thirty-foot long process maps that take up the entire wall in a conference room. While developing such a large map is a great achievement and may even impress management, the value of showing the process at a single abstraction level, especially a detailed abstraction level, is limited. Processes are hierarchical by nature and should be represented that way in process maps. Forming hierarchies enables us to better understand complex systems [8]. The problem with a single thirty-foot long process map is it does not let viewers have the end-to-end perspective that is important. Moreover, it makes it difficult to understand the natural divisions of the work. The value of process decomposition is it enables us to focus on the individual activities in the process as "black boxes" and how the activities are related without needing to understand the detailed workings of each activity at the same time. Process decomposition reduces the complexity of the overall process; it helps viewers focus on the different requirements of each subprocess; it helps viewers understand the process at different levels according to the needs of their analysis; and it makes explicit different management needs, resource needs, and the interdependencies between the subprocesses.

TABLE 10.2
Process Performance Measures with an Example Measure for Each

Performance measure	Definition	Example
Quality	Conformance to specifications and meeting of customer expectations. Measured at input, transformation, and output.	Rejected goods/total goods (manufacturing).
Efficiency	How well the resources are utilized to produce the intended results. A ratio of actual output / standard output.	Scheduled FTE nurse-hours / Actual FTE nurse-hours (healthcare).
Effectiveness	Accomplishment of the right things on time, within specifications, or expectations. Measured as the actual output / expected output.	Number of patients served / planned number of patients served (healthcare).
Profitability	The ability of the process to generate profit. Measured as revenue / cost.	Revenue / labor cost (healthcare).
Budgetability	The degree to which the process stays within budget. Actual costs / budgeted costs.	Actual outlays / budgeted outlays (insurance).
Productivity	The ratio of outputs to inputs.	Number of insurance claims processed / total person-hours (insurance).
Innovation	The degree to which innovation improves the process.	Number of employee process improvement suggestions (manufacturing).

The level of detail in the model should match the detail required for the intended analysis. For example, to specify a workflow that will be automated by a computer system requires a high level of detail. All possible actions and how the system should respond need to be modeled. If the purpose of the model is to train employees so that they have an understanding of how their work fits into the larger picture, then much less detail is warranted. In fact, too much detail can obscure the important aspects of the process that the modeler wants to convey. Recall in Chapter 3 on modeling that modeling is the art of deciding how much detail to include and how much detail to leave out. This fundamental modeling principle applies to process modeling as well.

Some process mapping techniques are structured to force this type of hierarchical thinking (e.g., IDEF0), but the other process mapping techniques can also support it even if they do not have explicit mechanisms to do so. There are many possible ways to decompose a process; here we define a generic process hierarchy that has proven to be useful. The process hierarchy is shown in Figure 10.2. At each level a transformation occurs. A *process* is a linked group of activities that transforms inputs into outputs to deliver a product or service to an internal or external customer. What distinguishes the process level from the other levels is the process output is always the final product or service. A process can be

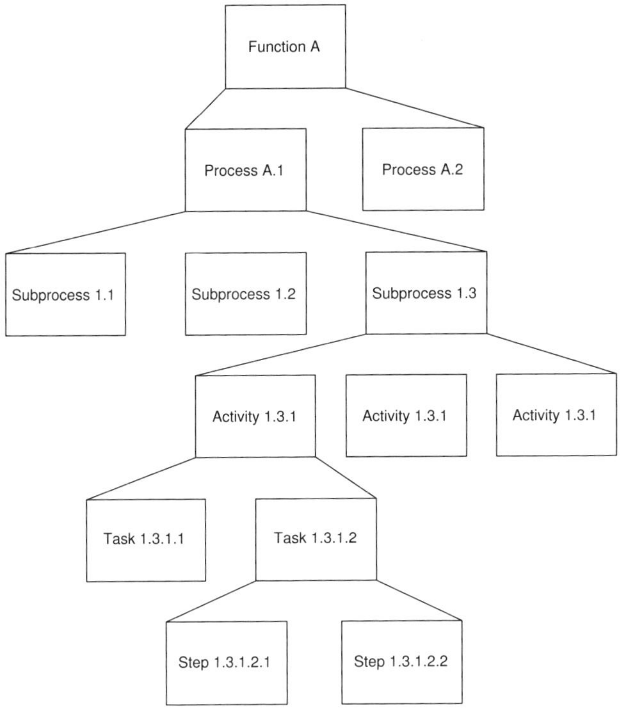

FIGURE 10.2
Process hierarchy.

decomposed into *subprocesses* to differentiate stages of the work. An *activity* is a middle aggregation to help the modeler understand what is done by a single resource, usually an organizational unit. An activity can be composed of two or more tasks. A *task* is defined as the lowest-level process-unit. A task is a complete, logical unit of work, with an input and an output, completed by a single resource. A *step* describes a single action or decision in the task that may not be a complete unit of work. In a model, one or more of the levels may be omitted depending on the needs of the analysis.

When defining a decomposition for a business process, several general rules should be followed. First, nothing is lost in the decomposition. All the inputs and outputs at the process level should appear somewhere as inputs and outputs at the subprocess level. Second, the activities should be semi-independent so that the activity design can be specified with some independence of the other activities. The interface between activities should be clear, usually it is an output-input interface.

There is no unique decomposition to a business process. Simon [8] notes that much of classical organization theory has examined how to decompose an organization and the relative merits of different decompositions. One of the proposed organizational decompositions is by business process. Similar to organizations, processes can be decomposed in many ways. While the one presented here is useful, it is not definitive, and other decompositions might be more appropriate depending on the particular situation.

To illustrate the process hierarchy, Figure 10.3 shows the decomposition of the function Provide Air Transportation in an airline (other functions including Reservations and Flight Schedule Planning are also shown for comparison). Provide air transportation takes as input a passenger with a ticket and his or her luggage, and transports that passenger to his or her destination. This is a core function of the airline. To provide air transportation, the airline

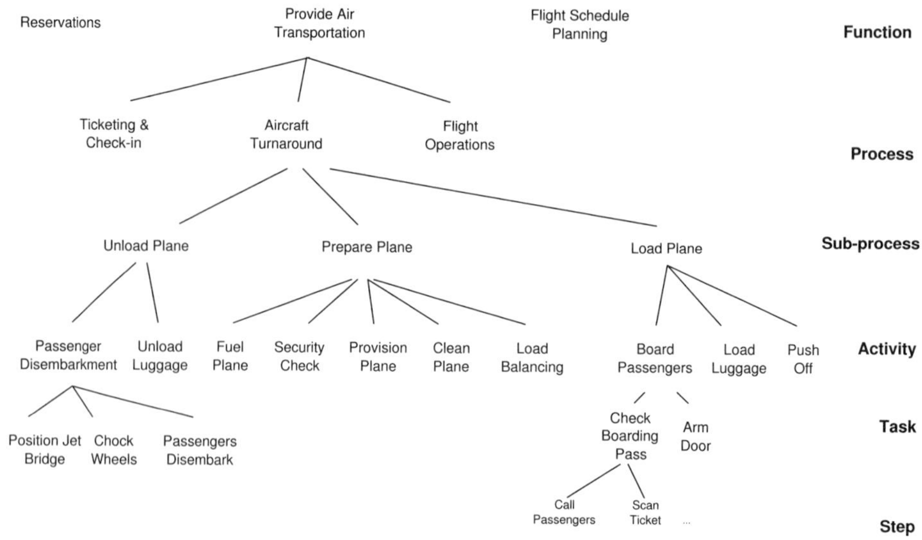

FIGURE 10.3
Airline process decomposition.

needs to do three processes of ticketing and checking in the passengers, turning around the aircraft at the gate, and operating the flight (flying the plane). Aircraft Turnaround is defined as when a plane lands until it takes off again. The Aircraft Turnaround process is decomposed into three subprocesses: unload plane, prepare plane, and load plane. The motivation for the decomposition is to accurately represent the sequence of activities. Each subprocess is decomposed into activities. The prepare plane subprocess includes five activities: fuel plane, security check, provision plane, clean plane, and load balancing. Each activity is performed by a separate organizational unit (e.g., fuel plane by the ground crew). The activities are decomposed into tasks (for clarity, not all decompositions are shown). The board passengers activity is decomposed into the tasks: check boarding pass and arm door. These tasks are further decomposed into the steps required. Check boarding pass is decomposed into the steps: call passengers, scan ticket, if passenger needs to gate check luggage get tag, complete tag, ... At the step level, the process describes detailed process steps done by a single person.

As mentioned earlier, how to model is somewhat of an art, and each process modeler given the same scenario will likely develop a different model. In the airline example, there are many alternative ways to decompose the Aircraft Turnaround process. Some of the participants on the modeling team argued for an alternative representation that has two subprocesses called Above-Wing Operations and Below-Wing Operations. Using this decomposition, the activities would be assigned differently to the subprocesses. Unload luggage and fuel plane would be together in Below-Wing Operations, while passenger disembarkment and clean plane would be part of Above-Wing Operations. This alternative model is equally valid for some context and purpose of the model.

The motivation for the second model decomposition is the above-wing operations are done by different staff than the below-wing operations. Moreover, the dependencies are stronger between above-wing operations than those between above-wing operations and below-wing operations. One goal of decomposition is to maximize cohesion in the group, and minimize the dependencies or coupling between different groups. It is possible the alternative decomposition does this better, but at the cost of not making the natural sequence of activities clear as is accomplished in the previous decomposition.

10.3 Classification of Business Processes

Not all processes are the same, and it is useful to classify processes so that we may treat each type of process differently. There are several different, useful classifications of processes. A useful classification is by Ould [5] who classifies processes into three classes:

1. Core processes – Have external customers and include the primary activities of the value chain. Frequently, these are also called "end-to-end business processes," because the process begins with a market or customer input (e.g., an order, a new product idea), and ends with an output that goes to the customer. Core processes have a bearing on the enterprise's ability to achieve its goals.
2. Support processes – Have internal customer and concern secondary activities in the value chain. Support processes are not essential to the enterprise.
3. Management processes – Manage core and support processes, set strategic direction for enterprise. Managerial processes are usually unstructured processes for making decisions. Sometimes these are called decision processes.

This classification is based on the process's role within the company and matches Mintzberg's [3] model for the structure of the organization (see Chapter 15). Another useful way to classify processes is based on their characteristics. Garvin [1] classifies three types of processes as work processes, behavioral processes, and change processes. Work processes describe the sequence of activities that transform inputs into outputs. Work processes may be core, support, or management processes. Behavioral processes describe widely shared patterns of acting, interacting, and decision-making. Change processes describe how the enterprise changes itself overtime.

In this book, we will focus primarily on work processes, whether they are core, support, or managerial processes. It makes sense that businesses, when analyzing processes, would first look at core processes, since the purpose of the business is to first satisfy external customer needs. Improvements in support processes only serve to improve the efficiency of the organization, but may not impact the competitiveness of the company. As such, core work processes are usually the main focus of business process reengineering and other business process improvement programs. A core process might entail many activities depending on how it is defined. For example, Delta Airline's management sees the company as having two core processes defined as customer experience (including reservations, gate check-in, seat assignment, baggage claim, boarding, and reward miles) and airline operations (allocation of resources, loading, flight departure, flight arrival, unloading aircraft, and cleaning aircraft) [6].

The use of information system to automate business processes has led to interest in understanding workflows. *Workflow* is the automation of a business process, in whole or part, during which documents, information, or tasks are passed from one participant to another according to a set of procedural rules. In our classification, a workflow could be either a core process or a support process. The emphasis of workflow is on the execution of the business process via an information system. What makes workflow unique is it is case-based such that an individual case, usually a document, can be tracked through the process. Thus, processing an insurance claim, opening a bank account, or making a plane

TABLE 10.3
Atypical Business Processes

Process	Input	Output
Order acquisition	Prospect or lead	Order
Order fulfillment	Order	Delivered product or service
Reservation	Reservation request	Reservation
Settle insurance claim	Insurance claim	Paid/unpaid claim
Provide customer support	Problem call	Resolved problem
Product design	Product concept	Product design specifications

reservation are all workflows because we can identify individual documents (respectively, an insurance claim, a bank account application, and a reservation request) that flow through the process.

The point of process classification is to understand the similarities of processes within each class so that general process modeling, analysis and design approaches can be used for each class. Furthermore, a process classification can help an enterprise create an inventory of all its processes, a first step in modeling the enterprise.

At a high level, the descriptions of processes within an industry are all very similar. Table 10.3 lists business processes that are typical in many industries. If we take the insurance industry, then all insurance companies will have a process to settle insurance claims. The differences between companies will manifest themselves at lower levels of the process decomposition.

10.4 Flowchart

The flowchart is a widely used technique to represent the activity flow of a process. Activity-flow oriented process models focus on the sequence of actions done to a work item in the process. This is the most common perception people have of a process.

A flowchart is a diagrammatic model of a process using symbols to represent operations, decisions, and documents, which are linked by arrows to show the sequence of activities. It is one of the earliest and most basic means to model a process. The main flowchart symbols are shown in Figure 10.4.

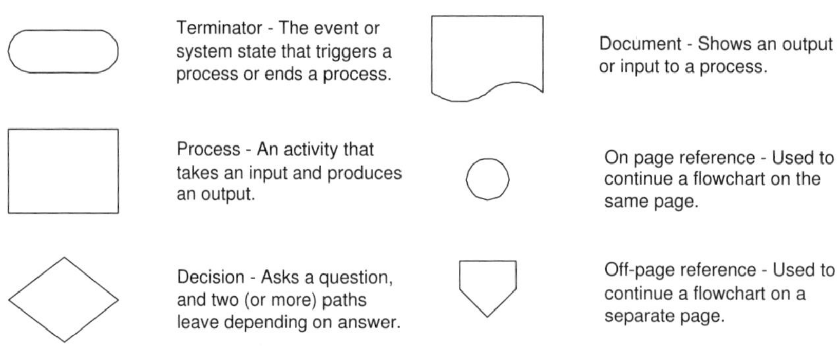

FIGURE 10.4
Flowchart symbols.

To create a flowchart, you first determine the start and end of the process, which are depicted with terminator symbols. The start of a process is either an event or state of the system that triggers the process. An example of an event is the arrival of a customer or the arrival of a part. A state of the system could be, for example, inventory drops below a critical value.

The main components of a flowchart are the process symbols. All the processes within a flowchart should be at the same abstraction level with the appropriate level depending on what the flowchart is intended to model. Commonly, a flowchart would depict a business process at the activity level. However, if you are modeling the steps of an algorithm, then the appropriate abstraction level is the step level. Process symbols should be named VERB-NOUN because processes represent work done in the process. Examples of names adhering to this advice are "Check Credit" instead of "Credit Check" or "Validate Order" instead of "Valid Order." This advice generally applies to all process modeling techniques.

The arrows between symbols in the flowchart depict the order of activities. A process can have multiple arrows enter it, but a process can only have one arrow exit it. Otherwise, if there were two or more exiting arrows, then there would be confusion on which arrow to follow. To show where activity flow can diverge, you use a decision node. The text in a decision node is a question and should have a question mark. A decision node has two or more output arrows, each of which is labeled to show when that arrow is taken. Frequently, decision nodes are answered in terms of "yes" or "no."

A flowchart can depict both sequential and parallel activities. In sequential flow, each activity is connected to the previous activity by a single arrow. To show parallel flow, a symbol of two horizontal, parallel lines is used at the beginning of the parallel activities. The parallel activities are shown side-by-side, and then when the synchronization ends, the two horizontal, parallel lines are used again. In this way, the flow splits into two parallel activity flows and then joins together again.

The flowchart shown in Figure 10.5 shows the cruise reservation process. It starts at the top when a guest inquires about a cruise. The first activity is to Search for a Cruise, which leads to a decision node. The decision node represents whether the guest finds a suitable cruise that he or she is interested in booking. If "yes," then the process continues as indicated. If "no," then the next process is to Save Search Results, and end the process. Assuming a "yes," the process continues with the activities shown. The process terminates with either the reservation placed on hold for 24 hours or the reservation is booked.

The cruise reservation flowchart depicts the sequence of activities necessary to create a cruise reservation. It does not show the information flow that occurs during the process nor the resources, both human and otherwise, necessary to do the activities. All the activities in the model are at the same abstraction level, some of which can be decomposed further, if needed. For instance, the Make Payment activity could be decomposed into smaller steps of calculating the total amount due, obtaining the credit card information, validating the credit card, and making the payment.

A variant of the flowchart is the cross-functional flowchart that adds swim-lanes horizontally or vertically on the drawing to show which organizational units do which activities (see Figure 10.6). A symbol appearing in a particular lane is within the control of that organizational unit. This technique allows the analyst to assign responsibility for performing an action and shows the relationship between different organizational units and their roles for a single process.

Interestingly, flowcharts do not necessarily show the inputs or outputs of each activity. They show the sequence of activities, but the data, which may be the inputs and outputs to the activities, are not shown.

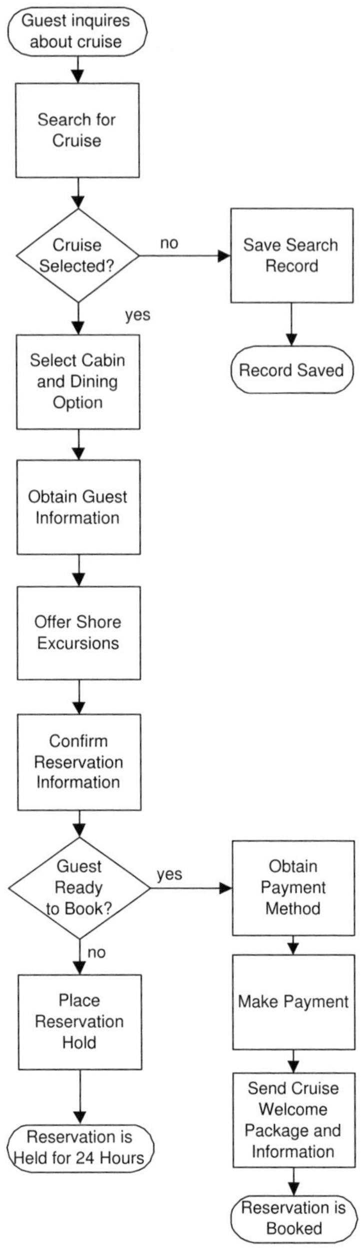

FIGURE 10.5
Flowchart of reservation process.

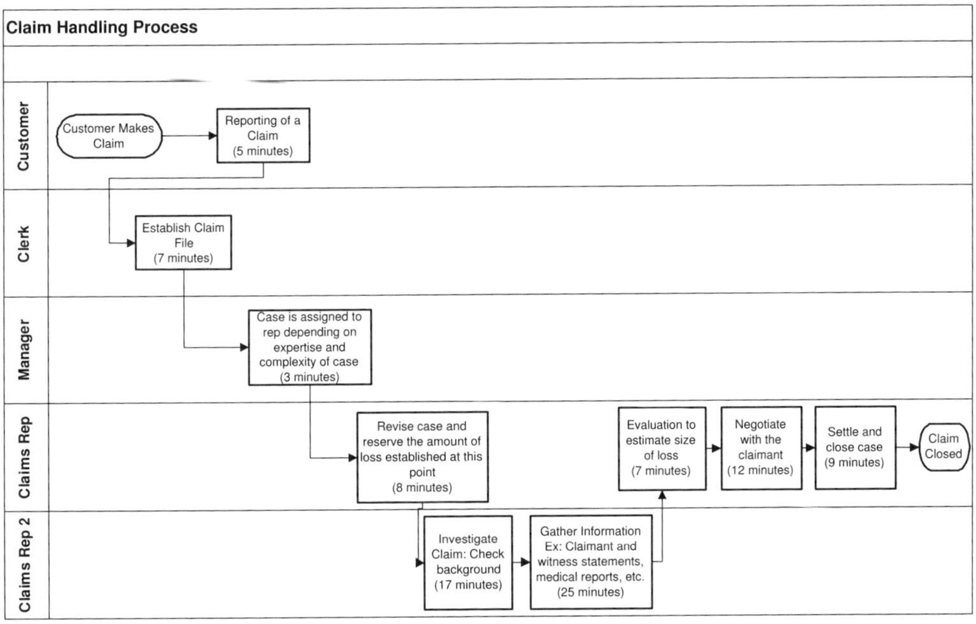

FIGURE 10.6
Cross-functional flowchart of insurance claims process.

10.5 Data Flow Diagram

A data flow diagram (DFD), as its name implies, depicts the flow of data in a system. Data flow diagrams are traditionally used in information systems analysis to show how information flows in the system. A DFD has three symbols and one connection (shown in Figure 10.7).

1. A *process* is represented by a rounded rectangle with a text description of the transformation. A process transforms one or more inputs into one or more outputs. In the process hierarchy; a DFD process can be an entire process, activity, or task.
2. An *external agent* is an interface and is represented by a square. External agents define the boundary of the system under study, whether a person, organizational unit, or other system.
3. A *data-store* is represented by an open-ended rectangle. A data-store can be a database, spreadsheet, file cabinet, or any other means used to store data.
4. *Data flows* are represented by arrows drawn between the other symbols.

A DFD shows the input and output relationships among the data in the system. The DFD does not show the activity flow or sequence of the processes. For this reason, a DFD has no start or end; all of the processes may operate in parallel or sequential – there is no way to tell from the DFD. The arrows should be interpreted as pipes or structural connections in which data can flow. The existence of a pipe between two processes says nothing about how frequently data flows, the time it takes, or any other attributes of the data flow. In a DFD each process may have substantially different durations.

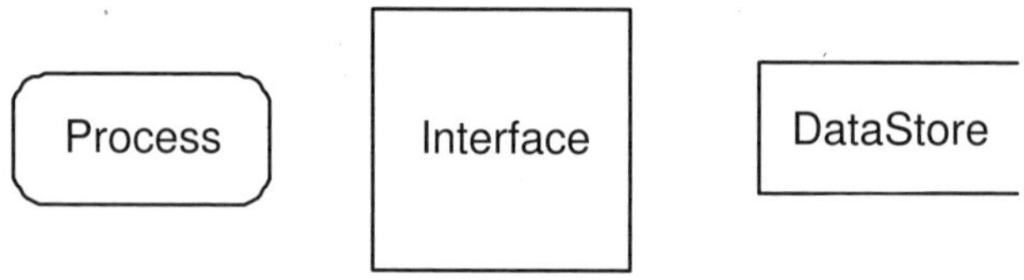

FIGURE 10.7
Data flow diagram symbols.

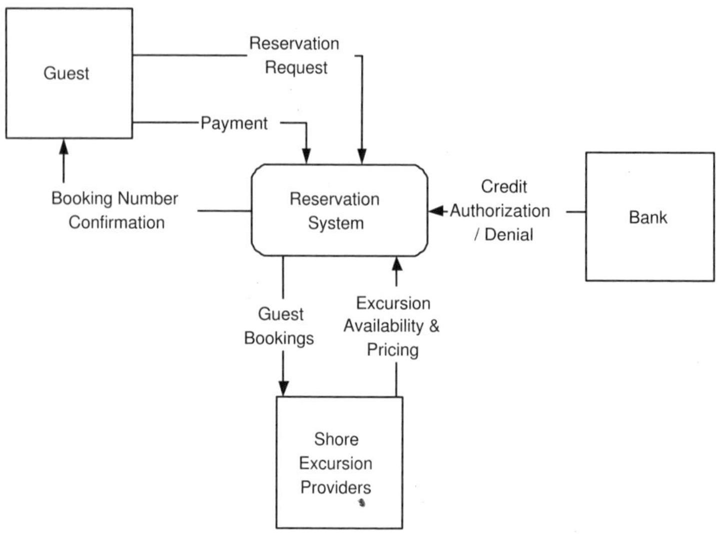

FIGURE 10.8
Reservation context diagram.

DFDs allow for the decomposition of higher level processes into more detailed processes. The highest level is called the *context data flow diagram*, which depicts the entire system as a single process and all the inputs and outputs between that system and its environment. In Figure 10.8 is the context diagram for the Reservation System. It shows the data flows between the Reservation System and three external agents. In the context diagram, each data flow represents a requirement for one or more system interfaces. The diagram is useful to understand the relationship of the system to its environment and is the top level in the data flow diagram hierarchy.

Figure 10.9 shows how a Level 0 process called Make Payment is decomposed into four Level 1 processes. To support decomposition the processes are numbered. The numbering convention is that the single process in the context diagram is numbered process 0, the next lower level is the Level 0 diagram and the processes at this level are numbered 1, 2, 3, and so forth. Any decompositions of Level 0 processes are numbered with the number of the parent process they are decomposed from with a dot and the number at that level. As shown in Figure 10.9, process number 6 is decomposed into processes 6.1, 6.2, 6.3, and 6.4 at Level 1.

All the data flows from the Level 0 diagram appear in the Level 1 diagram, which may have additional data flows.The data flow from the Bank, shown as Credit Authorization/Denial at Level 0, diverges into two separate data flows in the Level 1 diagram. Likewise, at Level 1 the Booking Number Confirmation and the Credit Denial Another Card Request data flows converge and are shown as a single data flow at Level 0. The showing of a single data flow at Level 0 is done to simplify the diagram at this abstraction level.

A process symbol in a data flow diagram can represent the entire system or process, a function of the system, an activity, or a task – it depends on the decomposition level diagram the process appears in. The process name is a verb phrase describing the transformation process. All processes must have at least one input and at least one output – otherwise there would be no transformation. Data flows from a process can be connected to another symbol, another process, an external agent, or a data store.

An external agent is named with a noun describing what it represents. An external agent must have at least one data flow, either input or output, and the data flow can only connect to a process.

The name of a data store corresponds to either a database name or an entity name in the information model (see Chapter 13). A data store must have at least one data flow, either input or output, and the data flow can only connect to a process. Notice the rules for data flows to an external agent and a data store means that these two symbols are never directly connected in a DFD. The reason is the external agent needs an intervening process to either enter data or retrieve data from the system.

Figure 10.10 shows the four types of data flows possible with a data store. These data flows are remembered by the acronym CRUD for **C**reate **R**ead **U**pdate and **D**elete, which are the only transactions possible with a data store:

1. Create - to create a new record in the data store.
2. Read - to read a set of records in the data store.
3. Update - to change an existing record in the data store.
4. Delete - to remove a record in the data store.

Data flows for Create, Update, and Delete make permanent changes to the data content of the data store and for this reason are shown as arrows into the data store. A Read data flow makes no changes to the data content of the data store, so it is shown as an arrow out of the data store.

A DFD cannot show decisions, yet they occur in processes. Two options to represent decisions in a DFDs are to either use composite data flows or a separate data flow for each possible outcome. In Figure 10.9, the data flow Credit Authorization/Denial is a composite data flow. Either the Bank decides to approve the transaction and provides a Credit Authorization, or the Bank decides to deny the transaction and provides a Denial. These are mutually exclusive options that can flow in the single data pipeline depicted.

The other way to depict a decision is two separate data flows as shown in Figure 10.11. In this DFD, the Guest can either confirm the reservation (the Confirmation data flow) or change the information (the Change Request data flow). Representing a decision this way is possible because data need not flow through each data pipe for every instantiation of the process. Also, a data object only flows through the Change Request pipe when the Guest requests a change.

In addition to this basic understanding of the DFD, several common modeling rules that you need to follow when creating a DFD are:

1. At least one data flow in and one data flow out is required for each process in the system.
2. The data output for each process should be modified from the data input.
3. Each data store must be involved with at least one data flow.
4. Each external entity must be involved with at least one data flow.
5. A data flow must be attached to at least one process.

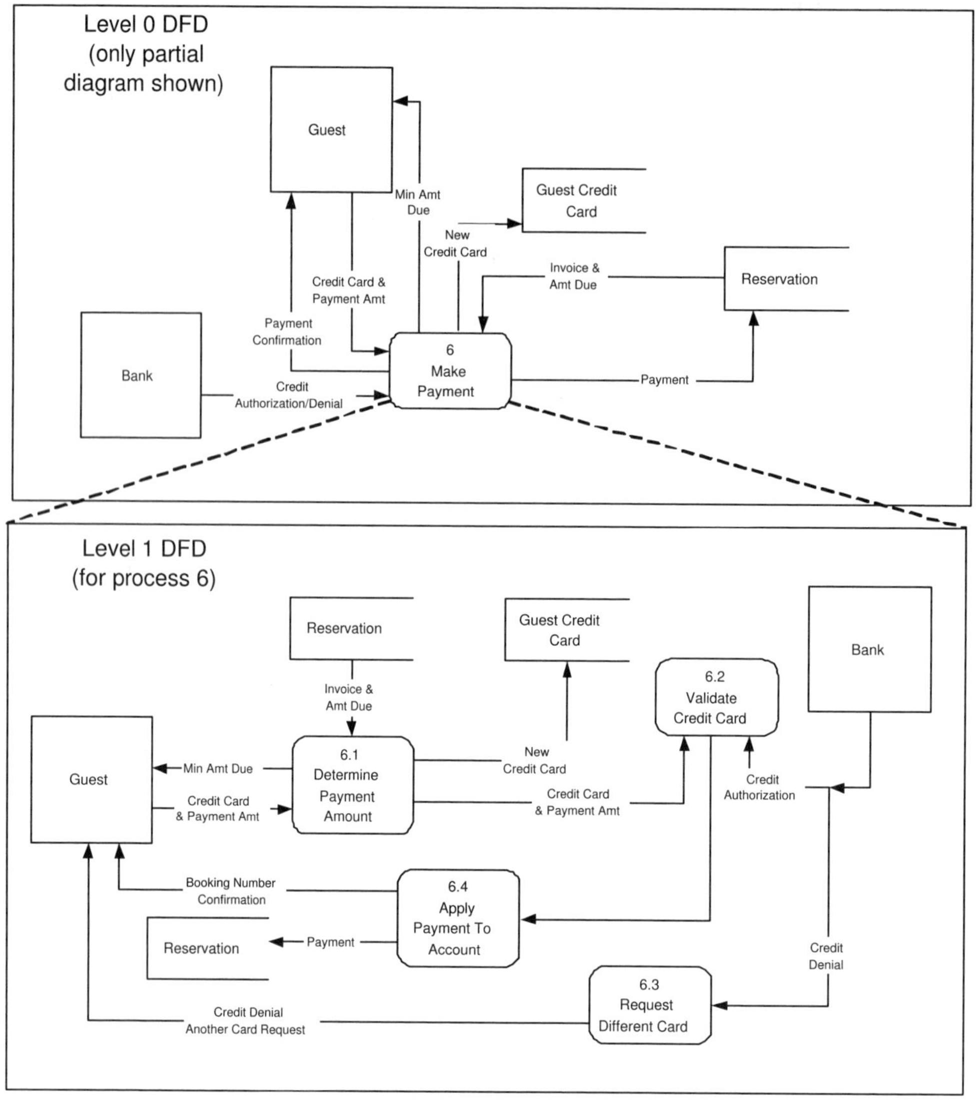

FIGURE 10.9
Decomposition of level 0 make payment process.

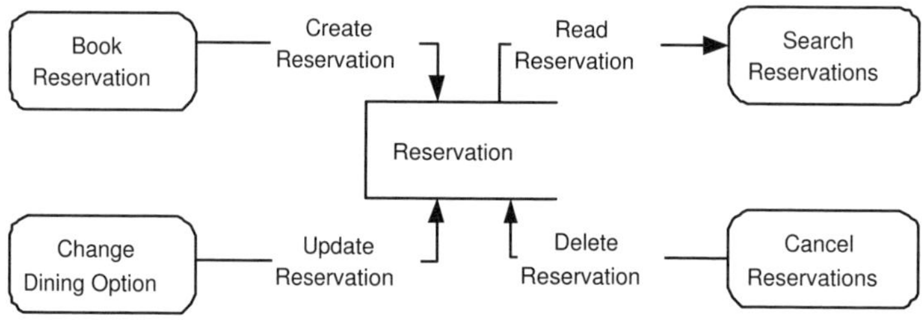

FIGURE 10.10
CRUD and data stores.

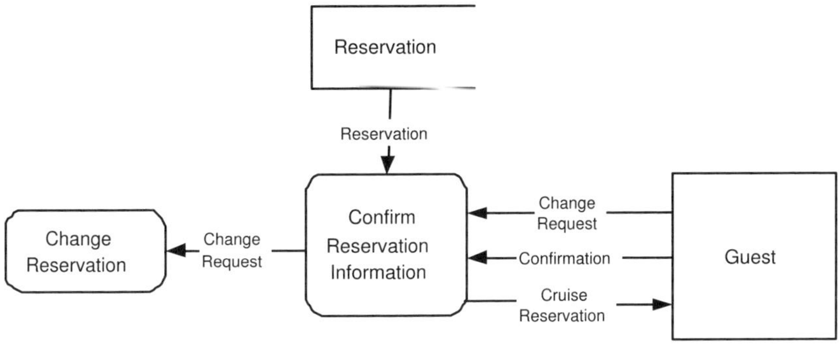

FIGURE 10.11
Modeling decisions in DFD.

6. Data can flow in one direction only.
7. A data flow cannot be returned to the process from which it originated.
8. Both branches of a joined data flow must have the same data type.

10.5.1 Data Structure

A *data structure* is a group of related data attributes that flow together through the data flows. Chapter 13 describes in detail the information model including definition and description of the data attributes. Yet, to properly specify data flows and make a data flow diagram, a basic knowledge of data attributes is needed. A data attribute is the smallest unit of data stored in a system. These data attributes are aggregated together to provide meaningful data structures that flow between processes, users, and data stores in the data flow diagram. Figure 10.12 shows a simple data flow diagram in which a customer submits a customer order that is recorded by the system and then confirmed to the customer. The three data flows: Order, Order Confirmation, and New Order represent data structures. The structures are:

```
Order =
CustomerName = FirstName, MiddleInitial, LastName
CustomerAddress = Street, City, State, ZipCode
CustomerTelephone = AreaCode, TelephoneNumber
Product = ProductNumber, Price, Quantity

New Order =
CustomerID, OrderID, OrderDate
Product = ProductNumber, Price, Quantity

Order Confirmation =
ConfirmationNumber, OrderDate
Product = ProductNumber, Price, Quantity
```

10.5.2 How To Construct a Data Flow Diagram

To construct a data flow diagram for a system, you start with the context diagram to define the boundaries of the system being modeled. In our running example of the reservation

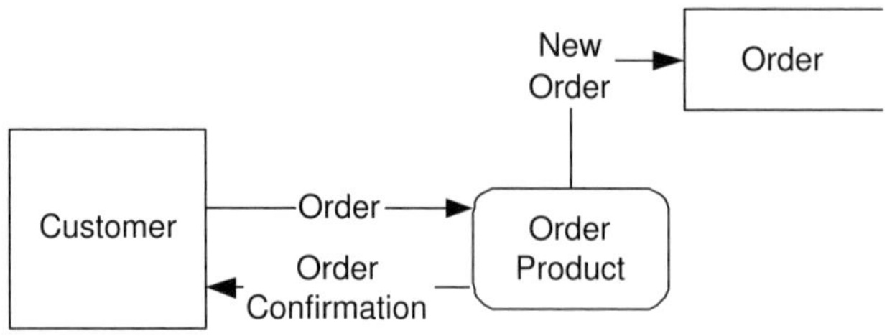

FIGURE 10.12
Data structure.

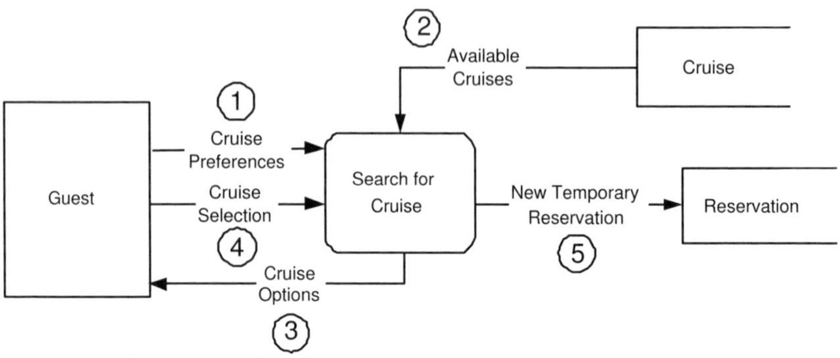

FIGURE 10.13
Event data flow diagram.

system, a single process of Reservation System is drawn in the context diagram. The next step is to identify all of the external agents: people, groups, organizations, and external systems that the system will interact with. For each external agent, identify the inputs and outputs remembering to combine data structures that flow together into a single packet. The resulting context diagram for the reservation system was shown in Figure 10.8.

Next, identify all the processes for the system from the functional requirements, and use them to create an event-response list (see Table 10.4). The event-response list aids the modeling team in identifying the inputs and outputs for each process. The trigger is a data input or a system state that starts the event. Response describes what the system does in response to the trigger. This includes obtaining further data (input data flows) and generating output data.

For each event in the event response list, you create an event DFD by making the event the process, the trigger an input data flow, and the responses are input and output data flows. To illustrate, the event DFD for the first event in the event-response list is shown in Figure 10.13 with numbers to indicate sequence (note, these numbers are not normally included in DFD). The trigger is Cruise Preferences, a data flow from the Guest to the Process. Then the process reads the Available Cruises from the Cruise datastore. These are processed and sent to the Guest as Cruise Options. The Guest will either select a cruise, abandon the search, or submit revised cruise preferences. Assuming the Guest makes a Cruise Selection, this is then used to create a New Temporary Reservation in the Reservation datastore.

Once a DFD is created for all the events, then you can combine the event DFDs into a *system data flow diagram*. In a system data flow diagram the symbols should be located so

TABLE 10.4
Event-Response List

External Agent	Event Process	Trigger	Response
Guest	Search for Cruise	Cruise Preferences	Obtain Available Cruises. Generate Cruise Options. Obtain Cruise Selection (if any). Create New Temporary Reservation.
Guest	Select State Room and Dining Option	Cabin Preferences	Obtain Available State Rooms. Obtain Available Meal Plans. Show Options to Guest. Update Temporary Reservation with Selections.
Guest	Obtain Guest Information	Request Reservation	Create New Guest. Obtain Guest Name and Address.
Guest	Offer Shore Excursions	Request Shore Excursion	Obtain Available Shore Excursions. Show Guest Shore Excursions. Obtain Guest Selections (if any). Create New Guest Shore Excursion.
Guest	Confirm Reservation Information	(End Reservation state)	Obtain Reservation. Show Cruise Reservation to Guest. Obtain Guest Confirmation.
Guest	Make Payment	Reservation Confirmation	Obtain Credit Card Number and Expiration date from Guest. Obtain Payment Amount. Obtain Credit Authorization or denial from Bank. Generate Payment Confirmation or Denial to Guest.
	Place Reservation on Hold	Hold Request	Create New Hold.
	Send Cruise Welcome Package	Time	Print and Send Welcome Package.

as to avoid too many overlaps of data flows. It is allowable to depict multiple times both external agents and data stores so as to avoid overlapping data flows.

At this point there is a context diagram and a single decomposition to the activity level. If further decomposition is called for by the modeling goals, then each activity is decomposed into steps. The data flows in the activity-level diagram may be composite flows that are deaggregated into two or more flows at the step-level diagram.

10.6 IDEF0 Function Model

IDEF0 is a function modeling approach and is part of a suite of tools for systems modeling that includes IDEF1x for information modeling. The United States Air Force commissioned the developers of SADT (Structured Analysis and Design Technique), a well-established

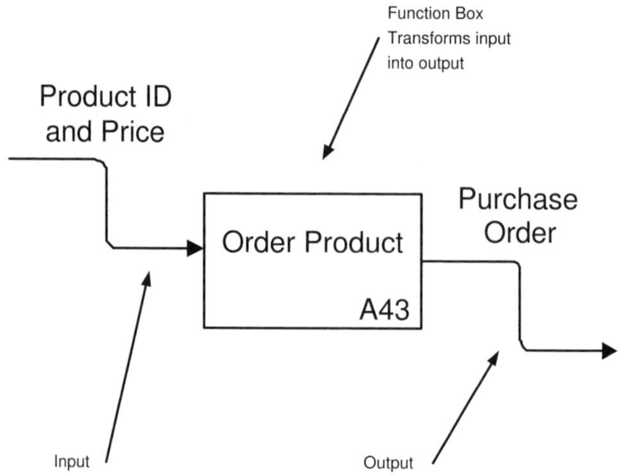

FIGURE 10.14
Function definition.

graphical language, to develop IDEF0 to represent the functions in a system. Originally, IDEF0 was defined in the context of manufacturing systems (a result of the Air Force Integrated Computer Aided Manufacturing ICAM project that it was developed for), but it has found wider use for all types of processes. IDEF0 models the functions or activities required by an enterprise and the functional relationships described by information and objects flows that support the integration of those functions [4].

10.6.1 IDEF0 Model Constructs

A rectangular box represents a function that transforms inputs into outputs. The name of the function is inside the box with its node designation. The function name should be a verb or verb phrase to convey the action performed by the function. In Figure 10.14 a clear action verb is used and the node number is A43. The numbering scheme indicates the position of the activity within the hierarchy, explained in detail later. In the figure, the function `Order Product` takes the input of the `ProductID` and `Price` and uses this information to create a `Purchase Order`.

In the IDEF0 diagrams, the arrows denote pipelines that either information or objects flow through. Thus, arrows are labeled with a noun or noun phrase to describe the information or objects that flow through them. The arrows assume different roles depending on where they enter or leave the box. The role the arrows fulfill is described by the acronym **ICOM**: - **I**nputs enter the left of the box, and are transformed or consumed by the function to produce outputs. - **C**ontrols enter the top of the box, and specify the conditions required for the function to produce correct outputs. Controls are usually information such as process instructions, inspection guidelines, etc. that guide the transformation process. - **O**utputs leave the right of the box, and are the data or objects produced by the function. - **M**echanisms enter the bottom of the box, and identify the means that support the execution of the function. Mechanisms are the resources used to effect the transformation, either human resources or machines. These ICOM objects are a distinctive feature of the IDEF0 model. The ICOM objects are shown in Figure 10.15.

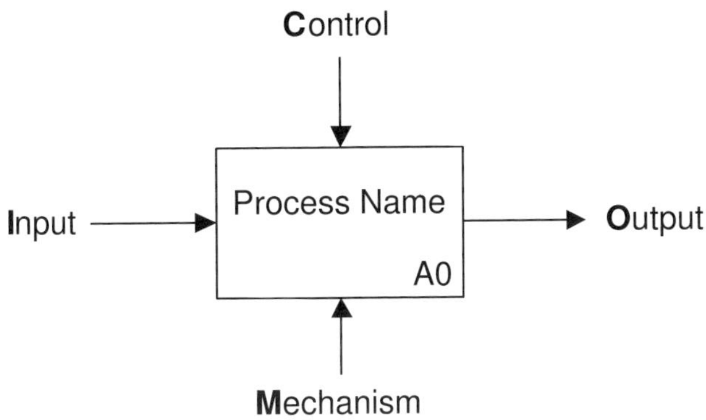

FIGURE 10.15
ICOM.

10.6.2 IDEF0 Decomposition Hierarchy

An IDEF0 function model is a collection of diagrams, organized in a hierarchy such that each diagram represents a decomposition of a higher-level diagram. The higher-level diagram is called the parent diagram and the lower-level decompositions from that diagram are each child diagrams.

Each individual IDEF0 diagram is drawn on a title block template and represents a single node in the hierarchy. The top level of the function hierarchy is the A-0 diagram (pronounced "A minus zero"). The A-0 level is a context diagram and represents the boundaries of the system, showing the entire system as a single function box numbered A0 and all the ICOM linkages with the external environment (see Figure 10.16). The A-0 level also includes statements of the model purpose and viewpoint. As indicated earlier in Chapter 3, all models are built for a particular purpose and from a particular viewpoint. IDEF0 provides a functional view of the system. At the A-0 level you provide a more specific description of the model's viewpoint. The purpose of an IDEF0 model embodies the reasons why the model is created – some common purposes are to create a functional specification for the system, to specify a design, and to describe the as-is system. A strength of IDEF0 is the explicit definition of the purpose and viewpoint at the highest level of the process model.

The A-0 level context diagram has a single function box that is decomposed on the A0 diagram. Each function on the A0 diagram is labeled A1, A2, ... Then each function can be decomposed further. The A1 function can be decomposed into functions A11, A12, ... ; and A2 can be decomposed into functions A21, A22, ... ; and so forth for each function. The function that is shown in the diagram is listed in the lower left corner of the title block. Figure 10.17 depicts the decomposition from the A-0 level to A0 and then further to A2. Higher level functions are very abstract and may be decomposed into activities of increasingly greater detail. The level of detail is determined by the modeling requirements. The function numbering is used to indicate a functions position in the hierarchy. For example, a function A213 is read backwards such that the number represents the third activity of the current diagram that is decomposed from the second activity of the next higher level diagram, which in turn is decomposed from the first activity of the diagram higher in the hierarchy.

A single IDEF0 diagram should have no fewer than three functions and no more than six functions. This rule essentially enforces decomposition by forcing the modeler not to crowd too many functions in a single diagram. By convention, the function boxes are arranged diagonally from the top left to the bottom right of the diagram.

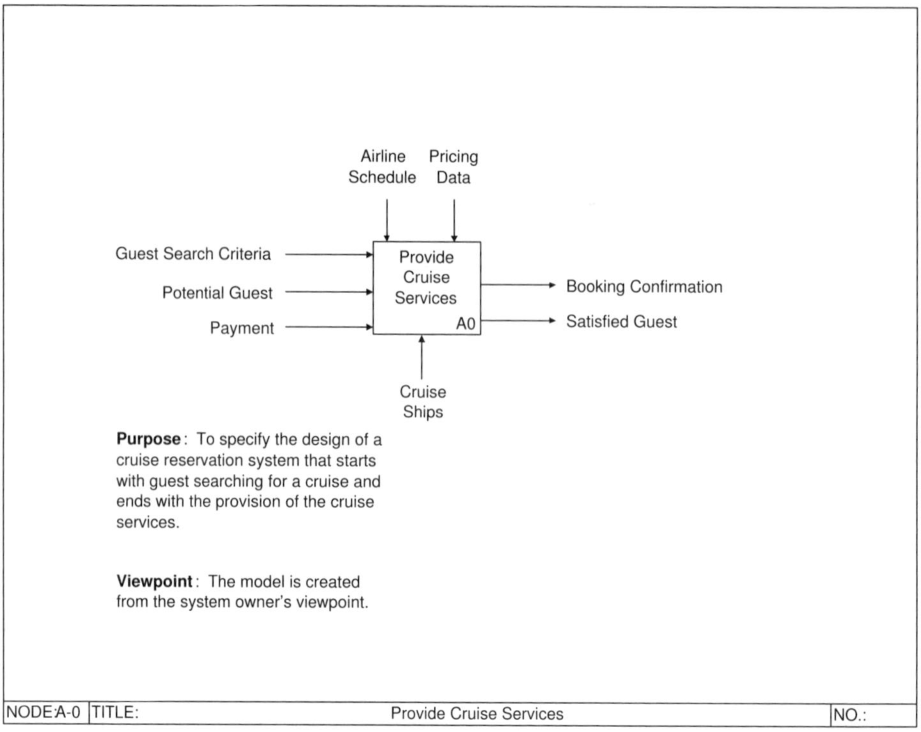

FIGURE 10.16
A-0 level.

In the IDEF0 decomposition, ICOM objects can neither be gained or lost. This means that every ICOM object in the parent diagram must appear in the children diagrams. Sometimes for purposes of clear exposition, the modeler might not want to show every ICOM at every decomposition level.

It is important that the ICOMs in each diagram are consistent with the ICOMs in higher level and lower level diagrams in the hierarchy. This imposes a constraint that ICOM objects can neither be gained nor lost in the decomposition. However, sometimes for clarity of exposition, a modeler might not want to show a particular ICOM object at every decomposition level. To suppress an ICOM object at a level, it can be "tunneled" by drawing parentheses around the end of the arrow to be suppressed. When an ICOM object is shown at the parent but not on the child diagrams, then the parentheses are around the arrow where it connects to the box. This means the ICOM object is not required for understanding the lower levels and will not be shown. To tunnel an ICOM object on the parent diagram then the parentheses are around the unconnected end of the arrow.

To illustrate how ICOM objects are handled in decomposition and how they can be tunneled, we use the example shown in Figure 10.18. The top of the Figure shows the partial A0 diagram. Function A3 has the mechanism object, MIS Department, tunneled at the box end. This denotes that the MIS Department does not appear in the child diagram. The bottom of the Figure shows the A3 diagram. From the parent diagram the single input `Verified Client` is shown as an input to function A31, and the output `ANI Activation/Cancellation File` is shown as an output from function A32. In the A3 diagram all the mechanisms are tunneled such that they do not appear in the parent diagram. Also the error handling outputs: `Recuperation Statistics Report` and `ANI Not`

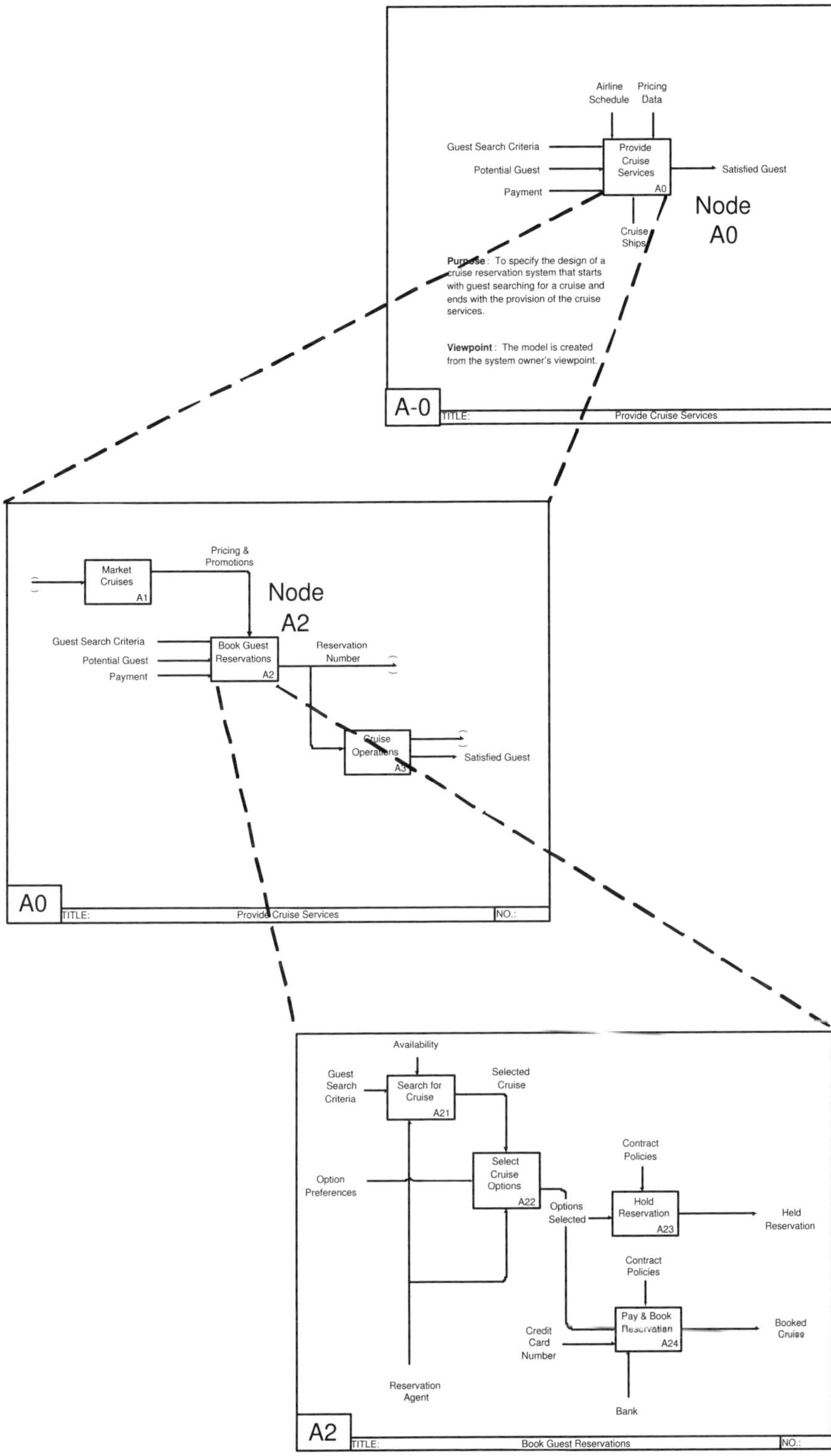

FIGURE 10.17
IDEF0 decomposition.

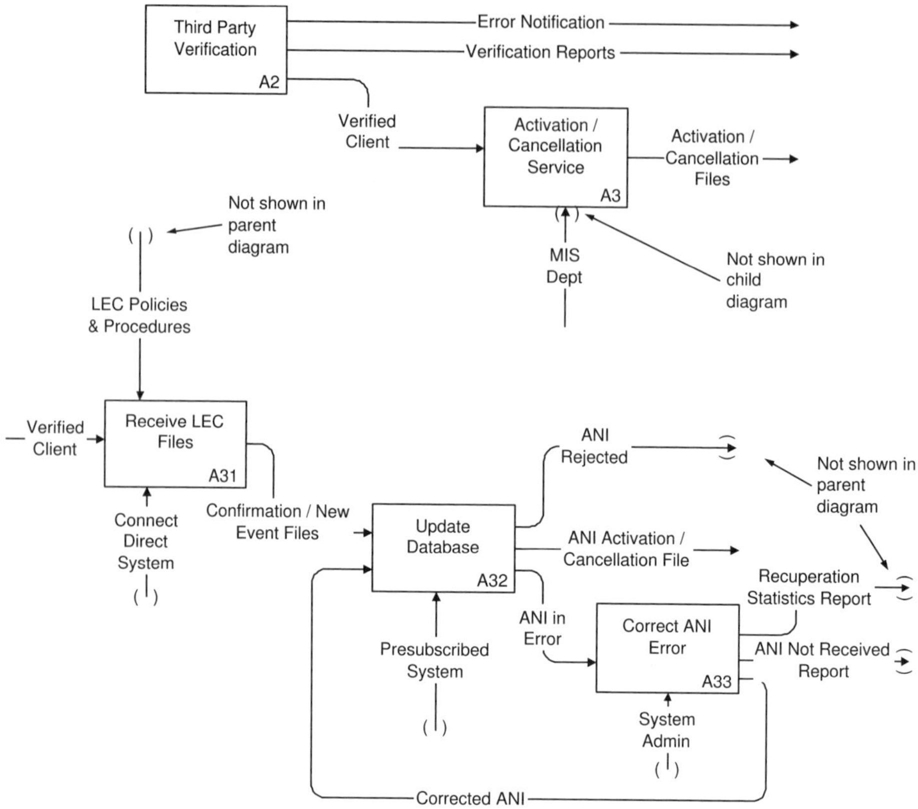

FIGURE 10.18
Tunneling at parent and child diagrams.

Received Report are tunneled and not shown in the parent diagram. Also note in the A3 diagram there are intermediate outputs and inputs that are wholly contained in the child diagram (e.g., `Confirmation/New Event Files`, `ANI in Error`, and `Corrected ANI`) that do not appear on the parent diagram.

10.6.3 IDEF0 Syntax and Semantics

An IDEF0 box may use any combination of the inputs to produce any combination of the outputs. In the example depicted in Figure 10.19, the Search for Cruise function uses a single input, the Guest Search Criteria, and can produce one of two outputs: either the Selected Cruise or No Cruise Available notification. The two possible outcomes are called the activations of the function. In this particular case, the activity never produces both of the outputs for a single input. The interpretation of the IDEF0 semantics depends on context, from the diagram alone it is not possible to determine if both outputs are generated from one input, or if only one of the outputs is generated from one input.

The arrows essentially denote constraints on the functions. In Figure 10.19 function box A22 requires both the control `Selected Cruise` and the input `Option Preferences` in order to activate.

Arrows can split or join in a number of ways. Figure 10.19 shows one example where the output `Reservation Information` splits and the entire output flows to the external user and part of the output, the `Confirmation Number` and `Address`, flows to function A26. In

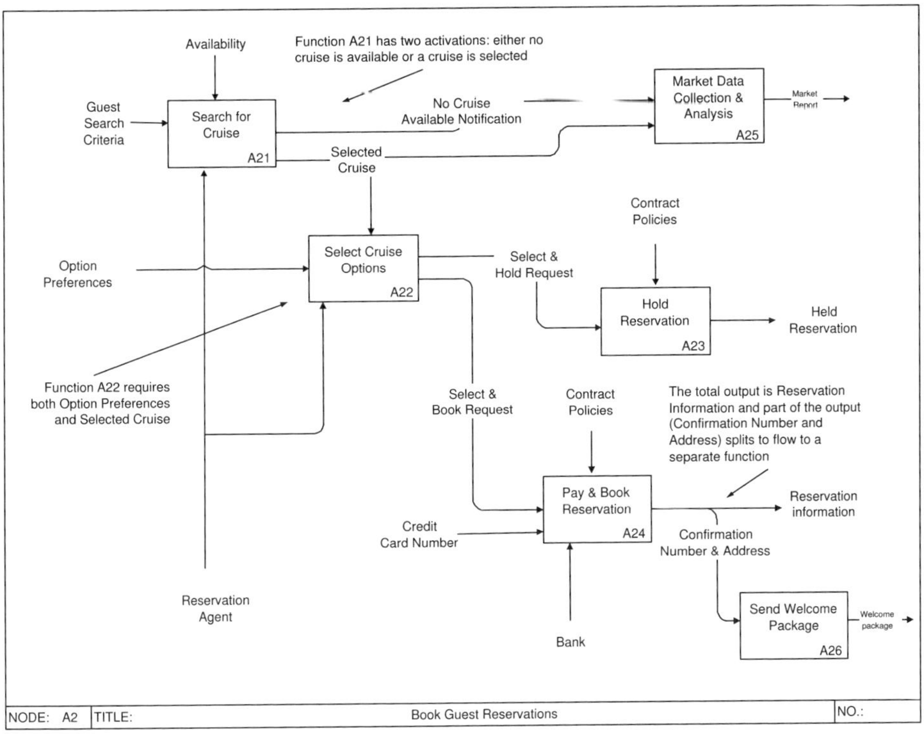

FIGURE 10.19
A2 diagram of reservation process.

the same figure, the output `Selected Cruise` from A21 splits but the full data object flows in both branches.

As a functional model it is important to remember that IDEF0 does not necessarily show sequence. In higher-level diagrams, the arrows denote constraints on activation, but the model is too abstract to infer a control sequence. In lower-level diagrams, there is usually sufficient detail that control sequence can be inferred.

Parallelism in IDEF0 is possible because a function box may be active if it has all the required inputs and controls. Figure 10.20 shows that function box A41 outputs object A that flows simultaneously to function boxes A42 and A43, both of which now have all required inputs and controls and therefore can activate in parallel.

An IDEF0 diagram depicts the functions of the enterprise, the input and output of each function, and how the functions interact through information flow. The primary strength if IDEF0 is its proven ability to model the functions of the system to any level of detail required by the modeler. It does function modeling well largely because the method ignores other system attributes and behavior common to process models. However, this can also be viewed as a limitation of IDEF0. IDEF0 does not explicitly model any timing, sequencing, or decision logic, which must be inferred, if possible, from the diagram. This is a problem when IDEF0 models are interpreted by people not expert in IDEF0, since there is a strong predilection to view the model as representing activity flow because of the modeling conventions of ordering activities from left-to-right and top-to-bottom. So, good advice when using IDEF0 is that the modeling experts provide sufficient supplementary documentation and instruction so that other stakeholders avoid misunderstanding the model.

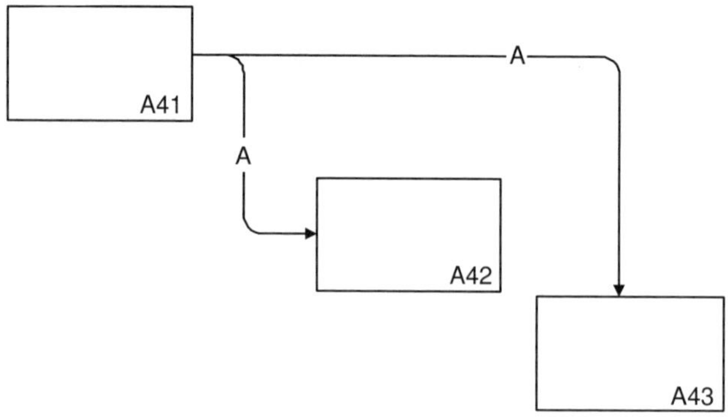

FIGURE 10.20
IDEF0 parallelism.

10.7 Other Process Modeling Techniques

Aside from the three modeling approaches detailed above, there are many other approaches to modeling processes. In this section we summarize some of the more common approaches.

Event-Driven Process Chains (EPCs) are one of the central components of the Architecture for Integrated Information Systems (ARIS) [7]. The ARIS architecture recognizes three views of a system: the data view, the function view, and the organization view. In earlier chapters, we stressed the importance that the views of an artifact must correspond, or in other words, be in agreement. ARIS provides the EPC diagram so that it can unite the organization, information, and function views defined by ARIS into a single diagram showing the process flow. EPC is different from the previous reviewed process modeling method because it emphasizes events as triggering functions. Figure 10.21 shows all the EPC symbols and how they are used. The function is triggered by an event and the function terminates with another event. The functions can be chained together in processes that include logical nodes (and, or, exclusive or). In the example shown, either of the input events can trigger the function. The EPC also shows the information input and output as well as the organizational unit responsible for the function. In this way, the EPC diagram shows how the three separate views of information, process, and organization come together.

EPC diagrams were adopted by SAP for the representation of the business processes in their ERP software version R/3 [2]. Because SAP is one of the largest software vendors in the world, their use of EPC models helped promote EPC as a modeling language for business processes, especially in Europe.

In academia, there continues to be great interest in Petri Nets to model business processes [10]. Petri Nets are graphs that have two types of nodes and model the state of the system. Petri Nets have a strong mathematical formalism that enables checking correctness of the model and other properties. This is useful for modeling workflows that will be executed by computers. However, Petri Nets are not easy for laymen to interpret, and their use for process modeling outside of academia seems limited.

There is a trend, mostly in the information technology community, to develop standards for a business process modeling language. The Business Process Modeling Language (BPML) is a meta-language for the modeling of business processes. The idea is to develop a process modeling standard similar to what has been achieved by the Unified Modeling Language (UML) for software design. The BPML while useful for its intended purpose (specification of computer-implemented processes) is not suitable for analysis nor is it eas-

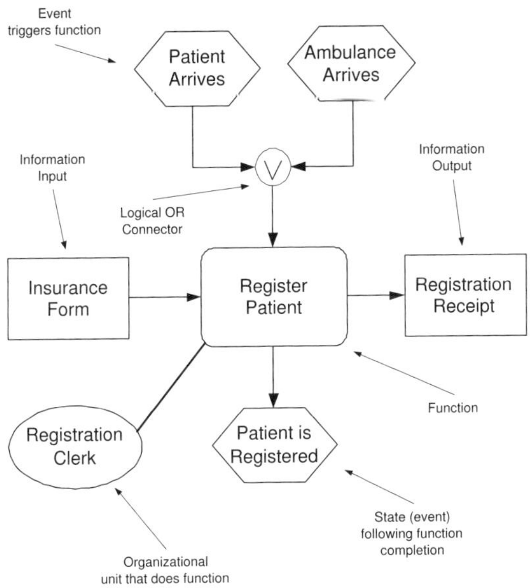

FIGURE 10.21
EPC diagram.

ily understood by laymen. UML itself has activity diagrams, collaboration diagrams, and state diagrams that to varying degrees can model a process. Again, the UML diagrams are mainly used in software engineering.

10.8 Summary

This chapter defined a business process as a set of a set of activities that generate a product or service of value to a customer. The chapter discussed the elements of a process as including a goal, inputs, outputs, resources, and measures. Processes can be modeled at different levels of abstraction, which leads to a process hierarchy. It is important that a single process diagram depicts the process at a consistent abstract level. Finally, processes can be classified as either core processes that serve external customers, support processes necessary for the enterprise's operation, and management processes that make decisions. In later chapters, the analysis focuses on core processes.

The chapter reviewed the process modeling techniques of flowcharts, data flow diagrams, and IDEF0 (strictly speaking, a function modeling approach and not a process modeling approach). Flowcharts depict the control flow of a process. Flowcharts are by far the most common process modeling techniques: they are popular in large part because they are simple to understand and draw. DFDs hail from the early era of information system analysis during the 1970s. They show the data flow in the process. IDEF0 has similar origins as DFD but is more formalized and was widely adopted by the U.S. government.

Both flowcharts and data flow diagrams are basic modeling techniques that have long been and still are used in process analysis and design. Together they represent the two types of process flows: activity flow and data flow. They only represent a static view of the process; there is no dynamic information in either model. Flowcharts model the process at a single decomposition level. Data flow diagrams can be used in a hierarchy of data flow diagrams in which higher-level processes are decomposed into lower-level activities and tasks.

Process Modeling Principles

- In a single diagram, all process activities should be at the same abstraction level.
- A process must have at least one input and one output. The output should logically flow from the input. The input and the output cannot be the same because that implies that no transformation took place.
- A process is described by a verb or verb phrase to denote the transformation it does.
- Input and output objects are described by nouns since they represent the objects that work by the process is done to.
- A process that shows interaction between a system and a user should provide feedback to the user in the form of a confirmation that the intended process was completed (e.g., reservation confirmation, sales confirmation, etc.)
- When multiple abstraction levels of a process are shown, then the inputs/outputs at parent levels should be consistent with the inputs/outputs at child levels.
- Models of core business processes need to show the end-to-end perspective from when the process is first triggered until the product or service is provided to the customer.

Enterprise system principles place emphasis on viewing the enterprise from a cross-functional, end-to-end perspective. As a process modeler, you should use a top-down methodology in which the top-level diagram shows the whole process on a single page, and use subprocesses to expand process detail at nested diagram levels. This allows readers of the model to zoom in and out of the model to describe any level of detail. The multi-diagram model may print out as multiple pages, but internally the integrity of a single model is maintained.

There are many process models and associated methodologies. There is no single, best way to model a process. Each modeler needs to consider the modeling requirements and determine what is the best modeling approach in that particular situation.

Review Questions

1. Explain why there are so many ways to model a business process.
2. Explain why decomposition is an important part of process modeling.
3. Describe the three different classes of processes.
4. Compare and contrast flowcharts and data flow diagrams.
5. List the main components of a data flow diagram.
6. Describe the difference between an input and a control in IDEF0.
7. Compare and contrast flowcharts and IDEF0 diagrams.

Exercises

1. Draw a data flow diagram for the warehouse control system (WCS) described below. You are modeling the system described. Remember to show: external agents, processes, data flows, and data stores.

   A warehouse distributes health food and related products. Customers order a particular product and quantity from the warehouse. The WCS saves the order and provides to the customer the order number. The WCS generates a pick list and shipping label, which tells the order-picker person how many of each item to pick to fulfill the order. The order-picker picks the items, places them in a box, and places the shipping label on it. The order-picker then uses the WCS to say the order is ready, and the WCS sends the order number, address, and payment data to the shipping company. At the end of the day, the shipping company arrives to pick up all the orders.

2. Draw a data flow diagram for the patient appointment scheduling system described below. You are modeling the system described. Remember to show: external agents, processes, data flows, and data stores.

   A patient contacts the healthcare clinic to request an appointment. The scheduler uses the system to check if the time slot requested and doctor are available, then the scheduler makes the appointment. The appointment is saved in the system and is confirmed with the patient. Additionally, a patient can contact the clinic to cancel an appointment. At the end of each day, the office manager prints the schedule for the next day.

3. The diagram below depicts a data flow diagram with the intent to represent the process whereby a student submits a change of address and the system updates the student's address. Identify what is wrong with the data flow diagram and create a correct version.

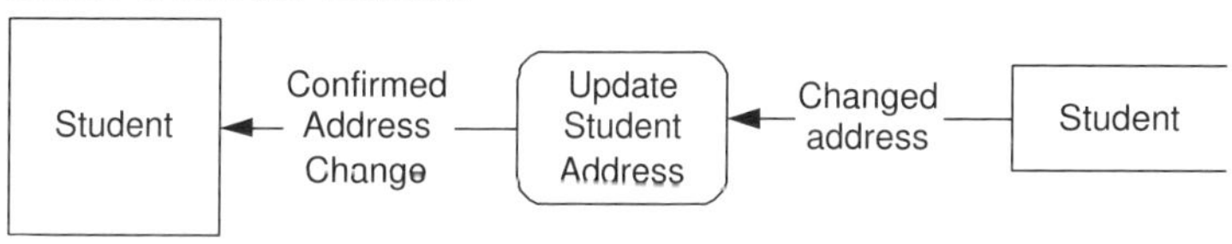

4. The diagram below depicts a data flow diagram with the intent to represent the process whereby management is alerted to low inventory levels that require replenishment. To check low inventory levels, the manager needs to know the current inventory levels and any outstanding purchase orders. Identify what is wrong with the data flow diagram and create a correct version.

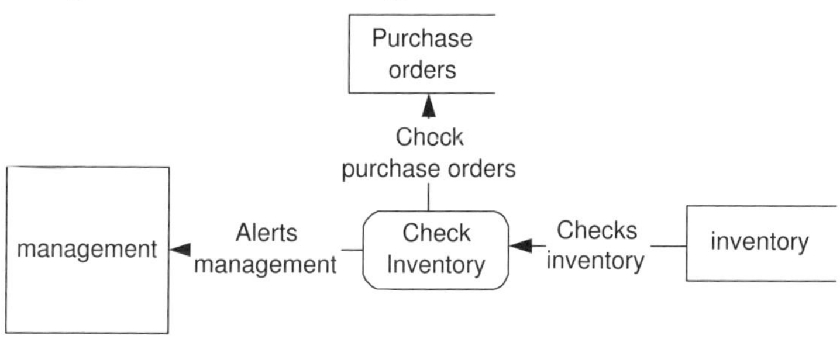

5. The diagram below depicts a data flow diagram with the intent to represent the order entry process in which a customer submits an order, the system saves the

order, the clerk reviews the order and if the order is acceptable, then the customer receives an order confirmation. Identify what is wrong with the data flow diagram and create a correct version.

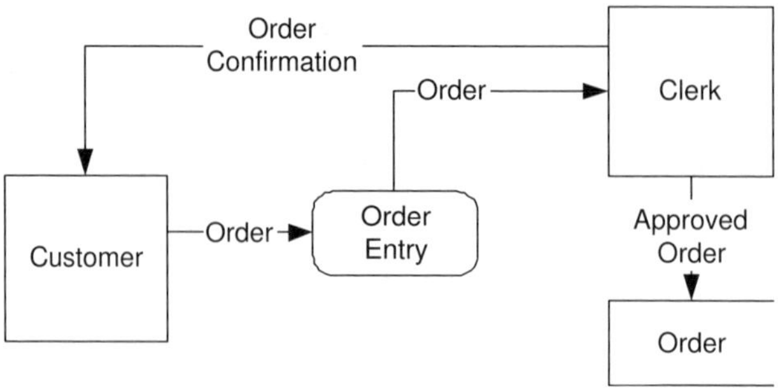

6. Draw a data flow diagram for the following event/response:

External Agent	Event	Trigger	Response
Customer	Address Change	New Address Form	Update address in customer database. Provide receipt to customer of address change.

7. Draw a data flow diagram for the following event/response:

External Agent	Event	Trigger	Response
Order Picker	Generate Pick List	New Order	Get pick locations from warehouse database. Output pick list. Update order status in order table.

8. Select a process you are familiar with (e.g., course registration at university or book loan process at library) and create an IDEF0 diagram of the process. Have at least two decomposition levels in your model.

9. Below is a single node of an IDEF0 diagram. Decompose node A33 into three functions of Calculate Sales Tax, and Calculate Shipping, and Calculate Order Total. Make sure the ICOM objects remain consistent in the decomposition.

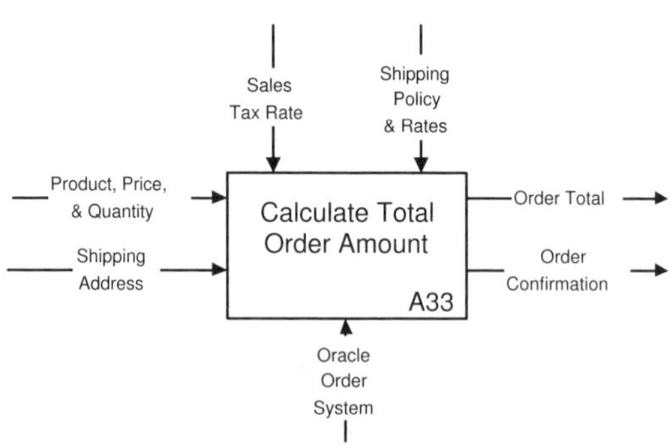

10. Select a process you are familiar with (e.g., course registration at university or book loan process at library) and identify a decomposition of the process from one activity to the lowest-level steps performed by a single person.

11. The Copper Canyon (*Baranca Cobre*) Tourist Agency in Mexico wants to develop a Website to promote the area. The Website must be in various languages (English, Spanish, French, and German) and enable the Website visitors to directly book reservations of hotels, tours and all other services provided by local vendors (e.g., rental car, train, etc.). Vendors will use the site to create an account to list the services they provide. The site must allow Travel Agencies to register and receive special promotions. The system will also be capable to produce reports to the Tourist Office of Copper Canyon (e.g., number of Website visitors, reservations made by seasons, etc.). The system also manages the yearly sweepstakes in which contestants can enter to win a week vacation package to the Copper Canyon.

    (a) Draw the context data flow diagram for this system. (Data flow diagram with only one process representing the entire system).
    (b) Create an event-response list.
    (c) Create data flow diagrams for each event.
    (d) List the attributes in each data structure in the diagram.

12. The master classroom schedule assigns each department time slots and classrooms they may use for the upcoming semester on a DEPARTMENT ROOM ASSIGNMENT LIST. Academic departments submit initial data about courses, instructors, times, and projected enrollment on a COURSE REQUEST LIST. The registration system generates a COURSE CATALOG, which is published and sent to all academic departments and students. Students select their courses on a ADD/DROP FORM which after being signed by their advisor is input into the registration system. Various systems, including the Library, Parking & Traffic Control, and Financial Aid can input a hold on a student's registration for unpaid bills. If no holds are in the system, then the student receives a confirmation called the STUDENT REGISTRATION from the registration system. The STUDENT REGISTRATION is also sent via the network to the Payment System, which will bill the student.

    (a) Draw the context data flow diagram for this system. (Data flow diagram with only one process representing the entire system).
    (b) Create an event-response list.
    (c) Create data flow diagrams for each event.
    (d) List the attributes in each data structure in the diagram.

13. Create a flowchart to model the registration process at your university.

14. Create a flowchart to model the reservation process used by a Website that does travel reservations. Indicate how your flowchart corresponds to each step in the Website's reservation process.

15. Create a data flow diagram to model the reservation process used by a Website that does travel reservations. Indicate how your data flow diagram corresponds to each step in the Website's reservation process.

16. A telephone company has a process called invoice customer. Two separate data structures flow to the customer: the invoice and the itemized list of telephone calls. These flow together at the same time. Create the data flow diagram to represent this process.

17. Create a flow chart for a process with activities A, B, C, D, E, F, and G. The first activity is activity A, then activities B, C, and D are performed in parallel. After the parallel activities, then activity E is done. Depending on the outcome of activity E, denoted as "yes" or "no", then activity F or activity G are done. The process then terminates.

18. A South Florida condo agency is developing a system to manage its listings (condos available for purchase). Below is an event-reponse list. Draw the data flow diagram for each event.

Event	**Trigger**	**Response**
Seller enters agreement to have agency sell house	Listing Request (from seller)	Create listing in database Create seller in database Provide seller with sales documents and contract
Buyer makes offer on condo	Offer (from buyer)	Create offer document Create offer in database Send offer to seller
Seller accepts buyer's offer	Purchase Agreement (from seller)	Create notarized purchase agreement Notify buyer of acceptance Confirm agreement with seller
Seller makes counteroffer	Counteroffer (from seller)	Create counteroffer in database Send counteroffer to buyer

Bibliography

[1] D.A. Garvin. The processes of organization and management. *Sloan Management Review*, Summer:33–50, 1998.

[2] G. Keller, M. Nüttgens, and A.W. Scheer. *Veröffentlichungen des Instituts für Wirtschaftsinformatik*, chapter Semantitische Procezeßmodellierung auf der Grundlage "Ereignisgesteuerter Prozeßketten (EPK). Saarbrücken, Germany, 1992.

[3] H. Mintzberg. *Structure in Fives: Designing Effective Organizations.* Prentice Hall, Englewood Cliffs, NJ, 1992.

[4] NIST. Integration definition for function modeling (IDEF0). Technical report, Federal Information Processing Standards Publication 183, 1993.

[5] M. Ould. *Business Processes: Modelling and Analysis for Re-engineering and Improvement.* John Wiley & Sons, Chichester, England, 1995.

[6] J.W. Ross. Creating a strategic IT architecture competency: learning in stages, CISR WP No. 335. Technical report, MIT, Cambridge, MA, 2003.

[7] A.W. Scheer. *Architecture of Integrated Information Systems: Principles of Enterprise Modeling.* Springer-Verlag, Berlin, Germany, 1992.

[8] H. Simon. *The Sciences of the Artificial.* 3rd ed., MIT Press, Cambridge, MA, 1996.

[9] D.S. Sink and T.C. Tuttle. *Planning and Measurement in Your Organization of the Future.* Industrial Engineering and Management Press, Norcross, GA, 1989.

[10] W. van der Aalst and K. van Hee. *Workflow Management: Models, Methods, and Systems.* MIT Press, Cambridge, MA, 2004.

11

Queueing Theory

"All human wisdom is summed up in two words - wait and hope" – Alexandre Dumas (1802-1870), French author of *The Three Musketeers* and *The Count of Monte Cristo.*

Queuing theory is the study of stochastic processes in which the arrival rates and service rates are random variables and the service capacity is constrained. As the arrival rate approaches the service rate, a queue will form – where a queue is a waiting line. Queueing theory provides analytical models to understand and predict the performance of systems with these characteristics.

Business processes can be modeled with queueing theory. In business processes, a job is routed from activity to activity, and at each activity, some transformation is done to the job, until the job finally departs the process. Each activity of the process is performed by a resource: either a human resource or machine resource. If the resource is busy when the job arrives, then the job will wait in a queue until the resource becomes available. To analyze and design a business process with these characteristics, you would typically want to know how long does it take the job to transit the process? How many jobs can the process do per hour? How busy are the resources? To answer these and other performance-related questions, we can use queueing theory.

The benefits of applying queueing theory to analyze business processes is first, they provide the analyst with insight into the performance of business processes, and second, the performance analysis can be conducted rapidly, allowing for fast generation of alternative process designs.

After completing this chapter, you should be able to:

- Describe the three main components of a queueing system.
- Characterize the variability of a system using the coefficient of variation.
- Understand the classification of queueing systems.
- Analyze the performance of single node queueing systems.
- Use Little's law to analyze processes.
- Explain how queueing theory can be used to analyze processes.

11.1 The Queueing System

A queueing system has the following structure (see Figure 11.1): jobs arrive and join a queue to wait for service provided by m servers. After receiving the service, the job exits the system. To understand the queueing model, we describe the arrival process, the queueing process, and the service process. Hereafter, we follow the convention that a job is the unit

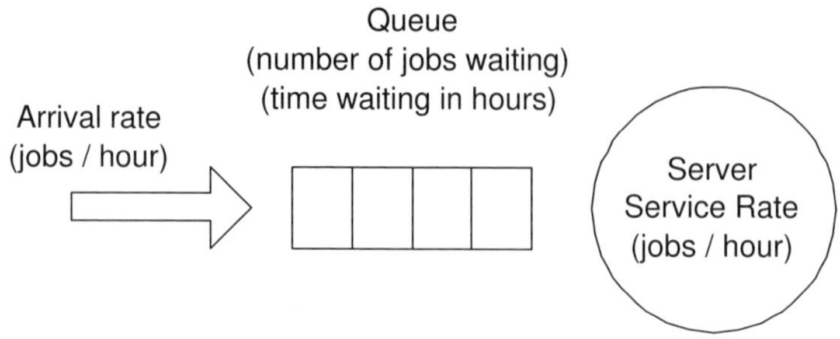

FIGURE 11.1
Queueing model.

demanding service, and the server provides the service. The terminology is used regardless of the scenario. Since queueing models can be applied to a wide range of scenarios, the interpretation of job or server changes in each scenario. For example,

- Patients arrive to a healthcare clinic for service by physicians.
- Internet data packets arrive to a router to be directed on their next hop.
- Job orders arrive to a machine shop to be processed by a piece of equipment.
- Queries arrive to a database to be processed.

There are many more such examples, which only serves to confirm the wide applicability of queueing models to many different systems.

The reason waiting lines or queues form is that the customers are competing for the same service. If we knew that customers arrived exactly every 5 minutes to a supermarket checkout counter and that the service time was exactly 4.9 minutes, then we could set up the supermarket so that a customer would never have to wait. As one customer was completing service, the next customer would arrive exactly 0.1 minutes later. This does not happen because both the customer arrival rate and the service time are random – we can specify their average and variance but we cannot predict with certainty any particular customer arrival. What happens in actuality is that sometimes customers arrive faster than the server can process them, and thus, these customers need to wait. It is the variability in both the arrival rate and the service rate that leads to queueing.

Figure 11.2 shows 10 customers who arrive to the supermarket with a mean interarrival time of 5 minutes and a mean service time of 4.9 minutes. At any time period where they overlap, this indicates the customer is waiting. For example, at 4 minutes, customer #2 is waiting until 6 minutes when customer #1 departs before customer #2 can be served. At 30 minutes customer #6 arrives while customer #5 is being served, and then customer #7 arrives too. Notice the server is idle from between 10 and 14 minutes and then again between 22 and 26 minutes. It is the randomness of arrivals and service that leads to the paradox that the system exhibits both idle time of the cashiers and waiting time of the customers.

One way to avoid queues is to increase the capacity of the servers so that it is much greater than customer demand. For example, if the supermarket could add more counters, then there would be no waits at the checkout counters. However, there are usually constraints on the availability of capacity. Constraints may be due to the physical layout or dimensions of where the service is provided – the supermarket only has space for so many checkout counters. Or, consider Disney's theme parks, where in order to shorten the line on

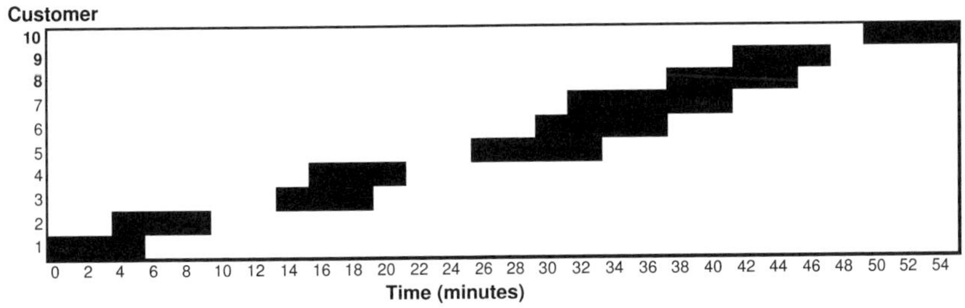

FIGURE 11.2
Effects of variability

an attraction, such as *Pirates of the Caribbean,* would require extensive, maybe not even possible, redesign of the current attraction or adding a whole new attraction. Another reason for limited capacity, may simply be the desire of management to minimize cost. Only by having capacity in large excess of the expected demand can waiting be eliminated. But then, the servers will experience high idle times when no customers are in the system. The design goal is to set capacity to balance cost and meet some service level for the customers. For example, the supermarket might want to have customers wait no more than five minutes to be served by the cashier. Queueing theory lets us design a system to meet these types of requirements.

Most of queueing theory's analytical equations are limited to stable and steady-state conditions only, so we need to understand what is meant by the terms stable and steady state. *Stable* means that the long-term average arrival rate of work to the queue is strictly less than the long-term average service rate.[1] Otherwise, if work arrives faster than it can be served, the queue will keep growing to infinity. *Steady state* means the system parameters: arrival rate, service rate, and number of servers never change. More formally, steady state means the probabilities that the various state variables will be repeated remains constant. So, if at time t the probability of the average queue length being five units is 50%, then at times $t+1$, $t+2$, and so forth the probability of the average queue length being five units remains 50%. In steady state, the initial conditions do not affect the current state variables. Hereafter, in the remainder of this chapter we assume steady-state conditions. At the end of the chapter, we discuss transient conditions. Transient conditions occur when the probabilities change with time.

11.1.1 Understanding Variability

Variability describes the differences in an attribute observed between members of a single class of entities. For example, given a group of people (the class of entities) their height (an attribute of the class) will differ between each individual. *Variance* measures the absolute variability and the *standard deviation* is the square root of the variance. Oftentimes, more useful is the *coefficient of variation* (CV), which defines relative variation as the standard deviation divided by the mean. Let c denote the coefficient of variation, and σ denote the standard deviation and t denote the mean, then $c = \sigma/t$. When the standard deviation is large relative to the mean, then there is significant variation. In queueing theory, instead of the CV, most approximations make use of the *squared coefficient of variation* (SCV), which is $c^2 = \sigma^2/t^2$.

[1] In small intervals of time, it is possible that the arrival rate is greater than the service rate – this occurs due to the variability of both arrivals and service and is the reason why queues form.

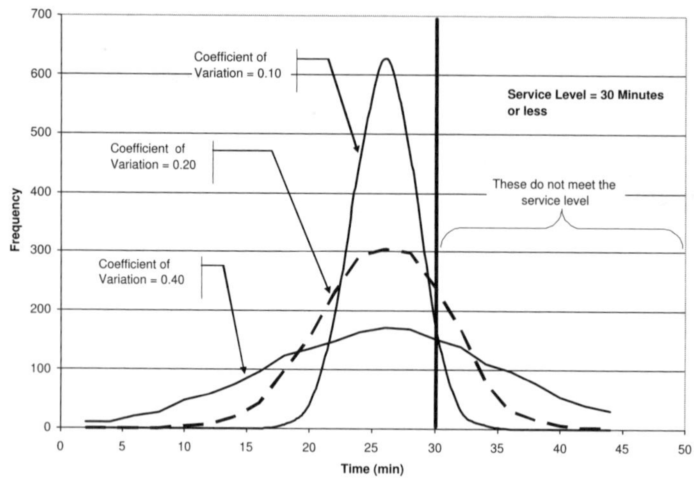

FIGURE 11.3
Understanding variability.

Figure 11.3 shows three processes with the same mean of 25 minutes but different CVs of 0.1, 0.2, and 0.4. Suppose the service level for the process requires that 90% of the jobs be completed in 30 minutes or less. Even though all the processes have a mean of 25 minutes (less than the 30-minute requirement), only the process with a $CV = 0.1$ will meet the service level criteria. For the process with $CV = 0.1$, 97.7% of the jobs will be completed in less than 30 minutes. When the $CV = 0.2$, then 84.1% of the jobs meet the service level, and when the $CV = 0.4$, then 69.1% of the jobs meet the service level agreement. The example illustrates that systems cannot be designed for the mean only, but system design should also take into consideration the entire distribution of events.

Variability is either special-cause variability or common-cause variability. *Special-cause variability* can be assigned to causes that can then be analyzed and reduced. *Common-cause variability* is essentially a catch-all term to describe system variability that cannot be assigned to any specific cause. It is the intrinsic variability of the system. Sources of variation come from inputs to the system, the transformation process, the resources that do the work, the information available for the system, and in service systems the interaction between the customer and the service provider.

In business processes, variability is also caused by the variety of inputs, variety of sequencing, or variety of tasks performed in the process (see Pentland [10]). For example, a hospital might design a system with a process to handle patients. The process is designed to handle a large variety of patients, each of whom is slightly different than the other patients, and will consequently require slightly different care. For this reason, we often see much higher variability in service industries than in manufacturing industries, which control the inputs, sequencing, and tasks used to produce goods.

11.1.2 Arrival Process

Customers, jobs, or any work unit requiring service enter the process. The arrival rate is frequently characterized by the interarrival time, defined as the time (in seconds, minutes, or hours) between subsequent arrivals. The interarrival times are assumed to be independent identically distributed (abbreviated *iid*). The term, *iid*, is used to describe random variables when each variable has the same probability distribution as the others and all are mutually independent. To illustrate the *iid* assumption, if the time between customers arriving to a bank is independent of any other customer arrival and through time studies we can show the distribution of the interarrival times is identical, then the *iid* assumption holds. If however, we have customers arriving to a fast-food restaurant in groups, then the interarrival times are no longer independent since the customers arrive as a group, and the assumption no longer holds.[2] It is always important to understand the assumptions underlying a mathematical model and to determine whether the assumptions hold in the scenario you wish to use the model.

To describe the arrival process in queueing theory, let's assume customers arrive at times $t_1, t_2, ..., t_n$. The time between subsequent arrivals is denoted as α, where $\alpha_j = t_j - t_{j-1}$. The parameter α is called the *interarrival time* and is assumed to be a sequence of independent and identically distributed (*iid*) random variables. Over a period of time from 0 to t, the total number of arrivals during the time period is denoted as $\alpha(t)$. The arrival rate λ is the number of jobs arriving per unit time, and can be calculated as $\lambda = \frac{\alpha(t)}{t}$. Furthermore, the arrival rate is the inverse of the interarrival time, so $\lambda = \frac{1}{\alpha}$.

Example

A nationwide insurer receives an insurance claim on average every 12 minutes. The average number of insurance claims received per week is:

$$\lambda = \frac{1}{\alpha} = \frac{1}{12\text{ mins}}\left(\frac{60\text{ mins}}{1\text{ hr}}\right)\left(\frac{24\text{ hrs}}{1\text{ day}}\right)\left(\frac{7\text{ days}}{1\text{ wk}}\right) = 840\ \frac{\text{claims}}{\text{wk}}$$

In many situations, the arrivals occur at a constant average rate and are independent of each other, which means the arrivals occur according to a Poisson process. The Poisson probability distribution is a discrete distribution of the form

$$f(n) = \frac{(\lambda t)^n e^{-\lambda t}}{n!} \tag{11.1}$$

where

$\lambda =$ mean arrival rate
$n =$ number of arrivals
$t =$time period

The Poisson distribution gives the probability of n arrivals during time period t. If the arrival rate of a process is 3 jobs per hour, then during a 1-hour time period the probability that there will be zero arrivals is

$$f(0) = \frac{(3*1)^0 e^{-3*1}}{0!} = 0.0498 \approx 5\%$$

The probability that there are more than 2 arrivals during the 1-hour period is

$$1 - f(2) = \frac{(3*1)^2 e^{-3*1}}{2!} = 1 - 0.2240 = 77.6\%$$

When the arrival rate is described by a Poisson distribution, then the interarrival time is

[2]An interesting example of testing *iid* is the paper by Klaassen and Magnus [5] that tests whether points in tennis are *iid*. The authors use data from Wimbledon, and show that winning the previous point has a positive effect on winning the current point. Consequently, the *iid* assumption does not hold for tennis.

described by an exponential probability distribution with mean $\frac{1}{\alpha}$. The exponential distribution is a continuous probability distribution, and has a useful property called the memoryless property. The memoryless property means the next arrival, measured from any instant in time, is expected to occur in time t regardless of when the last arrival occurred. Additionally, the standard deviation of the exponential distribution is equal to the mean, which means the coefficient of variation (defined as the variance divided by the mean) is 1. The properties of independence, memoryless, and the standard deviation equaling the mean make the calculations in queueing theory much easier. For this reason, it is often assumed that both the interarrival time and the service time are exponentially distributed. In effect, the exponential distribution represents complete randomness in the arrival rate. Interestingly, many times the interarrival time can be accurately modeled as being exponential.

Example

It is known that the interarrival time is 2 minutes and it is independent and exponentially distributed. The last arrival to the system occurred at 10:30 AM. It is now 10:34 AM. What is the expected time until the next arrival?

Given an independent and exponential distribution then we have the memoryless property; consequently, the next arrival does not depend on the last arrival. The expected waiting time to the next arrival is 2 minutes, which means we should expect the next arrival at the current time (10:34 AM) + 2 minutes or 10:36 AM. The term "expected" is used interchangeably for the mean value; in other words, the next arrival is equally likely to be less than 2 minutes or greater than 2 minutes.

In the example above, it is more likely you would want to know the probability that the time until the next arrival will be below or equal to a certain value. To calculate the probability that the time to the next arrival x will be less than or equal to time t the equation is:

$$P(x \leq t) = 1 - e^{-\lambda t} \tag{11.2}$$

Continuing the example above, given the interarrival rate is exponential with mean of 2 minutes. What is the probability the next arrival will be in: (i) 1 minutes? (ii) 2 minutes? (iii) 10 minutes?

Using Equation 11.2 we calculate:

$$P(x \leq t) = 1 - e^{-\lambda t} = 1 - e^{-\frac{1}{2}(2)} = 0.393$$

$$P(x \leq t) = 1 - e^{-\lambda t} = 1 - e^{-\frac{1}{2}(3)} = 0.632$$

$$P(x \leq t) = 1 - e^{-\lambda t} = 1 - e^{-\frac{1}{2}(10)} = 0.993$$

The analysis indicates that at 10 minutes we are almost, surely guaranteed to have an arrival.

11.1.3 The Queue

The queue is where work is held temporarily until the server is available to process it. A queue may be finite or infinite. Finite queues have a maximum capacity, for example a waiting room that can only hold 20 people. An infinite queue has no capacity constraint. Queues are managed by what is called the queueing discipline, which is the business policy used by the server to remove work from the queue. A simple, yet common queue discipline is First-In-First-Out (FIFO). Other queue disciplines are random order, last-come-first-served, use of priorities, process sharing, and others. Some systems will have more complex queueing disciplines, for example emergency departments employ triage, in which the severity of the patients' problem determines their position in the queue. In general, the analysis becomes very complicated if the queue discipline is other than FIFO. Other issues involving the queue is whether customers renege or balk from the queue. Reneging is when a person joins a queue only to leave before ever reaching service. Balking is when a person observes a queue and decides not to join the queue. The basic queueing equations assume no balking and no reneging, although both are observed human behavior, especially when the queues are long.

Common Queueing Terminology

Most queueing equations use the service rate, measured as jobs/minute or customers/minute. Yet, it is common to measure the service time. For this reason, the service time will often be written as the inverse of the service rate, i.e., $\frac{1}{\mu}$. For example, the service rate is 8 customers per hour. What is the average time spent serving each individual customer? The service time is $\frac{1}{8} = 0.125$ hours or 7.5 minutes.

11.1.4 The Server

The server provides the service to the work items or job. A process will have one or more servers, and each server is characterized by the service time, defined as the amount of time needed to process a single item of work. Service time is assumed to be *iid*. Essentially, this means the service time for each work unit is independent of every other work unit, but the distribution is the same for all work units. The server is characterized by the parameters

m = number of servers

t_s= average service time

μ = average service rate, which is the inverse of the service time, $\mu = \frac{1}{t_s}$.

For arrival rates it is common and valid to assume a Poisson arrival process. However, for service rates it is not uncommon to encounter wider deviations from the exponential distribution, where the $CV = 1$. Service rates are under greater control of the enterprise, which would naturally strive to remove randomness. Buzacott [1] published data, shown in Table 11.1, on the service time variability. Note that blue collar, or manufacturing work, is characterized by low variability while life insurance sales, or a white collar job, is characterized by high variability. Manufacturing work has been extensively studied for a long time, and companies have managed to standardize the work, train workers, and design the jobs so that there is little variability. This smooths out the production workflow and leads to improvements in throughput and productivity as we have seen. Service-oriented work exhibits greater variability because the service needs of customers vary more, and the performance of the work is more dependent on the individual doing the work. Also, a life insurance salesperson likely works on a wide variety of different types of life insurance, which adds to the variability.

TABLE 11.1
Service Time Coefficient of Variation for Various Types of Industries [1]

Blue collar	Crafts	Professional	Life insurance sales
0.20	0.32	0.50	1.2

11.1.5 Queueing System Performance

The entire queueing system is characterized by the following performance parameters

N_q= average number of jobs in queue

W = average time in queue

N = average number of jobs in process

The traffic intensity, ρ, also called the server utilization, is the ratio of the arrival rate to the effective service rate such that

$$\rho = \frac{\lambda}{m\mu} \tag{11.3}$$

The server utilization ranges from 0 to ∞; however, unless the server utilization is less than 1, then the queue in front of the server would grow to infinity under steady-state conditions. This is called the stability condition (i.e., $\rho < 1$). The server utilization is the main parameter affecting waiting time. As server utilization increases, the waiting time increases exponentially.

Example

A physician can service a patient in 30 minutes, the clinic has an arrival rate of 6 patients per hour, and there are four physicians. What is the average utilization of the physicians?

$$\rho = \frac{(6)}{\frac{1}{(30)}\left(\frac{60}{1}\right)4} = 0.75$$

11.1.6 Queueing Model Assumptions

Queueing models make two essential assumptions. First, we assume the queueing system is stable, which means the arrival rate is less than the service rate, shown mathematically as $\lambda < \mu$ (or as shown above $\rho < 1$). Under this condition, the departure rate from the queueing system is equal to the arrival rate, which is equal to the throughput rate. A useful analogy is to think of water flowing into a container. If the water flows into the container at a rate of 10 gallons/minute (the arrival rate), and the container has a hole that allows water to exit at a rate of 15 gallons/minute (the service rate), then the average rate of water leaving the container will be 10 gallons/minute. You must remember that we are considering an inflow that varies over time with an average rate of 10 gal/min. When the inflow is temporarily greater than the outflow then water will accumulate in the container, and when the inflow rate is less than the outflow rate water will leave the container. If the container is empty, then the outflow rate cannot be greater than the inflow rate. Over time, it can be seen that the time average outflow rate will be equal to the inflow rate.

11.1.7 Queueing System Notation

The field of queueing theory has developed a taxonomy to describe systems based on their arrival process, service process, and number of servers written as Arrival/Service/Number Servers. The basic notation, widely used in queueing theory, is composed of three symbols separate by forward slashes. The symbols are M for Poisson or exponential distributions, D for deterministic (constant) distributions, E for Erlang distributions, G for general distributions (any arbitrary distribution) and GI for general independent in the case of arrival rates. To illustrate we could write:

- M/M/1 for Poisson arrivals, Poisson service rates, and a single server.
- GI/G/1 for general independent arrivals, general service rates, and a single server.
- M/G/m for Poisson arrivals, general service rates, and m servers.

11.2 Little's Law

Little's Law says the average number of jobs in a system is equal to the average rate jobs go through the system multiplied by the average waiting time in the system. Let N denote the average number of jobs and W denote the average waiting time in the system. The parameter λ not only denotes the arrival rate, but also it is equal to the throughput rate since in a stable system in steady state, over the long-term, the flow in must equal the flow out (assuming there is no loss in the system and nothing is added). Little's Law[3] is

$$N = \lambda W \tag{11.4}$$

What makes Little's Law so powerful is that it holds regardless of the distribution assumed for the arrival rate or the queue discipline used. Little's law relates the long-term, average values of three important system parameters. The important implication of Little's Law is that if you know any two of the three parameters, then the third parameter is restricted by the equation. This limits your design freedom when specifying system performance.

You can apply Little's Law to a single server or to the entire system, it is just a matter of defining the boundaries. Consequently, Little's Law may be stated in many different forms. In Figure 11.4, we can draw the boundaries around only the queue and derive the equation $N_q = \lambda W_q$. If we draw the boundaries around the entire system consisting of the server and the queue, then we can derive the equation $N_s = \lambda W_s$. In both equations, we use λ without a subscript because the flow into the queue equals the flow out of the queue, which is the flow into the server.

[3]In manufacturing, it is usually written in manufacturing-specific terms as $WIP = TH * CT$ where WIP is work in process, TH is throughput rate, and CT is cycle time. The only difference is a matter of terminology.

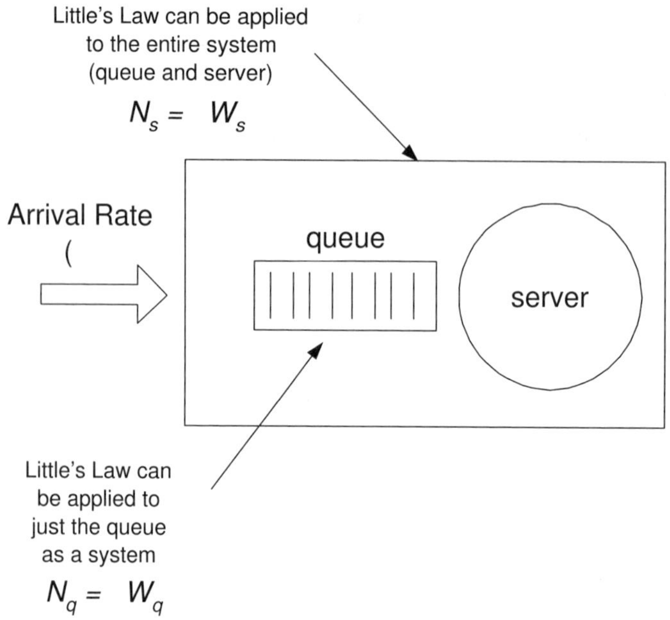

FIGURE 11.4
Little's law.

> **Example**
>
> Suppose 10 patients arrive every hour to an emergency department. Data gathering determines that on average 8 patients are either waiting or being served by the emergency department. What is the average time each patient spends in the emergency department?
>
> The arrival rate is 10 patients/hour and the number in the system is 8 patients. The waiting time is $W = \frac{N}{\lambda} = \frac{8}{10} = 0.8$ hours.

11.3 Queueing System Performance

There are three performance measures of queueing systems that are calculated: number in queue, waiting time in queue, and waiting time in the system. If you know the waiting time in the queue (or the waiting time in the system) and the service time, then you can calculate the other. They are related by the equation

$$W_s = W_q + t_s \tag{11.5}$$

11.3.1 M/M/1 Queue

In an M/M/1 queueing system, both the interarrival times and service times are exponentially distributed, which means the standard deviation equals the mean. For a M/M/1 queue, the estimated average number of customers in the system is given by the queueing equation

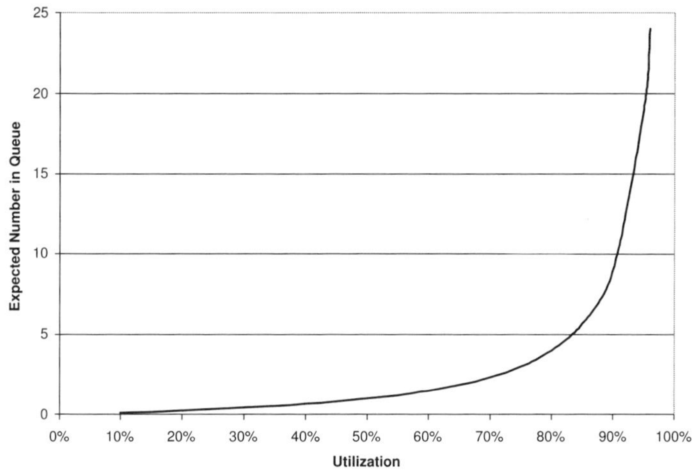

FIGURE 11.5
Graph of queue length for M/M/1.

$$N_s = \frac{\lambda}{\mu - \lambda} \tag{11.6}$$

Frequently, using the relationship $\rho = \frac{\lambda}{m\mu}$ where m is the number of servers (in this case $m = 1$), Equation 11.6 is rewritten as

$$N_s = \frac{\rho}{(1 - \rho)} \tag{11.7}$$

If the server utilization is 1, then the denominator of Equation 11.7 is 0, and division of any number by zero leads to infinity. At this point, the queue would keep growing until infinity. Figure 11.5 shows that the queue length increases exponentially to infinity as utilization increases. The trend is especially observable at $\rho = 0.7$ and above. The implication is that systems cannot operate in the steady state at server utilizations of 1 or greater. While $\rho = 1$ is not possible in stochastic systems, if the arrivals and service was deterministic, then it is theoretically possible to have server utilization equal to 1. This is an important concept to understand; in the stochastic system, the variability of the process creates waiting whereas in a deterministic system it would be possible to synchronize arrivals with the server to have no waiting.

The expected waiting time in the system (queue plus server) is calculated by Little's Law as

$$W_s = \frac{N_s}{\lambda} = \frac{1}{(\mu - \lambda)} \tag{11.8}$$

The total time in the system is the waiting time in the queue and the time to be served. Consequently, we have the relationship

$$W_s = W_q + t_s \tag{11.9}$$

To find the average waiting time spent in only the queue then subtract the service time, where $t_s = \frac{1}{\mu}$, from W_s. The equation for waiting time in the queue is then

$$W_q = W_s - \frac{1}{\mu} = \frac{\lambda}{\mu\left(\mu - \lambda\right)} \tag{11.10}$$

Given knowledge of the average number of customers in the system, it is possible to determine the average number of customers in the queue by subtracting $\frac{\lambda}{\mu}$ from the average number in the system. Consequently, the average number in the queue is

$$N_q = N_s - \frac{\lambda}{\mu} \tag{11.11}$$

and we obtain

$$N_q = \frac{\lambda^2}{\mu\left(\mu - \lambda\right)} \tag{11.12}$$

11.3.2 M/M/m Queue

The M/M/m queue has the same exponential interarrival times and exponential service times as the M/M/1 queue except it has m parallel servers with a single queue. All customers enter the queue and wait for the next available server. The steady-state properties of the M/M/m queue can be calculated exactly, but the ensuing equations are very cumbersome to use. Instead, we use an approximation [11]. The average time spent in the system (queue plus server) is estimated as

$$W_s = \frac{\rho^{\left(\sqrt{2(m+1)}-1\right)}}{\mu m\left(1 - \rho\right)} + \frac{1}{\mu} \tag{11.13}$$

As was done for the M/M/1 queue, the other performance parameters of W_q, N_q, and N_s can be determined by using Little's Law (Equation 11.4) and the relationship between the service time and the queue time (Equation 11.9).

Example

Students are starting to complain about the slow response of the Web-based course registration system. The IT manager is considering to add a second application server in parallel to shorten the waiting time. Student requests arrive at a rate of 300 per hour and wait in a single queue to be processed by the next available application server. The application servers process requests at a rate of 350 per hour. If we assume a M/M/n system, then calculate the average time in system.

In the current as-is system with a single application server, we have a M/M/1 queue and the average time in the system is 1.2 minutes. By adding a second application server in parallel, then the utilization $\rho = 300/(350 * 2) = 0.43$ and the time in system is approximately

$$W_s = \frac{0.43^{\left(\sqrt{2(2+1)}-1\right)}}{350(2)(1 - 0.43)} + \frac{1}{350} = 0.22 \,\text{minutes}$$

So an additional server decreases the time in system from 1.2 minutes to 0.22 minutes for about an 80% reduction.

The above example demonstrates a disproportionate improvement in service by adding an additional server. This insight is the basis for pooling servers discussed in the next chapter.

11.3.3 GI/G/1 Queue

If the interarrival time is described by a general distribution and the service time is also described by a general distribution, then there is no exact analytical equation for the performance measures. To calculate the performance of a GI/G/1 queue, we use approximations based on the first two moments of the interarrival time and the service time. The first moment is the expected value (mean), and the second moment is the squared coefficient of variation (SCV). Let c_a^2 denote the SCV of the interarrival times and c_s^2 denote the SCV of the service times. To calculate c_a^2 the equation is

$$c_a^2 = \frac{\sigma_a^2}{t_a^2}$$

where t_a is the expected interarrival time and σ_a is the standard deviation of the interarrival times. Likewise, to calculate c_s^2 the equation is

$$c_s^2 = \frac{\sigma_s^2}{t_s^2}$$

The approximated expected waiting time in the queue is given by [4][4]

$$W_q = \left(\frac{c_a^2 + c_s^2}{2}\right)\left(\frac{\rho}{1-\rho}\right) t_s \tag{11.14}$$

In Equation 11.14, the first term is the stochastic variation, the second term is the utilization, and the last term is the expected service time. Notice that the approximation is exact for a M/M/1 queue because in a M/M/1 queue, the coefficients of variation each equal 1. Consequently, the equation could be rewritten as

$$W_q = \left(\frac{c_a^2 + c_s^2}{2}\right) W_q\,(M/M/1) \tag{11.15}$$

The approximations are found to be good for most scenarios. The approximations are good when the utilization is between 0.1 and 0.95 and when the coefficients of variation are not too much larger than 1 [3]. If accuracy greater than 10% is desired, then other analysis means such as simulation should be used. To obtain the total waiting time in the queue plus the server you add the service time to Equation 11.14.

11.3.4 GI/G/m Queue

The approximation for the GI/G/m queue follows from the observation made to develop Equation 11.15 except now we use W_q(M/M/m) instead of W_q(M/M/1). The approximation for waiting time in the system is

$$W_q = \left(\frac{c_a^2 + c_s^2}{2}\right) W_q(\mathrm{M/M/m}) \tag{11.16}$$

11.3.5 Fork/Join Queue

A fork/join queue with synchronization (shown in Figure 11.6) is when a job arrives to a fork station where it splits and is then processed in parallel by two or more sibling servers. As the sub-jobs are completed by the siblings, they wait at the join node until all the other

[4]In the queueing network analyzer, we use a version of this equation improved by Whitt [12].

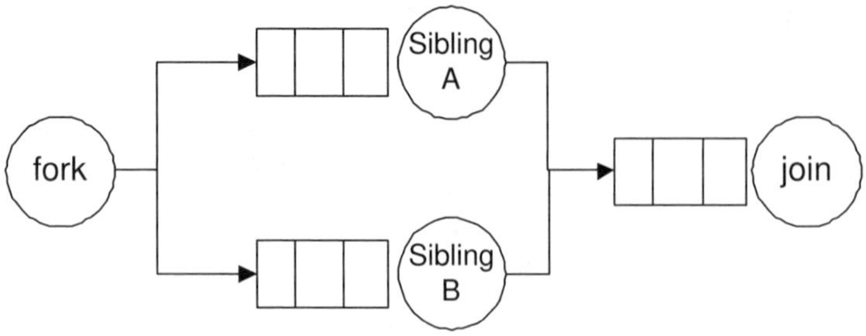

FIGURE 11.6
Fork/join queue.

siblings are done. Then they are joined together at the join node. Besides this definition of a fork/join queue, there are many other variations on the definition. For example, in some definitions of fork/join queues there is a queue in front of the fork and the siblings have no queues. Instead, jobs wait at the fork until all siblings are available. Other definitions change the types of distributions assumed and whether there is synchronization or not. Except for the simplest fork/join scenarios, there are no exact solutions, so most research has been on developing approximations or bounds on the performance of the fork/join queue.

Why are approximations necessary? It seems that the sojourn time of the fork/join queue should be the maximum time of the sibling queues, t_a and t_b, respectively. In many business process redesign projects, this is the conclusion reached by the engineering team. The problem with this analysis is it ignores the stochastic nature of the process. The simplistic, deterministic approximation (which is what $\mathrm{Max}\,(t_a,\, t_b)$represents) will provide a lower bound on the stochastic fork/join process. Unfortunately, it is not a consistently good lower bound because it can be much lower than the actual sojourn time.

For the simple case of two homogeneous M/M/1 siblings, Flatto and Hahn [2] provide an exact solution of

$$W_s(2) = \left(\frac{12-\rho}{8}\right)\left(\frac{1}{\mu-\lambda}\right) \tag{11.17}$$

The second term is the waiting time in the system for a M/M/1 queue. The equation says that the fork/join queue will have a time in system greater than that for a M/M/1 queue by a factor of $\frac{(12-\rho)}{8}$. Figure 11.7 shows the comparison between the two queues in sequence, the fork/join queue, and a M/M/1 queue. The time in system for the tandem (two queues in sequence) is 60 minutes; by parallelizing the servers as shown in the fork/join queue the time drops to 42.19 minutes. Because many people estimate the fork/join as equivalent to a single M/M/1 queue it is shown for comparison, and has a time of 30 minutes. This shows that parallelization does reduce the flow time, but it will not reduce it by half because of the stochastic behavior of the two parallel servers.

In Figure 11.8, we show the performance of the fork/join with two parallel M/M/1 queues compared to the lower bound set by the M/M/1 queue.

When all the sibling queues are homogeneous M/M/1 queues, it has been observed that the bounds on the cycle time grow at the same rate as the harmonic sum $H(m)$, where $H(m) = \sum_{k=1}^{m} \frac{1}{k}$ and m is the number of siblings [8]. Using this idea, Narahari and Sundarrajan [7] reported good results with an approximation for m siblings as

$$W_s = \left[\frac{H(m)}{H(2)} + \frac{4}{11}\left(1 - \frac{H(m)\rho}{H(2)}\right)\right] W_s(2) \tag{11.18}$$

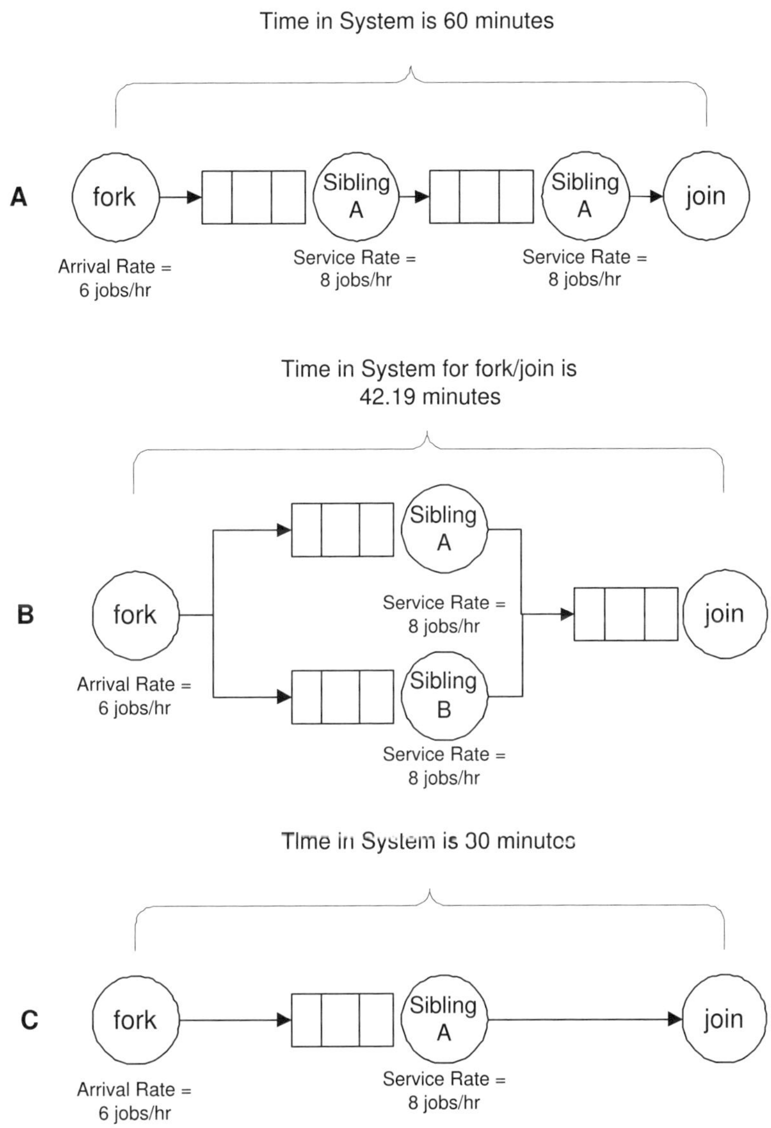

FIGURE 11.7
Fork/Join queue compared to a M/M/1 queue.

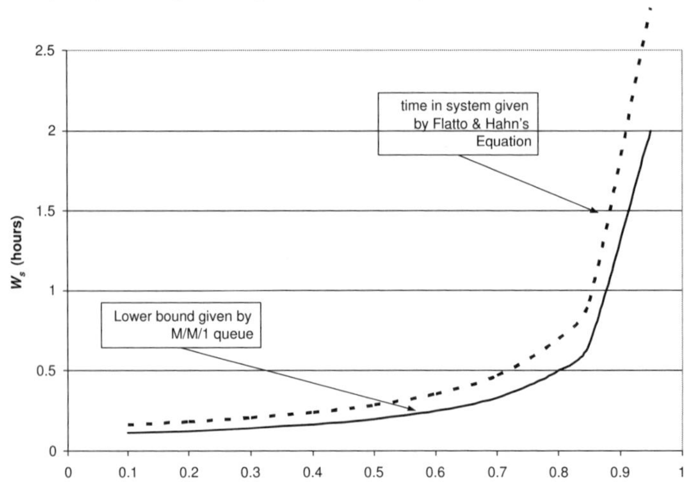

FIGURE 11.8
Max(t_a, t_b) as lower bound for fork/join.

Figure 11.9 used Equation 11.18 to show that as the number of siblings increases the expected waiting time in the fork/join system increases.

The equations for the homogeneous fork/join queues provides some insight into the performance of fork/join queues in general. When all the siblings have the same service rate distribution (in this case exponential) with the same mean, then the maximum of those siblings will be greater than their mean. There are other approximations if the siblings are normally distributed or for any arbitrary distribution (i.e., general).

What happens if the siblings are heterogeneous, in other words their service time means are not the same? Then depending on how far apart their means are, the maximum service time sibling will dictate the overall performance. Because there is no single approximation that fits the many different types of scenarios that may be encountered, instead of using closed-form approximations, we turn to Monte Carlo simulation. A Monte Carlo simulation takes a large number of random samples from the random variables and computes the results. In our example, the random variables are the interarrival times and the service rates. To generate the random variables we can use the random number generator in a spreadsheet and then perform the caculations for total waiting time in the fork/join queue.

To illustrate, we present a fictional example. In the example, the customer is a patient who arrives to a healthcare clinic for diagnostic tests. The diagnostic tests of laboratory and x-ray are conducted in parallel. In this example, the customer does not split, but the laboratory is done on the person's blood that was taken earlier while the patient is simultaneously being x-rayed. Once both tests are completed, the patient can be seen by the doctor in the next activity.

To model the fork/join queue in a spreadsheet using Monte Carlo simulation, we need to know the arrival rate of patients to the fork queue and the service rates of the sibling servers (x-ray and laboratory). In the spreadsheet there are various functions to model distributions. If we assume the interarrival time and service times are exponentially distributed, then if the arrival rate is put in a cell labeled `ArrRate`, it can be simulated in Excel as

```
=-1*(1/ArrRate)*LN(RAND())*60
```

and if the service rate is put in a cell labeled `SerRate`, it can be simulated in Excel as

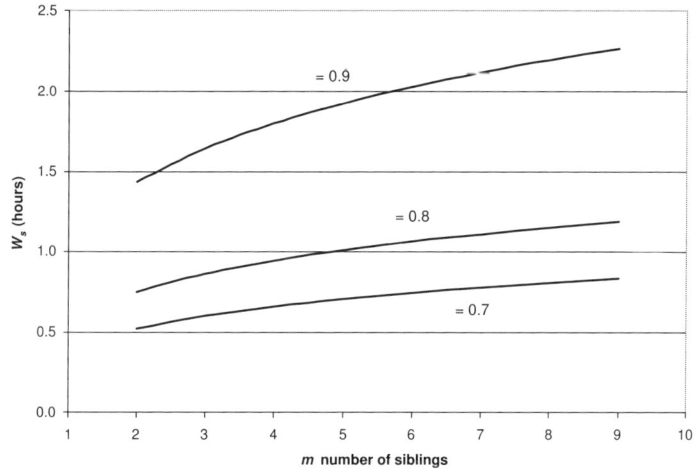

FIGURE 11.9
Fork/join *with m* homogeneous siblings.

patient number	Arrival to fork	diagnostic start	service	departure	x-ray start	service	departure	join arrival	pt total time	wait time
1	0.0	0.0	5.58	5.58	0.0	39.53	39.53	39.53	39.53	0.00
2	10.6	10.57	19.66	30.23	39.5	8.56	48.10	48.10	37.53	17.87
3	14.3	30.23	0.68	30.91	48.1	17.69	65.78	65.78	51.52	33.83
4	25.6	30.91	0.32	31.23	65.8	18.25	84.03	84.03	58.40	40.16
5	27.1	31.23	10.42	41.64	84.0	6.79	90.82	90.82	63.74	53.32
6	36.5	41.64	5.89	47.54	90.8	2.76	93.58	93.58	57.05	51.16
7	38.9	47.54	4.53	52.07	93.6	7.94	101.53	101.53	62.67	54.72
8	49.8	52.07	10.23	62.29	101.5	4.83	106.36	106.36	56.54	46.31
9	55.8	62.29	2.78	65.07	106.4	2.79	109.15	109.15	53.34	50.54
10	56.2	65.07	0.10	65.17	109.2	17.35	126.50	126.50	70.29	52.94

FIGURE 11.10
Spreadsheet for Monte Carlo simulation.

```
=-60/SerRate*LN(RAND())
```

In these spreadsheet formulas, the function `RND()` generates pseudo-random numbers between 0 and 1.

The spreadsheet is set up as shown in Figures 11.10 and 11.11. In the first row of the spreadsheet, the arrival time, the service times, and the departure times are created for the first patient using the above formulas. The departure time is the arrival time plus the service time. The arrival to the join is the maximum of the departure times from diagnostics or x-ray. The patient's total time is the duration from arrival to the fork to arrival to the join. The time the patient waits is the total time minus the maximum service time of the siblings. The next patient arrives to the fork the time of the previous patient plus the randomly generated interarrival time for patient #2. The start time for patient #2 at the diagnostic node is the maximum of the departure of the previous patient from diagnostics or the arrival to the fork of patient #2. Likewise, the same calculation of the start time at x-ray is performed. Afterwards, each patient is coded the same as patient #2. A Monte Carlo simulation must do many experiments, maybe 1000 separate rows, in order to obtain valid results.

B	C	D	E	F
patient number	Arrival to fork	diagnostic start	service	departure
1	=0	=C9	=-1*(1/DSerRate)*LN(RAND())*60	=D9+E9
=B9+1	=C9+-1*(1/ArrRate)*LN(RAND())*60	=MAX(F9,C10)	=-1*(1/DSerRate)*LN(RAND())*60	=D10+E10
=B10+1	=C10+-1*(1/ArrRate)*LN(RAND())*60	=MAX(F10,C11)	=-1*(1/DSerRate)*LN(RAND())*60	=D11+E11
=B11+1	=C11+-1*(1/ArrRate)*LN(RAND())*60	=MAX(F11,C12)	=-1*(1/DSerRate)*LN(RAND())*60	=D12+E12
=B12+1	=C12+-1*(1/ArrRate)*LN(RAND())*60	=MAX(F12,C13)	=-1*(1/DSerRate)*LN(RAND())*60	=D13+E13
=B13+1	=C13+-1*(1/ArrRate)*LN(RAND())*60	=MAX(F13,C14)	=-1*(1/DSerRate)*LN(RAND())*60	=D14+E14
=B14+1	=C14+-1*(1/ArrRate)*LN(RAND())*60	=MAX(F14,C15)	=-1*(1/DSerRate)*LN(RAND())*60	=D15+E15
=B15+1	=C15+-1*(1/ArrRate)*LN(RAND())*60	=MAX(F15,C16)	=-1*(1/DSerRate)*LN(RAND())*60	=D16+E16
=B16+1	=C16+-1*(1/ArrRate)*LN(RAND())*60	=MAX(F16,C17)	=-1*(1/DSerRate)*LN(RAND())*60	=D17+E17
=B17+1	=C17+-1*(1/ArrRate)*LN(RAND())*60	=MAX(F17,C18)	=-1*(1/DSerRate)*LN(RAND())*60	=D18+E18

Next column continues below

continues from column above

G	H	I	J	K	L
start	service	departure	arrival	time	wait time
=C9	=-60/XSerRate*LN(RAND())	=G9+H9	=MAX(F9,I9)	=J9-C9	=K9-MAX(E9,H9)
=MAX(C10,I9)	=-60/XSerRate*LN(RAND())	=G10+H10	=MAX(F10,I10)	=J10-C10	=K10-MAX(E10,H10)
=MAX(C11,I10)	=-60/XSerRate*LN(RAND())	=G11+H11	=MAX(F11,I11)	=J11-C11	=K11-MAX(E11,H11)
=MAX(C12,I11)	=-60/XSerRate*LN(RAND())	=G12+H12	=MAX(F12,I12)	=J12-C12	=K12-MAX(E12,H12)
=MAX(C13,I12)	=-60/XSerRate*LN(RAND())	=G13+H13	=MAX(F13,I13)	=J13-C13	=K13-MAX(E13,H13)
=MAX(C14,I13)	=-60/XSerRate*LN(RAND())	=G14+H14	=MAX(F14,I14)	=J14-C14	=K14-MAX(E14,H14)
=MAX(C15,I14)	=-60/XSerRate*LN(RAND())	=G15+H15	=MAX(F15,I15)	=J15-C15	=K15-MAX(E15,H15)
=MAX(C16,I15)	=-60/XSerRate*LN(RAND())	=G16+H16	=MAX(F16,I16)	=J16-C16	=K16-MAX(E16,H16)
=MAX(C17,I16)	=-60/XSerRate*LN(RAND())	=G17+H17	=MAX(F17,I17)	=J17-C17	=K17-MAX(E17,H17)
=MAX(C18,I17)	=-60/XSerRate*LN(RAND())	=G18+H18	=MAX(F18,I18)	=J18-C18	=K18-MAX(E18,H18)

FIGURE 11.11
Spreadsheet formulas for Monte Carlo simulation.

TABLE 11.2
Monte Carlo Results

Arrival Rate (pts/hr)	Service Rate (pts/hr)	Cycle Time (min)	Fork/Join Time (min)	Lower Bound (min)
6	8	8	10.8	7.5
6	8	10	7.8	7.5

In the spreadsheet, every time you press the **F9** key it recalculates all the formulas in the spreadsheet, which generates new random numbers. In this way, you can do many experiments to obtain multiple samples.

Using the spreadsheet with the assumptions described above we explore two scenarios. We first examine the case when both siblings have the same service time, and then we examine the case when one sibling is faster than the other sibling. Table 11.2 shows the input data and the Monte Carlo results. The lower bound is the deterministic case where the fork/join time is the max(sibling1, sibling2). When the siblings are equal, the simulated fork/join time is 10.8 minutes, which is 44% greater than the lower bound. However, if one sibling is improved by processing 2 more patients per hour, then the unbalanced fork/join time is 7.8 minute, which is only 4% greater than the lower bound. The reason this happens is because when both siblings have the same mean service time, then each sibling has a 1/2 probability of being greater than the mean service time. The probability of max(sibling1, sibling2) being greater than the mean is 75% when they are equal. Improving one sibling sufficiently such that the probability of it being greater than the other sibling means that only the slower sibling determines the fork/join time. This is why in the second case, the simulated value is close to the lower bound. The process design insight is that the overall process performance can be improved just by improving one of the siblings.

	SERVICE TIMES triage	register
expected (minutes)	6.00	8.00
std dev (minutes)	1.25	1.75

patient number	patient arrival time	triage start	triage end	registration start	registration end	taken to room	wait time
1	10:45	10:45	10:50	10:50	10:57	10:57	0:12
2	10:45	10:50	10:56	10:56	11:03	11:03	0:18
3	10:45	10:56	11:02	11:02	11:08	11:08	0:23
4	10:45	11:02	11:10	11:10	11:19	11:19	0:34
5	10:45	11:10	11:17	11:17	11:22	11:22	0:37
6	10:45	11:17	11:23	11:23	11:29	11:29	0:44
7	10:45	11:23	11:27	11:27	11:35	11:35	0:50
8	10:45	11:27	11:34	11:34	11:37	11:37	0:52
9	10:45	11:34	11:39	11:39	11:50	11:50	1:05
10	10:45	11:39	11:45	11:45	11:52	11:52	1:07
						mean	0:40
						max	1:07

FIGURE 11.12
Spreadsheet to do Monte Carlo analysis.

11.3.6 Transient Behavior of Queues

Most queueing equations deal with the expected value of stable, steady-state queues. However, it is often of interest to understand the transient performance of a queue. For example, many systems start/stop each day, consequently for at least part of the day, the system is not in steady-state. How do we analyze these system? There are some analytical results for these queues. However, a far easier and more versatile analysis method is by Monte Carlo simulation.

To illustrate the use of Monte Carlo to analyze the transient behavior of a queue, we will use the healthcare clinic example. The healthcare clinic starts each day at 11:00 AM, but it opens the doors at 10:45 AM to triage and register the first group of patients. The clinic has observed that a large number of patients are always waiting to enter at 10:45. So, in the spreadsheet model we assume there are ten patients arriving at 10:45. During this start-up period, the clinic is not in steady state and the analytical equations for queue performance cannot be used. To determine how much improvement can be expected from having two nurses do triage, called double triage, instead of one nurse, called single triage, a simulation model is used.

Figure 11.12 shows the spreadsheet to do the Monte Carlo analysis.We assume there are 10 patients who are waiting at 10:45 AM to enter the clinic. We know the expected service time and its standard deviation for triage and registration. The spreadsheet calculates the start and end time for each task, for each patient. The mean waiting time for all 10 patients and the maximum waiting time is calculated across all simulations (only the first simulation is shown, which explains why the mean and max waiting times do not match the values for the simulation shown).

When comparing the single triage and double triage policies we see a large reduction in the mean cycle-time and an even larger reduction in the maximum cycle-time.

Note, it is not possible to use queueing analysis with the equations presented earlier in the chapter for studying the triage policy decision. The queueing equations require input

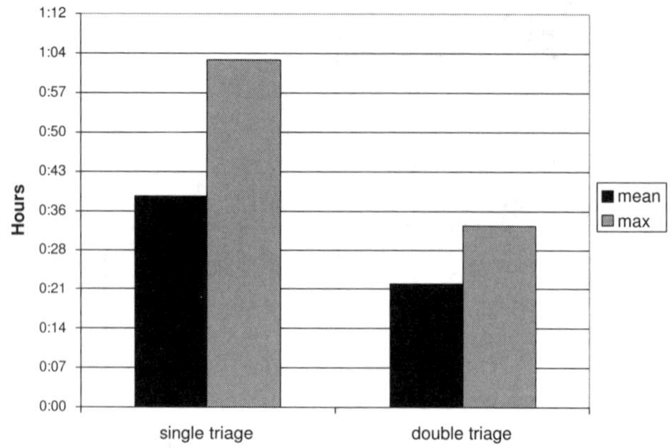

FIGURE 11.13
Comparison of triage policies.

parameters of arrival rate and service rate. There is no arrival rate for the triage start-up and there is no accommodation in the queueing equations for specifying a pre-existing queue. For this reason the Monte Carlo analysis presented here is an appropriate analysis technique. The alternative is to do a discrete-event process simulation, which should obtain similar results.

11.4 Queueing Networks

Until now, our discussion of queueing theory has been limited to systems with a single queue. The system may have had multiple servers in the queue, but it was only a single queue. Many enterprise systems are better described as a queueing network. Queueing networks describe a system with multiple queues in which the departure from one server becomes the arrival to another server in the same network. The dependencies created by the customer flow in the network creates complexity that is not easy to handle analytically. Jackson did show for a queueing network with certain properties a closed-form solution could be found. However, for general queueing networks no closed-form solutions can be found. Instead, most researchers have turned to various approximation techniques. Here we describe different classes of queueing networks and then we briefly summarize what is called the two-moment parameter decomposition approach that has been developed by Whitt and others.

For queueing networks we can draw a distinction between open and closed queueing networks. Open queueing networks have customer arrivals from outside of the system (coming from a conceptually infinite calling-population), and then they later leave the system. In closed queueing networks, the number of customers is constant, and no customer enters or leaves the system. Many manufacturing systems, such as just-in-time production systems, are modeled as closed queueing networks because the same number of parts is always in the system. Most service systems are better modeled as open queueing networks because the customers arrive to the system, are served, and then depart the system.

A second way to classify queueing networks is whether they serve a single customer

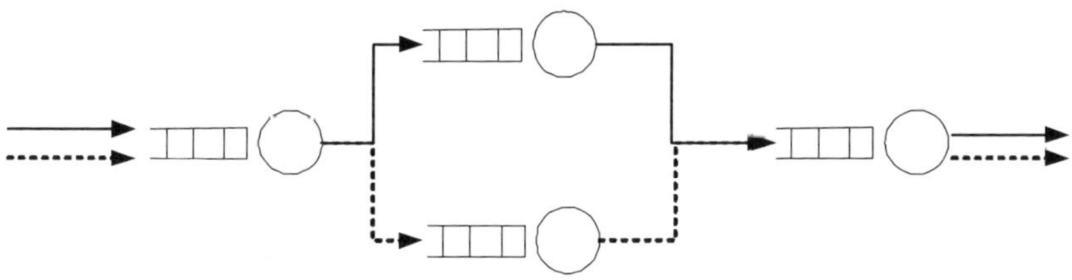

FIGURE 11.14
Queueing network.

class or multiple customer classes. In a single class queueing network, there is only one customer class such that all the customers share the same characteristics, importantly the same route and service times. The single class has a single external arrival rate to the system. In multiple customer class queueing networks, there are many different types of customers, with different external arrival rates, and different service rates at each node. Figure 11.14 shows an example multi-class open queueing network. Here there are two customer classes that arrive to the queueing network at the first node with different arrival rates. Each customer class follows a different route through the network. The customers have different service times at each node they visit, which depends on the customer class they come from. Finally, after being served the customers depart from the queueing network.

11.5 Psychology of Waiting

Nobody likes to wait for service. Queueing theory as explained above assumes customers are patient and wait in the queue regardless of the queue length. Actual customers do not behave this way. Customers might *renege*, which means they refuse to join the queue. Customers who *balk* join the queue, but then become impatient and leave the queue before receiving service. Understanding how customers react to queues can help in the design of the system. For example, Pazgal and Radas [9] conducted experiments that found that customers balked when the queue length was greater than some critical value. If the queue was less, then they joined the queue, otherwise they balked. The critical value depends on the customers' estimate of the waiting time, estimate of switching cost (defined as going to a competitor), and the value they place on both. Knowing this could help determine what maximum queue length to allow when designing the system.

Table 11.3 lists behavioral responses to waiting. Studies have shown that anxiety creates dissatisfaction. A major source of anxiety is uncertainty. Customers want to know how long the wait is expected to be. Amusement parks seem to have found good results by posting at the beginning of queues the expected waiting time.[5] Customers then know how long the wait is, and can decide whether they wish to join the queue or not. Customers also accept fair waits. So, first-come-first-serve and other similar queue disciplines are readily accepted. Another strategy to reduce the perceived length of the wait, and the negative feelings that come with it is to occupy the person while he or she are waiting. There has been much made about using mirrors next to elevators in tall buildings, having magazines in

[5]One engineer at Disney said they also manage queue length with the use of characters. If an attraction's queue is longer than desired, they send Mickey Mouse or some other character to that area. Many children with their families in tow will renege from the line to meet and greet the character, resulting in a shorter queue for those who remain in the line.

TABLE 11.3
Human Reaction To Waiting [6]

Human Reaction	How To Mitigate
Occupied time feels shorter than unoccupied time	Mirrors next to elevators; music, magazines or TVs in waiting room; entertainers at Disney.
People want to get started	Have patients fill out forms while in waiting room; provide menus, drinks at restaurants.
Anxiety makes waits seem longer	Keep people informed about wait, system conditions; let people stay in charge.
Unfair waits are longer than equitable waits	Generally, first-in first-out (FIFO) is accepted as fair; have a single line instead of multiple lines; economy passengers accept that first-class passengers who pay more receive faster service; emergency room patients understand more serious cases get treated first.
Unexplained waits are longer than explained waits	Airline pilots tell passengers reason for wait; customers can see that restaurant is full.
The more valuable the service, the longer the customer will wait	Express checkout counters for supermarket customers with 10 or fewer items; express check-in for passengers without luggage.
Solo waits feel longer than group waits	Have a group waiting room; promote sense of group waiting.
Uncertain waits are longer than known, finite waits	Disney posts estimate waiting times at entrance to attractions; telephone call lines provide estimates of waiting time.

doctor's offices, TVs in airport lounges, and other strategies to entertain or at least occupy a person's attention while he or she waits. Other strategies to mitigate the effects of waiting is to split the wait into stages, called in-process waits. For example, a dentist will have you wait a short period in the waiting room, then you are taken to the examination room, where you wait some more. This is the in-process wait. After a while the hygienist will see you and set up. Then a further wait before the dentist sees you. The process is designed such that the wait is broken up into several smaller waits separated by processes. The overall psychological effect is the client perceives better service than if they had a single long wait prior to any service.

How to mitigate the negative effects of waiting is mostly an operational decision. However, the enterprise engineer needs to be aware of the need to design system structure, processes, and policies so that waiting is minimized and so the operational policies to mitigate waiting can be implemented. This might include proper facility design to allow for in-process waits, designing the process to allow for breaking up the work into tasks that can be separated, designing systems to specify expected waiting times, and so forth.

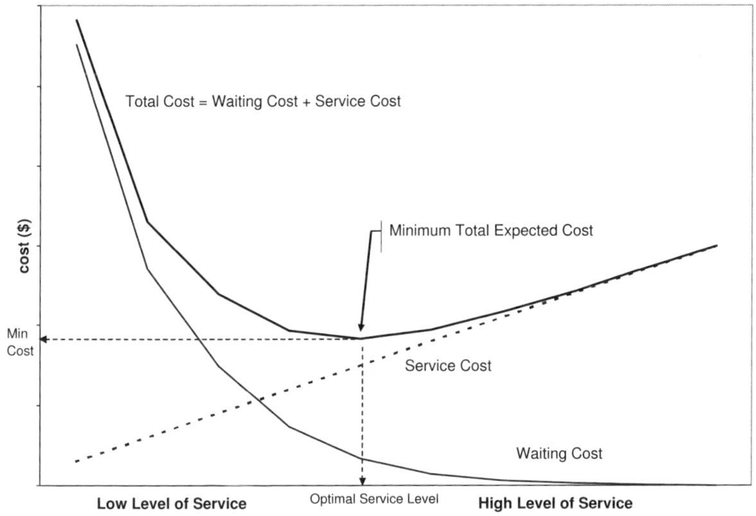

FIGURE 11.15
Queueing cost.

11.6 Queueing Costs

In the design of queueing systems, there are two dominant costs that must be traded off: the cost of service and the cost of customer waiting time. The cost of capacity is the cost of the servers, whether human or machine. If waiting time is high, then by adding servers the waiting time can be decreased. The cost of customer waiting time is an opportunity cost. If the waiting lines are too long, customers will be unsatisfied, may not come again, might balk, or renege from the queue. For this reason, enterprises wish to keep the waiting time as low as possible. Of the two costs, the capacity cost is the easier one to measure. Enterprises know how much each server cost to employ. The waiting costs are much more difficult to measure.

Figure 11.15 shows the service cost, waiting cost, and the total expected cost. The service cost is linear with respect to the level of service provided. For example, a grocery store can add checkout clerks, and each additional clerk costs a constant amount per hour. The waiting time decreases exponentially as the service level increases. The total queueing system cost is the summation of these two costs. Figure 11.15 shows the minimum point to the total cost curve, which would be the optimal system design to minimize total cost.

11.7 What Queueing Theory Tells Us about Business Processes

In this section, we discuss how queueing theory provides a foundation on which to examine and understand business processes.

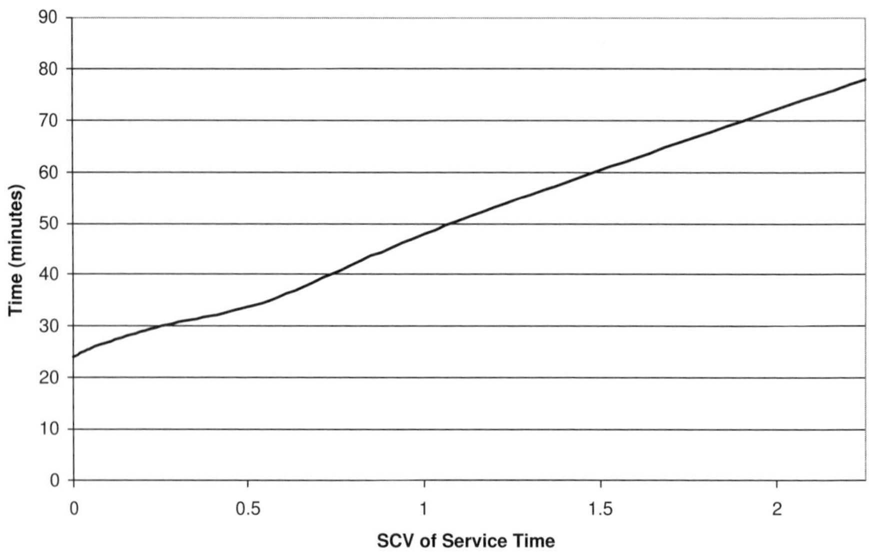

FIGURE 11.16
Effect of SCV on waiting time.

11.7.1 Service Time Variation

The larger the service time variation then the longer the waiting time, the greater the average number of customers waiting, and the lower the throughput rate. To illustrate, consider a simple process that has a single physician who examines patients. The average exam time is 12 minutes. The mean arrival rate is 4 patients per hour, and we assume the arrivals are exponential. This process has a utilization of $4/5 = 0.8$. The system can be analyzed as a M/G/1 queue, and the results of varying only the service-time variation are shown in Figure 11.16. The results demonstrate that as service-time variation increases, the average number of patients waiting increases and the average waiting time increases. This relationship holds in general. Given a process, as the service-time variation increases the average waiting-time increases. Since cycle time is the service time plus the waiting time, then the cycle time also increases. Also, the throughput rate decreases. What this analysis informs us is that as service time variation increases, then the performance of the system will get worse (cycle time increases and throughput decreases).

11.7.2 Capacity Utilization

Queueing theory demonstrates irrevocably that systems cannot operate over the long-term at 100% capacity. The stability condition states that $\rho < 1$ otherwise the queue length grows to infinity. Figures 11.5 and 11.8 show that as the traffic intensity approaches 1 the waiting time grows exponentially. Most enterprises would design systems to operate in the range of $0.7 \leq \rho \leq 0.9$ to provide good service while maintaining high capacity utilization. An article in the *Wall Street Journal* illustrates this concept, which is counter-intuitive to many managers who naturally desire to maximize utilization. The article describes how Avery Dennison Corporation, an adhesive-label maker, has been aggressively innovating new products to expand business. The Avery executives became upset at the time it took to bring these innovations to market – the product development schedules were slipping. A consultant group studied their processes and came to the counter-intuitive conclusion that Avery was trying to develop too many new ideas simultaneously. Their processes were being forced to operate at close to 100% capacity with the result, as can be predicted by queueing theory, that waiting time increased exponentially. The consultant's recommendation was to reduce the number of new products to create slack time so that they could more rapidly bring products to market.[6]

To Improve Process Performance

- Decrease variability in customer interarrival times and service times.
- Decrease capacity utilization by decreasing arrival rate, increasing service rate, or increasing the number of servers.
- Do activities in parallel.

What queueing theory tells us is that we cannot ignore variation. Optimal system designs and operating policies can be vastly different under the deterministic and stochastic models. In a deterministic model (which simplifies the actual performance of enterprise systems), resource utilization can be 100% without any deleterious effects. In fact, if you optimized a deterministic model to maximize throughput it would result in a design with 100% utilization (i.e., no slack in the resource). We know 100% utilization of a resource is a route to failure (as Avery Dennison found). The expected waiting time in queue approaches infinity in a stochastic model. Consequently, queueing theory shows that all systems should be designed with some slack to handle the natural variations in arrivals and service.

11.8 Summary

This chapter presented queueing theory, which provides an analytical basis to understand and design systems. Queueing theory can be used to model the performance of many systems as consisting of arrivals, queues, and servers. To use queueing theory to analyze enterprise system performance, we need to transform our process models into queueing models, collect data on the arrival and service rates of the process, and then determine the appropriate queueing model to apply. For some queues such as the M/M/1, the analytical results are exact. For others such as GI/G/1, the equations are approximations. If the squared coefficients of variation c_a^2 and c_s^2 are less than one, then it is possible to use the M/M/1 equations to provide a conservative estimate of performance. Whenever applying an analytical model, the modeling team needs to check the validity of the model assumptions against the actual system being modeled. If the assumptions are satisfied, then queueing theory can provide reasonably accurate results (<10% error) in a short period of time using standard software such as spreadsheets to either implement the queueing equations or to do Monte Carlo simulations.

[6] How innovation can be too much of a good thing: reducing the number can speed introduction of new product lines, Anders, G., *Wall Street Journal*, June 11, 2007.

Queueing theory highlights the importance of process variability on process performance. As variability in arrivals or service rates increases, then performance decreases. Little's law shows us the relationship between three important performance variables: time in system, throughput rate, and work-in-process. Because these variables are related, changing one will change the others.

In the following chapters, we use queueing theory to develop models to analyze and design business processes.

Review Questions

1. Describe the main assumptions made by queueing models for M/M/1 queues.
2. What causes variability in a process?
3. What does the term queue discipline mean, and what is a common queue discipline?
4. Why do many service-oriented processes have greater variability than manufacturing processes?
5. Explain what would happen in a queueing system if the steady-state arrival rate is greater than the service rate.
6. What scenarios are fork/join queues used to model in business processes?
7. Explain how a Monte Carlo simulation can be conducted.
8. Describe the difference between transient conditions and steady-state conditions.
9. What is the difference between a customer who balks and a customer who reneges?
10. Explain how increasing process variability would affect system performance.
11. What are the two major costs in a queueing study?

Exercises

1. The mean service rate for a database server is 8 transactions per minute. The calculated standard deviation is 3 transactions per minute. What is the coefficient of variation? What is the squared coefficient of variation? What is the mean service time in seconds?
2. The Casa Manufacturing Company makes partially pre-fabricated homes that are assembled on the construction site. Their manufacturing facility is manufacture-to-order. They receive, on average, about 25 orders per five-day work-week. To finish a single order takes on average 7 days. What is the average daily inventory you would expect in this facility?
3. In the Casa Manufacturing Company they have an autoclave to cure the concrete castings they make. This piece of equipment has been identified as a bottleneck. The industrial engineer has studied this work center and obtained the following data. On average, approximately 10 new parts per day arrive to the autoclave. The amount of work in progress in front of the autoclave is on average about 11

parts. What is the expected waiting time in the queue? (Hint: use Little's Law but now the system is defined as the queue only).

4. The arrival rate of customers to an ATM is 15 per hour. If it is observed to be a Poisson process, then what is the interarrival time in minutes? If a customer arrives at 9:30 AM, what is the probability that a customer arrives before 9:35 AM?

5. A review of Silvia's online jewlery Website log reveals an average load of 15 simultaneous transactions being processed. The average arrival rate is 20 transactions per minute. What is the average time in system? Does this include network latency?

6. In the Port of Miami a tugboat takes ships in and out their berths. The average number of requests for tugboat services is 1 per hour to either be taken to, or from a berth. The requests have been found to follow a Poisson process. The average time required to tow a ship in or out of its berth is 45 minutes and is exponentially distributed.

   (a) What is the expected waiting time of a ship waiting for a tow?
   (b) If a second tugboat is put into operation, then what is the waiting time for a tow?
   (c) It is estimated that a ship's waiting cost is $100 per hour. If a new tugboat costs $250,000, then would it pay for itself in the first year of operation?

7. An insurance company collects data for a month and determines the average number of insurance claims being processed on a given day is 1200. The average number of new claims that arrive each day is 20. What is the average flow time to process an insurance claim? If the company wants the average to be 30 days to process a claim, then what average work-in-process will they have, assuming the arrival rate does not change?

8. An IT department receives service calls from the other organization units to repair IT problems. Suppose the IT department has a single secretary who receives the service calls and generates a work order. Calls arrive in a Poisson process with a mean interarrival time of 20 minutes. Time studies indicate the mean service time is 12 minutes with an SCV of 1.5.

   (a) Use a GI/G/1 model to estimate the expected waiting time.
   (b) Use a M/M/1 model to estimate the expected waiting time.
   (c) Compare your answers and discuss the accuracy of making an exponential assumption in the M/M/1 model instead of using the actual SCV of 1.5 in the GI/G/1 model.

9. An engineering firm has a single plotter for color engineering drawings. Requests for plots arrive in a Poisson process at a rate of 20 per hour. Plotting takes an average of 80 seconds, and has been found to follow an exponential distribution. Calculate the following performance parameters:

   (a) The percentage time the plotter is used.
   (b) The average number of jobs waiting in the queue.
   (c) The average number of jobs in the system.
   (d) The average time a job spends waiting in the queue.
   (e) The average time a job spends in the system.

10. Miguel's Machine Shop retains a service crew to repair airplanes at Orlando International Airport. Repair requests occur with an average rate of 3 per 12-hour workday and are approximately Poisson in nature. The crew can service an average of 8 airplane repairs per workday. The service rate is approximately exponential.
    (a) What is the utilization rate of the repair crew?
    (b) What is the average down-time of a broken airplane?
    (c) On average, how many airplanes are waiting for repair?
    (d) If to save money, Miguel cuts his crew in half, and they can then service an average of 4 airplanes per workday, then what is the new utilization rate and average down-time of a broken airplane? Do you think his airline customers would tolerate this down-time?
11. A system architecture has two application servers in parallel to process customer orders and is served by a single queue. Orders arrive according to a Poisson process at an average rate of 100 per hour. The service rate for each application server is exponentially distributed with a mean of 80 per hour. What is the utilization of the system architecture? What is the average time in system for a customer order?
12. Pick orders arrive to a warehouse and are split into two sub-orders that are picked separately by the order-pickers. Assume the arrival of orders is described by a Poisson process with a mean rate of 20 per hour, and the order picking rate is exponentially distributed with a mean rate of 18 per hour.
    (a) Use Flatto-Hahn's equation to calculate the time in system.
    (b) How does this time compare to the situation where instead of the sub-orders being picked in parallel, they were picked in series?
13. The processing of a syringe is highly automated and the production rate is 100 per hour with a standard deviation of 10. The materials arrive to the station at a rate of 75 per hour with a standard deviation of 30. Assume a GI/G/1 queue and calculate:
    (a) The percentage time the production machine is used.
    (b) The average number of jobs waiting in the queue.
    (c) The average number of jobs in the system.
    (d) The average time a job spends waiting in the queue.
    (e) The average time a job spends in the system.

Bibliography

[1] J.A. Buzacott. Modeling teams and workgroups in manufacturing. *Annals of Operations Research*, 126:215–230, 2004.

[2] L. Flatto and S. Hahn. Two parallel queues created by arrivals with two demands. *SIAM Journal of Applied Mathematics*, 44:1041–1053, 1984.

[3] W.J. Hopp and M.L. Spearman. *Factory Physics*. Irwin McGraw-Hill, Boston, MA, 1996.

[4] J.F.C. Kingman. On queues in heavy traffic. *Journal of Royal Statistical Society*, Series B, 24:383–392, 1962.

[5] F.J.G.M. Klassen and J.R. Magnus. Are points in tennis independent and identically distributed? Evidence from a dynamic binary panel data model. *Journal of the American Statistical Association*, 96(454):500–509, 2001.

[6] D.H. Maister. *The Service Encounter*, chapter The psychology of waiting lines. Heath and Company, Lexington Books, 1985.

[7] Y. Narahari and P. Sundarrajan. Performability analysis of fork-join queueing systems. *Journal of the Operational Research Society*, 46:1237–1251, 1995.

[8] R. Nelson and A.N. Tantawi. Approximate analysis of fork/join synchronization in parallel queues. *IEEE Transactions on Computers*, 37(6):739–743, 1988.

[9] A.I. Pazgal and S. Radas. Comparison of customer balking and reneging behavior to queueing theory predictions: An experimental study. *Computers and Operations Research*, 35(8):2537–2548, 2008.

[10] B.T. Pentland. Sequential variety in work processes. *Organization Sciences*, 14(5):528–540, 2003.

[11] H. Sakasegawa. An approximation formula $l_q = \alpha\beta^\rho (1 - \rho)$. *Annals of the Institute of Statistical Mathematics*, 29:67–75, 1977.

[12] W. Whitt. Approximations for the GI/G/m queue. *Production and Operations Management*, 2(2):114–151, 1993.

12

Process Design

> *"The more important the subject and the closer it cuts to the bone of our hopes and needs, the more we are likely to err in establishing a framework for analysis."* – Stephen J. Gould (1941-2002), American evolutionary biologist.

This chapter describes how to analyze and design business processes. The design methodology is based on queueing theory. We emphasize analysis and improvement of core business processes because they deal with the external customers of the enterprise and they often have the most room for improvement. The goals of core business processes are a combination of satisfying customer needs for timeliness, cost, and quality. This chapter focuses on minimizing process cycle-time as a means to delivering timeliness. To keep the process costs low, the process needs to be efficient by having a high throughput-rate. The chapter shows how to quickly determine bounds on both these performance measures: a lower bound on cycle time and an upper bound on throughput rate. Quality refers to both the process and the product or service the process outputs. Quality has many meanings and interpretations, of primary importance to process design is customer-perceived quality and manufacturing quality. The chapter defines these terms and describes how to establish a quality control process to maintain and improve process quality. After completing this chapter, you should be able to:

- Describe the process analysis method.
- Compare and contrast the process analysis techniques.
- Describe four goals of a process.
- Determine the critical path in a process.
- Calculate the theoretical minimum cycle-time of a process.
- Calculate the theoretical maximum throughput-rate of a process.
- Estimate the cycle time of a process using queueing theory.
- Estimate the throughput rate of a process using queueing theory.
- Describe how to design quality into a process.

12.1 Process Analysis Method

The first step to design a process is to understand the activities, their relationships, and the relevant performance metrics. Process design generally involves the following tasks:

1. Define the process boundaries that mark the entry points of the process inputs and the exit points of the process outputs.
2. Construct a process model of the process activities, their inputs, their outputs, and their interrelationships.
3. Define the goals of the process; answer the question, "What makes a good process?".
4. Define measures for process performance based on the process goals.
5. Analyze the process with respect to the desired process performance.
6. Use the analysis to make design decisions to improve the process. Iterate with the previous step until a satisfactory process design emerges.

In this chapter we assume steps 1, 2, and 3 have been completed. We focus on the analysis and design steps. There are three general approaches to process analysis and design: an analytical approach, a computational approach, and a qualitative or knowledge-based approach.

- **Analytical approaches** – An analytical approach models the process using mathematics, and then by solving the mathematical system, you can obtain performance results or other process characteristics of interest. Queueing theory provides a framework to analyze business processes. As an analysis tool, it has the benefits that an analysis team can quickly build a model and obtain results. The drawbacks are that queueing models can be mathematically complex, are approximate for more complicated systems, and the results are only valid if the actual system being studied matches the underlying assumptions of the queueing model. The mathematical complexity is, for many, no longer a major issue since software is available to do the calculations and the analyst does not need to write the equations every time they do an analysis. However, the other limitations may be too restrictive such that a queueing analysis is not recommendable.

- **Simulation approaches** – A simulation approach exploits the ability of computers to perform thousands, if not millions, of computations per second to simulate the behavior of a process. Within computational approaches there are three main simulation models: continuous simulation, discrete-event simulation, and agent-based simulation. The benefits of simulation are that a system of almost any complexity could, in theory, be modeled accurately. A drawback is that simulation modeling often requires significant expertise and time to both develop a model and to analyze it. Also, to obtain greater accuracy of the results requires obtaining more accurate data than what might be called for in analytical approaches.

- **Knowledge-based approaches** – A knowledge-based approach uses rules or heuristics based on best practices to guide the analysis. There are many heuristics that can be applied; it is through the knowledgeable application of these rules that you can improve a process.

Analysts often debate whether to use analytical models or simulation models. The conclusion you reach depends on a trade-off of several factor, primarily speed of analysis, system complexity, and desired accuracy. If a queueing model can be built in a matter of days, then it would take a matter of weeks to build a simulation model of the same system. In general it holds that queueing models can be built much more quickly than simulation models. However, the queueing network model is limited in what it can validly represent. If the process is exceedingly complex, has many iterations, or other features that deviate from the underlying assumptions of the queueing model, then it cannot be used to model the process.

TABLE 12.1
Common Process Goals

Goal Type	Examples
Quality	High customer satisfaction
Cost	High resource utilization High throughput rate Low production cost
Timeliness	Short cycle-time Short customer waiting time

Lastly, even when the process under study does fit the queueing model assumptions, it is not as accurate as a simulation model. Generally, a queueing model is considered a valid model of the process if its predictions are within 10% of the actual system. Simulation models can achieve greater accuracy than this.

A process analyst should really be using all three approaches concurrently. To use analytical or simulation models, you should also be applying heuristics to understand problems in the process design and how the process can be improved. Likewise, you should not just apply heuristics without some type of analysis, whether by queueing models or simulation models, of how the heuristic will affect the process performance.

12.1.1 Process Goals

Process design starts with determining the goals of the process. This chapter focuses on core processes, which need to provide value to the customer. Value is usually defined in terms of the characteristics of the product or service: (1) quality, (2) cost, and (3) timeliness (see Table 12.1 for some examples). The analysis starts by determining what makes a process valuable to the customer. By identifying the process customers and goals, the analyst will stay focused on attaining the process goals in the most effective and efficient manner.

Timeliness depends on the process and its context. Timeliness of an online retail store includes the response time to a customer request. Timeliness in a make-to-order manufacturing system is mostly a matter of the lead-time between the customer order and the fulfillment of that order. Timeliness in many service processes is the cycle time, where *cycle time* is the total time to process a single work item from start to finish. Actually, in many processes cycle time is often the most important performance measure of a process because it measures the speed with which a customer order can be satisfied. The objective is to minimize the cycle time.

Cost as a process goal refers to the process output, either a product or service. To reduce costs of the output, the process designers can reduce the costs of the input, which reduces the costs of the resources to execute the process, and improves the efficiency of the process. An important measure related to cost is the throughput rate of the process. The *throughput rate* is the number of work units the process completes per unit time. The objective for throughput rate is to maximize it. The more output a process can generate per unit time and holding everything else constant (in other words, not increasing throughput by adding more resources), then the cost per work unit will decrease.

Quality refers to the quality of the product or service generated by the process. There are many views of quality, but two are important in this context. Quality is the conformance to specification such that process generates output exactly how it was intended, and quality is perceived by the customer such that certain product features, performance, and aesthetic qualities contribute to customer's recognition of product quality. Conformance to specification can be measured using statistical tools such as the number of defects per million work units produced or using statistical quality control. Perceived quality is measured by customer satisfaction surveys and related methods. Process quality is achieved by embedding quality consciousness into the organization, developing feedback loops in the process so that the workers know what type of quality is being generated, and instituting continuous quality improvement programs to constantly measure and improve process quality.

There are other important measures of process performance, but in this chapter we will focus on cycle time, throughput rate, and quality. The rationale for focusing on these three is that they are common across many industries and are frequently the most important competitive factors in many markets. We will see that there is a trade-off between cycle time, throughput rate, and quality.

Example

To illustrate the determination of process goals, consider the analysis of an outpatient clinic. The core business process is to treat patients. The clinic has multiple goals, of primary importance are two: first, to remain profitable, and second, to provide superior patient services. The latter goal is based on the clinic's strategy of differentiating itself by providing superior service to its patients.

To track attainment of profitability they need to be able to treat as many patients as possible in a given time frame (e.g., a day), and they must efficiently utilize their resources in providing patient services. To track attainment of providing superior patient services, they measure patient satisfaction. Patient satisfaction is measured by a questionnaire that is done one to three days after the patient visits the outpatient clinic. The questionnaire helps identify those activities or process characteristics that are value-adding from the perspective of the patients. In this case, the questionnaires indicated that patients were primarily dissatisfied with two process features. The first was that patients were most dissatisfied with the long waiting times in the clinic, and second by the courtesy of the phlebotomist (person who draws blood).

The first issue deals with timeliness and the second with quality. To improve customer value the outpatient clinic should focus the process analysis on ways to reduce waiting time. Additionally, studies and training should be conducted to improve the courtesy of the phlebotomist or the perception of courtesy (it is possible it is the nature of the task, drawing blood, and not the person that causes dissatisfaction).

Similar to many situations, there is a trade-off between meeting the process objectives. To reduce waiting times, queueing theory tells us we can increase the number of servers (resources such as doctors, nurses, clerks, and equipment). However, resources cost money and the clinic must also remain profitable by keeping costs low.

The outpatient clinic has two options for the analysis. It can analyze and design the process to minimize waiting time subject to a maximum number of nurse full-time equivalents (FTEs). Alternately, it can establish a minimum service-level and then determine the minimum number of nurse FTEs to provide this service level, where a service level is a statement such as, "90% of all patients will wait less than 15 minutes to be taken to the examination room."

The insight is that process performance is defined firstly according to the customer's definition of value. Improving a performance measure not valued by the customer does not increase the customer's perception of the value the process provides. Secondly, a process must contribute to the organization's overall goals for that process, in this case profitability.

12.2 Cycle Time and Throughput Rate

A process consists of many activities arranged in a network. Within the activity network, the work unit might split up to have separate parts processed in parallel, or the work unit might take alternative paths depending on the outcome of a decision node. Of all the possible paths, the *critical path* is the longest path, in terms of activity duration, from start to finish in the process network. The critical path can often be identified by inspecting the flowchart for the process and generating every possible path. For each possible path calculate the cycle time, and the path with the greatest cycle time is the critical path. If it is not possible through inspection, then there are algorithms to help identify the critical path. In general, to improve a process the analysis must focus on those activities that comprise the critical path.

To calculate the duration of a path you:

1. Add all the activity times that are in sequence.
2. For activities that split depending on a decision, multiply the probability by the activity duration.
3. For activities in parallel calculate the maximum of the parallel activity durations.

To illustrate the identification of the critical path, we use the process flowchart shown in Figure 12.1, which contains two paths as follows:

Path 1

$$A + B + 0.4C + 0.6(D + E) + F + Max\,(G, H)$$
$$= 7 + 10 + 0.4(16) + 0.6(5 + 4) + 7 + Max\,(9, 5) = 44.8$$

Path 2

$$A + B + I + J + K + F + Max(G, H)$$
$$= 7 + 10 + 8 + 4 + 1 + 7 + Max\,(9, 5) = 46$$

Path 2 has the longest duration and is the critical path.

12.2.1 Activity Components

The duration of an activity is decomposed into component times. Let i denote the index of each activity in the process route for a customer. The cycle time for the route is the summation of four components: the activity time (t_i^a), coordination time (t_i^c), movement time (t_i^m), and waiting time (t_i^w). The cycle time is

$$t = \sum_{i=1}^{N} (t_i^a + t_i^c + t_i^m + t_i^w) \tag{12.1}$$

The *activity time* is the time required by the worker to do the task-specific work. The *coordination time* is the time the worker spends doing coordination tasks. Coordination is the work done to integrate one task with another task in the process. Coordination work tasks include communication between workers, preparing work for the next step, delays while waiting for needed input or supervisory direction, and planning work tasks. The *movement time* is the time to move the work between work stations. The *waiting time* is the time

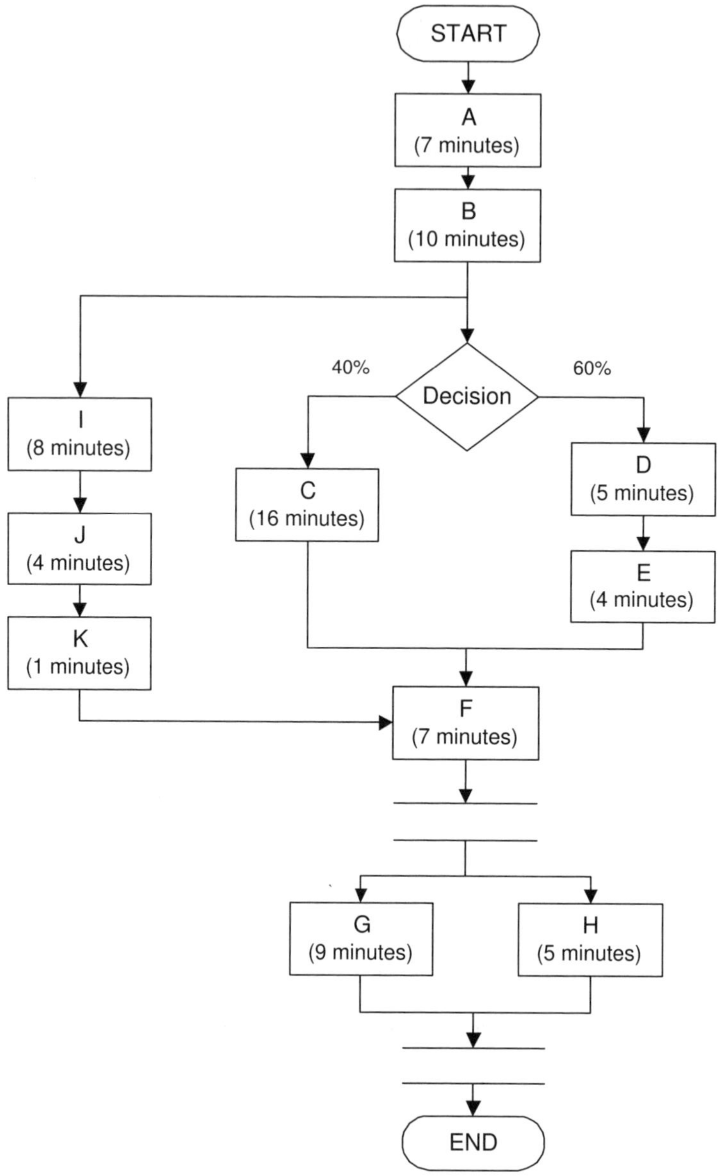

FIGURE 12.1
Critical path calculation.

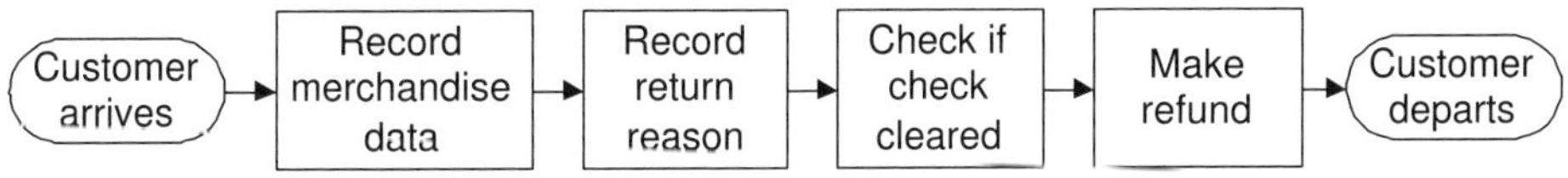

FIGURE 12.2
Check clearing process.

TABLE 12.2
Single Clerk (No Coordination Work)

Activity	Activity Time	Coordination Time
Record merchandise data	2.5	0
Record return reason	0.5	0
Check if check cleared	1.0	0
Make refund	1.25	0
Total times	5.25	0
Total time		5.25

the work unit waits because the worker is occupied doing some other task or working on another work unit.

To illustrate our characterization of work, consider the process of refunding a customer for returned merchandise that was purchased by check[1] (shown in Figure 12.2). The process involves recording the data on the returned merchandise, recording the reason for the return, verifying if the check cleared, and if the check clears, making the refund. The mean times for the activities are shown in Table 12.2. Note that the coordination time is 0 for all activities. If a process is done completely by a single employee, there is no need for coordination.

Now, suppose verification of the check was done remotely by another clerk as shown in Table 12.3. Then the clerk processing the return must coordinate with this clerk. They must communicate with the remote clerk by telephone, wait if not available, and then take down the information. The additional work encountered is what we term coordination work. Clearly, in this simple hypothetical example, the total amount of work depends on the process structure with the difference mainly being in the coordination work required.

We assume that, in general, the activity work is invariant or changes little with respect to process structure; it is the coordination work that changes when the process structure changes – where activity structure is the arrangement of activities. Referring back to Table 12.2, the activity of verifying if a check cleared involves matching the merchandise receipt to a check number and then searching a database for the check and verifying its status. The time to complete the activity is the same regardless of the process structures (see Table 12.3). It is the coordination time that changes when the process structure changes. However, the activity time would change if the nature of the activity changes. For example, if the verification process was automated with the use of information technology.

In some business processes, the movement time is essentially zero and may be ignored. If the process is an insurance claim that is routed electronically between employees, then the few seconds it takes the information system to move the claim are insignificant compared to the minutes of each activity. So, in this instance and others like it, you can safely assume movement time is zero. In production processes, the movement time is to move the work between workstations. In some processes, such as healthcare, the movement time is for the customer to move between workstations.

In manufacturing processes, they usually ignore the coordination time [2]. The reason

[1]Example adapted from Nuñez et al. [6].

TABLE 12.3
Two Clerks (with Coordination Work)

Activity	Activity time	Coordination time
Record merchandise data	2.5	0
Record return reason	0.5	0
Check if check cleared	1.0	1.5
Make refund	1.25	0
Total times	5.25	1.5
Total time		6.75

is coordination in most production processes, other than the movement of the work-in-progress from workstation to workstation, is negligible. Why is this so? Production lines are designed to standardize procedures for what is a repetitive process. They are designed to minimize the need for coordination. Consequently, coordination is generally negligible and not included in the analysis of production processes. When conducting a process analysis, the analysis team needs to determine which elements are significant and must be included in the modeling effort.

When deterministic analysis of the process is performed, the waiting time is frequently left out of the analysis. However, waiting time frequently accounts for a large proportion of the cycle time so not including it will lead to results that greatly underestimate the cycle time. Chapter 12 showed how waiting time can be estimated using queueing theory. To do this, the analyst needs to know the arrival rate and service rate for the activities.

12.2.2 Bounding Process Performance

To calculate the cycle time given by Equation 12.1, the analyst needs to understand each activity along the critical path in terms of the activity time, coordination time, movement time, and waiting time. For each activity, you need to know the expected service time and its variance. To collect this data involves observing the process and using stop watches to measure the duration of each activity component. In general, 20-30 data points should be sufficient to support an analysis. These times should be taken over a range of days, times, and people to obtain a representative sample. To estimate the waiting time requires application of the appropriate queueing equations. Once all the data are collected, then the expected cycle-time can be calculated.

In many situations it is useful to do a quick analysis to estimate the process performance. This is possible by calculating a lower bound on the cycle time and an upper bound on the throughput rate. Together, these two values bound the process performance. The advantage is that both bounds can be calculated quickly without the need of a thorough, complete process analysis.

The *theoretical minimum cycle-time* is the shortest possible time for a work unit to be processed. It ignores all waiting time, assuming that only a single work unit is being processed. Consequently, the theoretical minimum cycle-time is

$$t = \sum_{i=1}^{N} (t_i^a + t_i^c + t_i^m) \tag{12.2}$$

The advantage of Equation 12.2 is that it can be calculated once you know the expected service times for each activity in the critical path. This equation ignores the variability in the process and any waiting by using only the expected values. Adding variability and waiting can only increase the cycle time, so Equation 12.2 is a lower bound on cycle time.

Each resource pool in the process has a throughput rate defined as the number of work units it can process per hour. To calculate the mean throughput rate (r_i) of resource pool p we use

$$r_i = \frac{60n_p}{t_p^a} \tag{12.3}$$

where n_p denotes the number of servers in the resource pool and t_p^a denotes the service time the resource pool spends per work unit. The value 60 is to convert the service time to hours because the throughput rate is given in work units per hour. It is important to understand the throughput rate is calculated for resource pools and not activities. The reason is a single resource pool may perform several activities in the process. In this case the service time the resource pool spends per single work unit is the sum of the service times for each activity it performs.

The *theoretical maximum throughput rate* is the maximum number of work units per hour the process can process. It depends on the process bottleneck. The process *bottleneck* is at the activity performed by the slowest resource pool in the process. As the slowest activity, the bottleneck sets the maximum throughput rate for the overall process because all work units must be processed by that resource pool. In front of a bottleneck, work units will have to wait because they are being processed by upstream activities faster than the bottleneck can process them. Downstream from the bottleneck, the resources will have to wait to receive new work, called starving, because they finish their work faster than the bottleneck. The maximum theoretical throughput rate is given by the bottleneck and is denoted as r^{max}.

Generally but not always, the greatest process improvements are had by improving the performance of the bottleneck. For this reason, it is advantageous to always identify the process bottleneck and focus redesign activities on removing the bottleneck (at which point another activity would become the bottleneck).

The actual throughput rate of a process is the minimum of either the maximum theoretical throughput rate or the arrival rate, and is given by

$$r = \min\left(r^{max}, \lambda\right) \tag{12.4}$$

If the arrival rate of work to the process is less than the bottleneck rate, then the arrival of work is the bottleneck and that dictates the actual throughput rate because a process cannot produce work items faster than the input arrives. So if the arrival rate is less than the maximum throughput rate, then the throughput rate equals the arrival rate.

As the arrival rate approaches the maximum throughput rate, then whatever the arrival rate represents (e.g., input materials, customers, etc.) will queue in front of the bottleneck. This queue would grow to infinity as the arrival rate gets closer and closer to the theoretical maximum throughput rate.

When cycle time is graphed against throughput rate, the minimum cycle time and the maximum throughput rate create bounds within which the actual process performance will fall. In the next section we use an example to illustrate the ideas presented here.

Process Design Example

An urgent-care healthcare clinic sees patients on a walk-in basis.[2] Patients are triaged by an RN, they then register, are examined by a doctor who will order various treatments,

[2]This is a simplified version of an actual healthcare clinic that was analyzed in Jiang and Giachetti [3].

TABLE 12.4
Activity Times

Task	Resource Pool	Number of Servers in Resource Pool	Mean Service Time (min)	SCV
Triage	Triage RN	1	5	0.5
Registration	Clerk I	2	5	0.3
Examination	Physician	3	12	0.6
Treatment	RN	3	16	0.5
Re-examination	Physician	3	9	0.6
Discharge	Clerk II	2	7	0.3
Minimum cycle-time			**54**	

TABLE 12.5
Resource Pool's Throughput Rates

Resource Pool	Number Servers	Pool Service Time	Throughput Rate
Triage RN	1	5	12.0
Clerk I	1	5	12.0
Physician	3	21	**8.6**
RN	3	16	11.3
Clerk II	2	7	17.1

they are treated, the doctor re-examines them, and then they are discharged. Table 12.4 provides the activity data.

The *theoretical minimum cycle-time* is the sum of all the task mean durations. In the example, the theoretical minimum cycle-time is calculated with Equation 12.2, and assuming movement time and coordination time are zero, the minimum theoretical cycle-time is 54 minutes. We would not expect to see this cycle time because it ignores the variation of the task times, variability of patient arrivals, availability of staff, and consequently, makes no accommodation for waiting in the process.

Table 12.5 shows the calculation of the throughput rates for each resource pool according to Equation 12.3. The physicians do two activities, so the service time shown is the addition of the examination (12 minutes) and re-examination after treatment (9 minutes). The maximum theoretical throughput rate of the process is given by the bottleneck, which are the physicians with the lowest throughput rate of 8.6 patients per hour. It is also observed that some of the non-bottleneck resources will have significant amounts of idle time because they have much higher throughput rates. Both types of clerks will be frequently idle.

The *bottleneck* is the resource pool that limits the performance of the overall process by reaching full (100%) utilization first. In the example presented, the physicians are the bottleneck, which is what a manager of a clinic would want because they are the most expensive resource, so you want to utilize them close to 100%. Suppose the clerks were the bottleneck instead, then this would imply the physicians would have idle time, which makes little sense given the salary difference between clerks and physicians.

Figure 12.3 shows that the minimum theoretical cycle-time and the maximum theoretical throughput rate bound the actual performance of the process. The graph is obtained by varying the arrival rate from 1 to 8.25 in the Business Analyzer Software, which implements a queueing network model of the process. The graph provides the following important insights. First, throughput rate and cycle-time cannot be determined independently. We know this

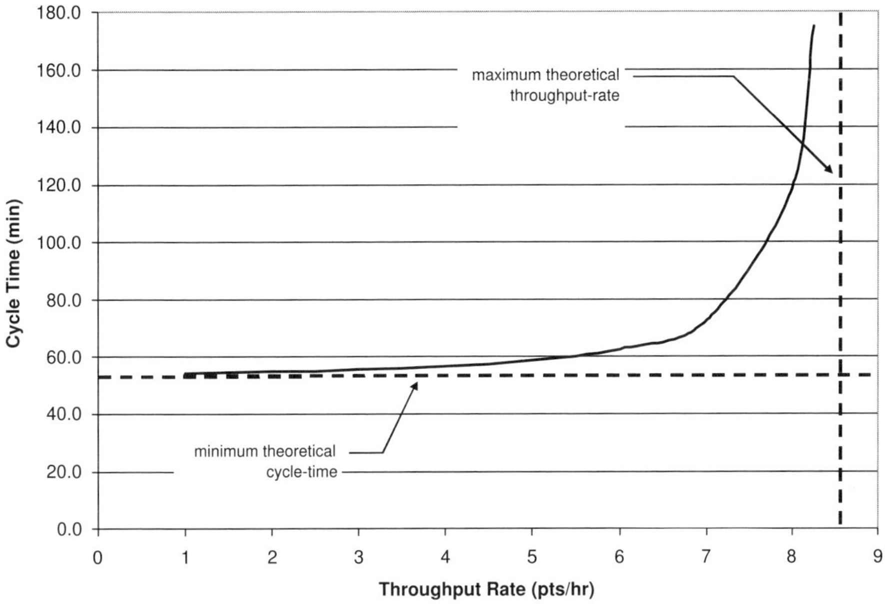

FIGURE 12.3
Cycle time versus throughput rate.

from Little's Law, and the graph illustrates it. To improve (decrease) cycle-time we must lower the throughput rate, which is undesirable. To increase the throughput rate and thus revenue, we must increase cycle-time, which is undesirable. The second insight, is that our simple analysis of the theoretical minimum cycle-time and the theoretical maximum throughput-rate provide bounds on the actual process performance. We cannot achieve these theoretical bounds, but we can approach them at the extreme values shown in the figure.

In a walk-in clinic, the clinic management would not have control over the arrival rate. In that case, the clinic would have to size the bottleneck resource pool to provide the desired service. If the arrival rate was between 7 and 8 patients per hour, then 3 physicians probably provide reasonable balance between the goals of a short cycle time and a high throughput rate. Otherwise, if the arrival rate is lower, they might consider having less physicians, and if the arrival rate is higher, then they will need more physicians. If the clinic scheduled patients, then they can exercise control over arrival rates, which adds another design variable to the problem. We will assume the arrival rate is 7.5 patients per hour for the following analyses. The cycle time of the as-is design is 87.0 minutes as estimated by the queueing network model.

Previously we had noted that the clerks have a high idle time. One improvement to the process is if instead of having two classifications of clerks, we cross-trained the clerks to do both activities. Then we could reduce the number of clerks and provide the same service level. Table 12.6 shows the effects of doing this. We are able to reduce the number of clerks by one and the bottleneck remains the physicians. The effect on cycle-time is minimal. If the arrival rate is 7.5 pts/hour, then the cycle time of the new design with cross-trained nurses is 90.2. The increase of 3.2 minutes over the as-is design is probably well worth the reduction of one clerk. The insight from this analysis is that changes to non-bottleneck resources will frequently have a negligible impact on process cycle-time or throughput rate.

TABLE 12.6
Resource Pool's Throughput Rates after Combining Clerks

Resource Pool	Number Servers	Pool Service Time	Throughput Rate
Triage RN	1	5	12.0
Clerk	2	12	10
Physician	3	21	**8.6**
RN	3	16	11.3

TABLE 12.7
Resource Pool's Throughput Rates after Reassignment of Work Activities

Resource Pool	Number Servers	Pool Service Time	Throughput Rate
Triage RN	1	5	12.0
Clerk	2	13	**9.2**
Physician	3	17	10.6
RN	3	19	9.5

So a reduction of non-bottleneck resources will reduce the cost of staffing the process with only a minor impact on process performance.

To illustrate the effects of changing non-bottleneck resources, suppose the variability of treatment is much higher with a SCV = 1.5. Making this change to the model, the cycle time increases from 87 minutes (at an arrival rate of 7.5 patients per hour) to a cycle time if 91.2 minutes. Again, this demonstrates that a large change in the variability of a non-bottleneck activity has minimal impact on the cycle time.

Given the bottleneck has such a large effect on process performance, it makes sense to concentrate efforts on improving the bottleneck. Suppose an examination of the work activities reveals that some of the work done by the physicians can be assigned to the RNs and a part of the RNs' work can be transferred to the discharge clerk. Table 12.7 shows the new data where the Physician's work decreases by 4 minutes, the RNs' work increases by 3 minutes, and the Clerk's work increases by 1 minute.

Shifting work from the bottleneck also shifts the bottleneck to a new resource pool. Now the clerks are the bottleneck resource. However, the process throughput rate will be greater than before, 9.2 patients per minute versus 8.6 patients per minute previously. The theoretical minimum cycle-time does not change in redistributing the work, it is still 54 minutes. The cycle time for the redesigned process is 76.1 minutes, a reduction of 12% over the original design with one less resource due to the cross-trained clerks. The redistribution of the work away from the bottleneck to other resources improves the process without reducing the total amount of work that is performed. It also changed the bottleneck, in this example to a less expensive resource. At this point the idle time of the physicians should be calculated to see if it is acceptable.

The insight gained from this healthcare example is that focusing redesign effort on identifying the bottleneck resource and ways to improve the bottleneck will have the greatest improvement for the process performance. However, improvements to non-bottleneck resources, such as the clerks here, can reduce costs with minimal impact on process performance. So these opportunities should not be overlooked, especially if easy, "low-hanging fruit" type of improvements are possible. The example also demonstrates that waiting time is often a large component of the cycle time. In the example, the activity time is 54 minutes

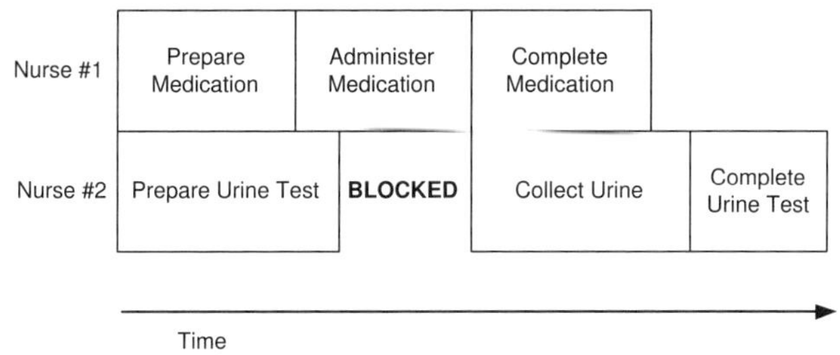

FIGURE 12.4
Coordination limits to parallelization.

and at a throughput rate of 7.5 patients/hour the cycle time was 87 minutes, the difference of 33 minutes due to waiting. The emphasis on the bottleneck should not be construed to mean that non-bottleneck resources can be ignored, as the example showed, cross-training of the clerks would allow for the reduction of one clerk with little change in cycle time. This change reduces the process cost while maintaining the same relative performance.

In many processes, the actual cycle-time is much greater than the theoretical minimum cycle-time; cycle-time efficiencies of less than 10% are not unusual. Part of the actual cycle-time will include down-time of the process – e.g., insurance agents work 9-5 so policies sit there overnight until the next day for processing.

12.2.3 Parallel Processes

Section 11.3.5 introduced fork/join queues to analyze parallel processing. It showed that the sojourn time of the parallelized activities would be greater than the maximum of the parallel activities, and an explanation of why this happens was provided. This section extends this analysis to include the consequences of changes in coordination time that occur when the structure of the work is changed from sequential to parallel.

Generally, when tasks are done in parallel they require more coordination than when the same tasks are done sequentially [4, 8]. This is one reason production lines usually do all the production tasks sequentially. Nuñez et al. [6] remark that a healthcare clinic that parallelized tasks only obtained a fraction of the expected gains a year later. The difference between the actual improvement realized compared to the expected improvement was attributed to the need for more coordination when nursing tasks were conducted in parallel. Figure 12.4 shows the tasks conducted by two different nurses (the first nurse is the top row and the second nurse is the bottom row). In this patient scenario, the patient needs to be administered a medication and to do a urine test. But, both cannot be done simultaneously, so one nurse is blocked until the other nurse finished. The period of being blocked is part of the coordination time. In this example, the parallelism did not save as much time as would be expected because of the blocking that occurs. In other scenarios, parallelization of work usually increases coordination work in terms of the need for greater communication and more preparation of activity outputs and inputs to mesh with the parallel activities. For these reasons, parallel processes tend to place greater coordination demands than equivalent sequential processes.

To illustrate how coordination can affect performance comparisons between sequential and parallel processes, Figure 12.5 shows two processes. Each process has five tasks. For the sake of comparison, we assume all task characteristics are the same in the two configurations except for the coordination work. The 5-minute activity duration in the sequential process is 4.5 minutes of activity time and 0.5 minutes of coordination time. The theoretical minimum

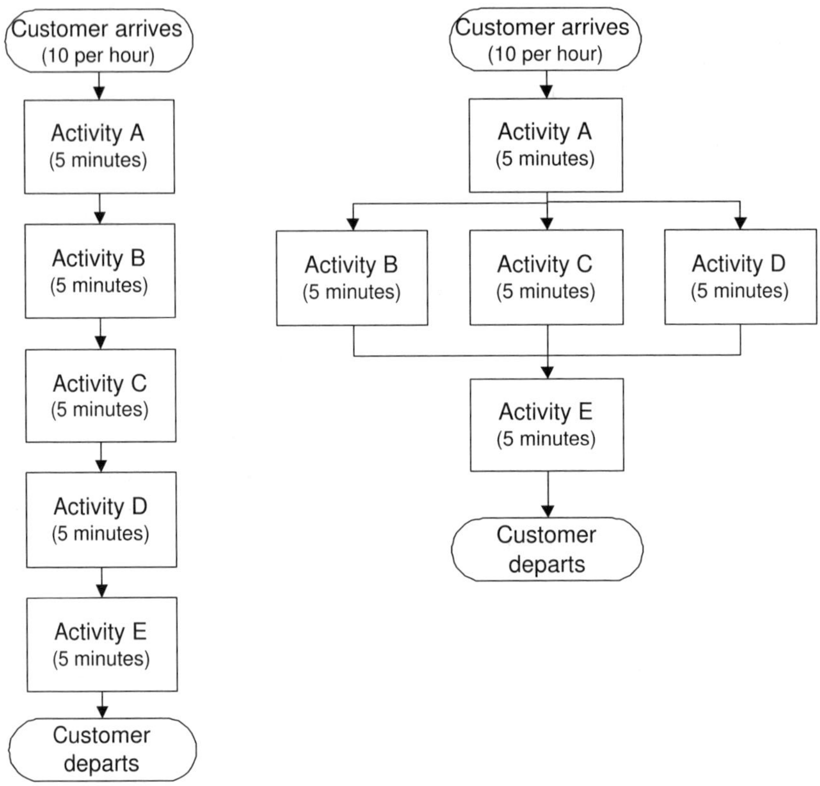

FIGURE 12.5
Sequential and parallel processes.

cycle-time for the sequential process is 25 minutes. If all activities are modeled as M/M/1 queues, then the actual cycle time when customer arrivals are 10 per hour is 150 minutes. In the parallel process we make the same assumptions, except we model the coordination time as being 20% greater than in the sequential process. Theoretically, the minimum cycle-time is 15 minutes for the parallel process, a 40% reduction. The model with the alteration of the coordination time estimates the cycle time to be 107.5 minutes, compared to the 150 minutes for a 28% reduction. The analysis helps to explain why there is a tendency to over-estimate the gains from parallelization of tasks.

The insight gained is that when a process is switched from a sequential structure to a parallel structure, then the coordination load also changes, and the coordination load in parallel may outweight any gains from parallalization. To assume that the coordination will remain the same regardless of the process structure is erroneous. For this reason, the time savings from parallelization will be less than what is indicated by a deterministic analysis due to two phenomena: first, queueing theory shows that the fork/join queue will perform slightly worse than the maximum of the parallel activities; and second, coordination theory finds that coordination load is greater for parallel activities.

12.3 Capacity

The determination of system capacity is a major enterprise design decision. System capacity is part of the operational strategy of the enterprise because to add large amounts of capacity takes significant amount of time and money. Additional capacity is usually added by acquiring more facilities, expanding existing facilities, partnering with others who can provide capacity, or outsourcing operations. For short periods of time, capacity can be increased by working overtime; however, this is not sustainable over the long term. Strategic capacity decisions were discussed in Chapter 7 on strategy; here we discuss capacity decisions at the process level.

Process capacity is how many products or customers the process can generate in a given time period. The *theoretical process capacity* is the maximum sustainable throughput rate for the process operating without interruption (no down-time due to maintenance, worker rest periods, and so on). To determine theoretical process capacity, you need to know the maximum sustainable throughput rate for the process, and the number of hours per day that process can be feasibly operated. In the example shown in Figure 12.3 the maximum throughput rate is 8.6 patients per hour. If the clinic operates 8 hours per day, then the theoretical process capacity is 68.8 patients per day. However, operating at 100% capacity is not sustainable because the waiting time would be enormous. So, the actual capacity is less, perhaps 80% of this number.

In service industries, capacity may be determined as much by the arrival pattern of customers as by the availability and capacity of resources. Consider a healthcare clinic that serves walk-in patients. Patients will not arrive uniformly throughout the day; instead there will be periods of high patient arrivals followed by lulls. Assume the clinic operates 12 hours per day because that is the longest shift a nurse can work straight. All the nurses work the entire day, so potential capacity will be constant throughout the day. Figure 12.6 shows the patient arrivals for each 1-hour period. If the clinic hires sufficient staff to process 400 pts/hour, then the performance shown on the left will be obtained. Notice, during the main part of the day a large number of patients will be waiting because the arrival rate in those periods exceeds the service rate of 400 pts/hour. If the clinic has staff to process 500 pts/hour, then most patients will not need to wait. Yet, in both cases the average

		400 pts/hr service rate		500 pts/hr service rate	
Time Period	Arrivals	Waiting	Served	Waiting	Served
1	100	0	100	0	100
2	200	0	200	0	200
3	500	0	400	0	500
4	725	100	400	0	500
5	400	425	400	225	500
6	300	425	400	125	425
7	350	325	400	0	350
8	475	275	400	0	475
9	500	350	400	0	500
10	225	450	400	0	225
11	50	275	325	0	50
12	50	0	50	0	50
	total		3875		3875
	Average per period		323		323

FIGURE 12.6
Clinic capacity analysis.

throughput rate is approximately the same and less than the available capacity in both cases. In this example, because the clinic has no control over patient arrivals the additional resource capacity (more nurses) does not translate into an increase in realized capacity because the bottleneck is the patient arrivals. Increased capacity translates into less waiting time for the patients.

The example illustrates several realities of designing service systems. The available capacity will likely have to be much more than the realized capacity because of the mismatch between demand and capacity. In this case, the service capacity is constrained by various work rules such that all staff work a full shift. More advantageous to the clinic management would be if they had flexibility to better match capacity with demand by modifying the shifts the staff work. They could, for example, hire part-time help during the peak period of the day. Another strategy is if they could manage demand, for example, schedule patients to arrive in off-peak hours or provide an incentive for patients to arrive in off-peak hours. Finally, it is possible the staff can use that time for other work that does not directly involve serving patients (updating paperwork, stocking examination rooms, telephoning patients, and so forth).

In service industries, or any industry where the product is perishable, the application of revenue management is appropriate to match demand and capacity. The healthcare clinic provides a perishable service if we define the service as the available capacity to serve patients. Capacity to serve patients in period one is lost if no patients are available because that capacity cannot be "saved" for later periods. Revenue management was pioneered in the airline industry but is now prevalent in car rental, hotel, cruise, and many other industries. A good treatment of revenue management can be found in [7].

Since 100% utilization is not feasible, the actual process capacity will always be less than the theoretical capacity. Reasons for lost capacity include:

- Resource breakdowns – A machine may become unavailable due to a breakdown or a human resource my be absent.
- Preventive maintenance – Machines require regular maintenance to operate at maximum

efficiency; this scheduled preventive maintenance makes the resource unavailable for processing. People may require periodic training (e.g., on an updated ERP system), which makes them unavailable.

- Process flow inefficiencies – A resource may become idle due to the unavailability of work. In a sequential line, if the work task in front of a task is slower, this unbalancing will cause starving at the task.
- Demand variation – As described above the mismatch between demand and capacity can cause under-utilization.

12.3.1 Pooling Capacity

A cost effective means to increase capacity without increasing cost is to pool either demand or resources. Pooling is when customer demands or resources are grouped together. To explain the difference between a pooled system and an un-pooled system consider two banks; let's call them Casa Rey Bank and Pobrecito Bank. Both banks serve commercial clients and individuals. Casa Rey Bank decides to have a teller dedicated to commercial clients and a teller dedicated to individuals. Pobrecito Bank also has two tellers, but decides to pool them and have a single line that both commercial and individual clients join for service. As a bank client, would you wait longer in Casa Rey Bank or Pobrecito Bank? The answer is that in most cases Pobrecito Bank with pooled resources will provide faster service. To illustrate the pooling effect, we will add numbers and use the queueing models described in Chapter 11.

Suppose the arrival rate of individuals is 7 per hour and the arrival rate of commercial clients is 7 per hour. The service rate for either customer is 10 per hour. There are two servers. In Casa Rey Bank, we model each dedicated teller as a M/M/1 queue. Using Equation 11.8, we calculate the total waiting time in the system for Casa Rey Bank as,

$$W_s = \frac{1}{(10-7)} = 20\,\text{minutes}$$

We model Pobrecito Bank as a M/M/n queue. The total arrival rate is 7 individuals per hour plus 7 commercial clients per hour for a total of 14 customers per hour. The number of servers is 2. Using Equation 11.13 we calculate the total waiting time in the system for Pobrecito Bank as,

$$W_s = \frac{0.7^{\left(\sqrt{2(2+1)}-1\right)}}{10(2)\,(1-0.7)} + \frac{1}{10} = 11.96\,\text{minutes}$$

The customers in Pobrecito Bank can expect to wait 40% less than those in Casa Rey Bank. The pooling of the customer demands provides a dramatic reduction of waiting time *without* the need for additional resources (tellers). It is rare that you can improve a system without additional cost.

Logically, it should be apparent that pooling can improve system performance. In the bank example, if we have three tellers serving three distinct customer types, then situations in which one teller has no customers waiting in line while customers are waiting in the other lines are likely to happen. However, if we pooled the tellers with a single line, then a teller would never be idle when there are customers waiting in line. The system would then make more efficient use of its resources and customers would, on average, wait less.

Pooling could be the key motivation behind an enterprise's design such as consolidation of sales regions, the cross-training of employees to do multiple tasks, or to have a single warehouse instead of many. Other examples of pooling are:

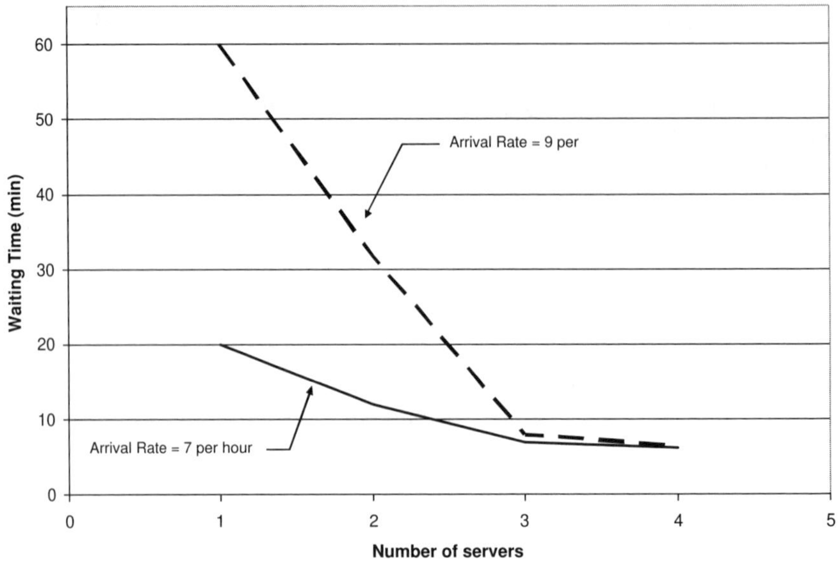

FIGURE 12.7
Diminishing returns on pooling capacity.

- Baptist Health of South Florida operates two hospitals and multiple outpatient clinics. Instead of having a radiologist in every facility, they pooled the radiologist in one location. At each facility when an x-ray is taken, it is sent electronically to the pooled radiologist who reviews the x-ray, writes the report, and sends it back to the facility. Here they pooled an expensive human resource, the radiologists.

- Call center agents are often pooled together to handle the multitude of different problem calls that might come into the company.

- A physician's office pools the physicians such that a patient can be seen by any physician in the clinic.

When pooling resources, there are diminishing returns. Figure 12.7 illustrates the reduction in waiting time when changing the system from one bank teller to two is larger than the later reductions.

The extent of improvement realized by pooling depends on the utilization of the servers, the variability in arrival and service rates, and the number of servers being pooled. The greatest benefits are when the utilization of the servers is high (because if utilization is low, then the servers are mostly idle and there will be little waiting anyway); when the variability of arrivals and service is high; and the more servers pooled the better, but as each successive server is added to the pool the incremental improvements diminish.

No resource can operate at 100% capacity (defined as theoretical capacity). Many factors cause resources to operate at much less than 100% capacity, and management continuously strives to design and manage the process so that close to 100% capacity is realized. Pooling is a powerful means to increase capacity without increasing resources.

TABLE 12.8
Process Cost Elements

Fixed Costs	Variable Costs
Facility	Consumable material
Machine	Labor
Labor	
Taxes, rent	
Utility	

12.4 Cost

Chapter 7 asserted the goal of an enterprise is to make long-term profits; therefore, understanding the cost of the process output is needed because profits equal revenue minus costs. Even for non-profit enterprises the cost to deliver a product or serve a customer is a major performance parameter of the process.

A *cost component* is any activity for which a separate cost measurement is desired. Example cost components are the materials consumed, inventory, labor, or overhead of the process. The process cost components can be classified as either fixed costs or variable costs. *Fixed costs* remain constant for all levels of output. So whether you serve a single customer per day or thousands, you incur the same fixed costs. Fixed costs are often set once a business is established. For example, once you purchase a building or sign a lease then the fixed costs of rent and property taxes are determined and are difficult to change in the short term.[3] *Variable costs* are the costs per work unit and therefore vary with the amount of sales. Table 12.8 lists examples of fixed and variable costs. Notice labor can be a fixed cost or a variable cost; it depends on how the employees are paid. An accountant is a fixed cost if paid a salary, whereas a truck driver is a variable cost because the truck driver is paid per hour spent driving.

Another way to classify costs are as direct costs and indirect costs. *Direct costs* are those costs that can be directly and exclusively attributed to a particular cost object. *Indirect costs* cannot be directly and exclusively attributed to a particular cost object. To illustrate the difference, consider the manufacturing production of a computer. Direct costs of the computer include the labor to assemble the computer and the components that go into the computer. The labor and components can be tied directly to a particular computer and they are exclusively done for that computer. Indirect costs in this example are the management salaries and rent for the factory. Neither of these two cost components can be directly and exclusively attributed to a particular computer.

Cost allocation is the assignment of costs to the work unit. Direct costs are traced to the work unit and assigned. Indirect costs cannot be traced to the work unit because they are common to many work units. Two methods are available to allocate indirect costs to the work unit. The first, more simplistic method is to determine a *cost driver* for the indirect costs. It remains common in many enterprises to aggregate all indirect costs into a single cost driver called overhead and change this as a percentage of the direct costs. For example, if the direct cost to produce a computer is \$100 and the overhead rate is 75%, then the indirect cost of 0.75 (\$100) = \$75 is allocated to the computer for a total cost of \$175. A single cost-driver for the entire enterprise is the most simplistic and least accurate means to allocate indirect costs. Greater precision can be obtained by differentiating the cost driver

[3] Not impossible to change; you can always renegotiate a lease or get tax breaks.

by work unit type, department, or by defining cost centers, which are groupings of resources that have a single cost driver.

A second means of indirect cost allocation is the activity-based costing method. The activity-based costing method attempts to allocate all indirect costs to activities based on the resources they consume. To illustrate, the computer manufacturer has the indirect cost of electricity. Normally, the cost of electricity would be aggregated with other indirect costs and uniformly allocated as described above. In activity-based costing, the enterprise determines how much electricity is consumed in the production of computers and assigns this number directly to the computer. Likewise, the area of factory floor used for computer production is determined and used to allocate the facility rental cost. Activity-based costing can provide more accurate cost allocation, the primary limitation to its adoption being the difficulty of establishing the cost accounting system to support it.

Illustration of Cost Estimate

A rapid prototyping shop has a stereolithography (SL) machine to make prototypes out of photo-polymer resin. The owner wants to determine the cost per hour of the process so that accurate estimates can be made for each order. The SL machine costs $250,000 and has an expected useful life of 3 years. It is estimated that the SL machine will run 3000 hours per year because a job can be started at the end of the day and the machine can continue through part of the night. We calculate the hourly cost of the machine as

$$c_m = \frac{\$250,000}{3\,(3000)} = \$27.78$$

The labor is done by a technician who earns a salary of $70,000 per year. The technician receives health and other benefits of 29% of his salary, which means the technician costs $\$70,000\,(1.29) = \$90,300$ per year. The technician works 2000 hours per year. We calculate the hourly cost of the technician as

$$c_l = \frac{\$90,300}{2000} = \$45.15$$

The facility is leased at $12,000 per month; the monthly utility, insurance, taxes, and other overhead costs are $4,000 per month. The facility operates 167 hours/month. We calculate the hourly facility costs as

$$c_f = \frac{\$16,000}{167} = \$95.81$$

The SL machine workroom only accounts for one third of the facility floor space, so the owner decides to allocate one third of the facility cost to the hourly operation of the SL machine, or $c_f = \$95.81/3 = \$31.94$.

The SL machine consumes a photopolymer resin that costs $400 per gallon. This is the material cost of the operation.

Table 12.9 shows the labor calculation and machine calculation for a sample order. The SL machine requires no operator, so no labor is associated with the run machine activity. In addition, the order will consume 2 gallons of resin, so the material cost is $800. There are several means to add the overhead to the estimate, here we add it as an hourly charge. The order consumes 13 hours at $27.78 per hour for an indirect charge of $361.11. The total cost is the cost of the labor, machine, consumable materials, and overhead for a total order cost of $1,643.83. This is the cost per work unit. The owner can then add a profit to the cost to arrive at the price to quote the customer.

If the customer ordered more of the same product, then the per work unit price would be

TABLE 12.9
Cost Estimation per Hour

Activity	Resource	Cost/hr	Time (hrs)	Hourly Cost
Generate CAD model	Technician	$45.15	3.0	$135.45
Generate production files	Technician	$45.15	1.5	$67.73
Set up machine	Technician	$45.15	1.0	$45.15
Run machine	SL machine	$27.78	6.0	$166.67
Post processing	Technician	$45.15	1.5	$67.73
	Total			$482.72

lower because the activities to generate the CAD model and the production files would only have to be done once for the entire batch. If there was a batch, then these costs ($135.45 and $67.73) would be spread out over all the units in the batch, lowering the per work unit cost.

A few comments are in order. This is a simplification to demonstrate the general approach of calculating cost elements and then calculating the cost per work unit. Many details were left out. The example showed the machine cost calculation for the LS machine, but what about the CAD software and computer? The example also bundled all the overhead into a single monthly charge of $4000. Better estimates can be made if more of this was calculated as a direct cost. For example, how much time did the salesperson spend to win the order? Also, what is the cost of machine maintenance? These costs could be applied directly to the order.

When the fixed costs account for the majority of the total cost, then management naturally tries to maximize the utilization of the resources that lead to the fixed costs. For example, once a cruise line decides to schedule a cruise, the costs are mostly fixed – whether a single passenger takes the cruise or two thousand passengers, the cost is essentially the same. For this reason, cruise lines practice revenue management to maximize revenue for the use of their fixed resources, the cruise ships. In fact, sometimes if a cruise is undersold, the cruise line will offer last-minute, low-cost fares because receiving some revenue and the possibility of more through onboard services (alcohol, shopping, gambling) is better than if the ship embarked with empty staterooms. The same logic holds for the healthcare clinic in the previous examples. They would naturally want to see as many patients as possible to reduce costs and improve profits (revenue minus cost). The number of customers that can be served is limited by capacity and/or limited by the desired service level the clinic wishes to provide. As previously demonstrated, if the number of patients is increased (equivalent to increasing the arrival rate), then the waiting time will increase. Increased waiting is undesirable, so a clinic may establish a service level, defined as the maximum waiting time experienced by the majority of its patients. A service level may be, "90% of all customers will be served in 1-hour or less." What this demonstrates is there is a trade-off between cost, which is correlated with capacity and the service provided.

To summarize, estimating the per work unit cost for a product or service involves first estimating the cost elements per unit time or unit product. Supposing the cost driver is measured as cost per hour, then the consumption of that cost element in hours is multiplied by the cost driver to obtain an estimate for that cost element. In the SL illustration above, we calculated the cost per hour of the technician, SL machine, and facility and multiplied that by the number of hours a work unit consumed each cost element. For the material cost we determined the cost per gallon as the cost driver and multiplied that by the number of gallons consumed. Using this approach, the cost per work unit can be estimated for any process.

12.5 Quality

Quality has two separate definitions, one from the customer's perspective and the other from the enterprise's perspective; both are important for the design of processes. The customer's perspective of quality is a product or service that satisfies his or her needs and expectations. This definition includes multiple dimensions of quality as performance, features, reliability, durability, serviceability, aesthetics, and perceived quality. The definition is also contextual because the satisfaction of needs and expectations implies a value-orientation. A customer has different expectations for a \$20-set of kitchen knives compared to a \$200-set of kitchen knives. Likely expectations for the cheaper knives are that they fulfill their basic function of reliably cutting food for a reasonable amount of time. The more expensive knives are expected to have greater aesthetic appeal, be made of more durable materials, have a sharper edge that remains longer, and include service or warranties beyond that provided by the cheaper knives.

The enterprise's view of quality is that the product or service conforms to the specifications designed for the product or service. This view of quality is that the process produces zero defects, and each product or service provided is consistent with little variability. This definition is a more objective engineering definition and provides a basis on which to control the processes generating the product or service. Returning to the knives's example, the cheaper knives could have better quality in this perspective because they conform to specifications better than the more expensive knives. This was the experience of Jaguar owners in the 1980s where the luxury car-maker became notorious for breaking down and needing repairs.[4] This was around the same period of time that Japanese automakers became associated with high quality but basic cars.

Customer Perspective of Quality – The degree to which the features and characteristics of a product or service meets the needs and expectations of the users.

Enterprise Perspective of Quality – The degree to which a product or service conforms to the specifications for the product or service.

The two notions of quality are related. A product or service cannot meet customer needs unless the enterprise determines those customer needs and designs a product or service to meet those needs. This is the quality by design. Then in the delivery of the product or service the enterprise must conform to the design specifications. Only by getting both the design and the delivery correct can an enterprise provide a quality product or service. Additionally, the enterprise needs to continuously work to improve the quality of its products and services. To institute quality processes an enterprise performs three functions:

1. Design quality products and services by understanding customer needs and translating them into product and service specifications.
2. Deliver quality products and services by having processes that produce no defects and have little variability.
3. Improve the quality of their processes by establishing continuous process improvement programs.

[4]Covert, C., Jaguar purring again after licking wounds, *Spokane Chronicle*, April 9, 1988, p. D1-D2.

Design Quality

The design of quality processes starts with understanding the customer needs and expectations. A thorough understanding of quality can only be achieved by asking the customer. Of course, sometimes, directly asking the customer will not elicit accurate information – the analysis team will need to use the data-gathering techniques discussed in Chapter 8. One caveat is that customers may not be able to express a need for an innovative product, service, or new technology. When Sony first developed the Walkman in 1979 there was no customer expectations for a portable music device. Sometimes, innovative ideas need to be introduced to customers first before they can understand and develop expectations for the new product or service. To do this, other market research tools are available. If the process output is primarily a service or has a large service component, then there are questionnaire instruments such as SERVQUAL designed to elicit customer perceptions of the service quality. Another commonly employed means to determine customer needs is via benchmarking with competing products/services. The benchmark lets the enterprise gauge itself against its competitors to see where it excels or where it falls behind. Using these approaches, an enterprise can determine the customer needs its products and services need to satisfy.

Patient Satisfaction Surveys

A healthcare clinic regularly measures and tracks patient satisfaction through the services of the Press-Ganey company that specializes in patient satisfaction surveys. The survey is sent to patients soon after they are discharged from the clinic. The survey has questions about waiting times, the facility, courtesy and competence of the staff, and other categories important to quality service in healthcare. The clinic regularly receives reports of the survey results with benchmarks against other healthcare providers that also utilize the survey. The survey enables the clinical management to understand how patients perceive the service and take actions to correct any shortfalls indicated by the survey.

The quality functional deployment methodology and tools translate the customer needs into the attributes of a product or service. The primary tool of quality functional deployment is the house of quality, which is a matrix that denotes the strength of the association between a product or service attribute and customer expectations in a range from 0 to 9, with 9 indicating a very strong association. Additionally, the matrix indicates the degree of correlation between product attributes as either weak, medium, or strong. To use the tool the enterprise must identify those product attributes that it can manipulate and then associate them with the customer expectations. Using benchmarking and other market analysis, the enterprise can prioritize the customer expectations that need improvement. The house of quality tells them what product attributes are most associated with those customer expectations. The quality functional deployment methodology and tools provide a systematic way to take customer needs and design a product or service to fulfill those needs.

Another approach to designing quality products is the robust design methods developed by Taguchi. Taguchi challenged the idea that a manufactured product meeting specifications was sufficient. He developed the quality loss function showing that any deviation from the target is undesirable (see Figure 12.8). Normally, any value that fell within the lower and upper specification was considered good – just like making the extra point in football where any ball between the goal posts gets 1 point. The quadratic loss function says that any deviation from the target value results in a loss. Taguchi's loss function is used in conjunction with design of experiments to remove as much variation from the process as

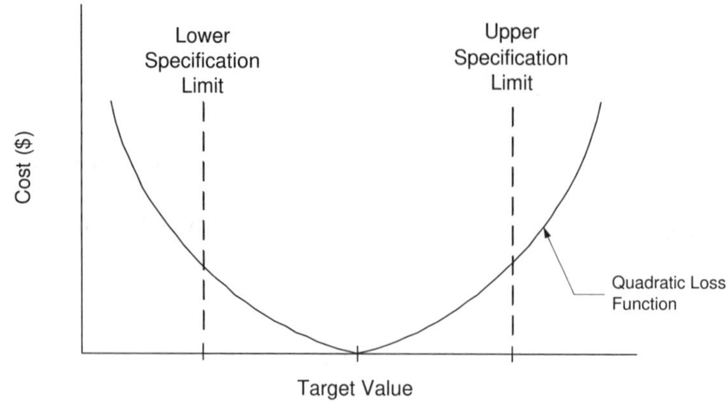

FIGURE 12.8
Taguchi's quadratic loss function.

possible. It is also used to design robust products that will maintain small variation even when the process inputs or process itself is subject to variation. Byrne and Taguchi [1] describe a case where a tile manufacturing was experiencing variation in tile dimensions because of uneven temperature in the kiln. Replacing the kiln was very expensive, but the company found a less expensive solution by changing the chemical composition of the tiles. The tiles became insensitive to the temperature differences so that regardless of where in the kiln the tile was, it adhered to the size specifications.

To design quality into the product, the process that produces the product should be fail-proof. A Japanese word that has come to embody this idea is *poka-yoke*, essentially the fool-proofing of a process. The observation is that process errors, mistakes, and variation can be attributed not to under-performing employees but to interruptions, fatigue, and other special causes. Using poka-yoke concepts, you design the process so that these errors cannot occur. In manufacturing, you might design parts so that they can only be assembled the correct way, thus removing the need for the worker to think about how the parts go together and possibly making an error. Enclosed car parking lots often have bars at the top indicating the maximum height clearance in the garage, which acts as a go/no-go gauge. In service industries such as hospitals or maintenance, checklists are used to remind the person to do all the procedures for every customer. Poka-yoke designs the process so as to eliminate the possibility of quality problems before they can happen.

Healthcare Patient Interaction Script

A script is a type of poka-yoke for dialog and behavior to standardize the quality of care when a staff member interacts with a patient. An example in healthcare is:

When first entering the examination room, greet the patient and identify yourself by saying, "Hello, my name is XXX and I will be your nurse today. Your doctor will be XXX..."

When leaving the patient say, "If you need anything, please let me know, you can always ring the buzzer and I'll come to the room."

Quality Control

The design of a quality process involves instituting the correct feedback loops to constantly monitor the process (see Section 2.1.6 for a discussion of feedback control). Statistical quality control uses control charts to monitor processes and determine when they control or when they need intervention to correct potential problems. Statistical process control samples the critical measures of output quality and charts them. The underlying principle of statistical quality control is that all processes exhibit variation in their output. Some of the variation is called *common cause* variation and is due to the inherent characteristics of the process. The common cause variation cannot be attributed to any particular problem or event. Other variation is the result of *special causes* that can be attributed to a specific problem such as a malfunctioning machine, fluctuations in temperature, or changes in the process inputs. Special causes typically cause greater variability than common causes and the statistical quality control method seeks to identify and eliminate special causes.

A control chart monitors the process and indicates whether it is in statistical control, meaning only common cause variation is present, or if there are any special causes that require process intervention. There are several different types of control charts available depending on the type of output variable being measured. Here we describe a typical scenario when the output variable being controlled is measured on a continuous scale. Suppose, a company manufactures rope for sailing applications and the rope is designed to be 0.25 inches in diameter. Too large or too small is a quality problem. The $\overline{X}$ control chart monitors the process mean and an R chart monitors the process range or variability. The upper and lower control limits are calculated based on the sample mean, sample standard deviation, and sample size. They are set to be three standard deviations above and below the mean using the generalized equations of

$$UCL = \mu + 3\sigma \tag{12.5}$$

$$LCL = \mu - 3\sigma \tag{12.6}$$

The above equations change depending on the size of the sample and underlying process. Figure 12.9 shows a typical control chart. When the line falls out of the limits or is on one side of the mean for a number of sequential samples, then it is considered out of control. The details on how to construct a control chart and interpret the results can be found in Montgomery and Runger [5].

The control chart closes the feedback loop between the process output and the process. If the rope is seen to trend out of statistical control, then the employees are alerted to a special cause problem that needs to be investigated. Other tools such as fishbone diagrams and so forth are used to isolate the special cause so that corrective action can be taken. Usage of control charts maintains process variability within a small range of values within which the variability is explained only by common cause variation.

12.5.1 Process Improvement

All enterprises should establish continuous process improvement programs that run parallel to the core business processes. Quality functional deployment, Taguchi methods, and statistical process control are all part of a continuous process improvement culture. In this section, we review methods and tools to take processes in statistical control and improve them even more. Six sigma is a continuous process improvement methodology that utilizes a data-driven approach to identifying quality problems and reducing process variation to make the process perfect. The term six sigma refers to a process that produces no more

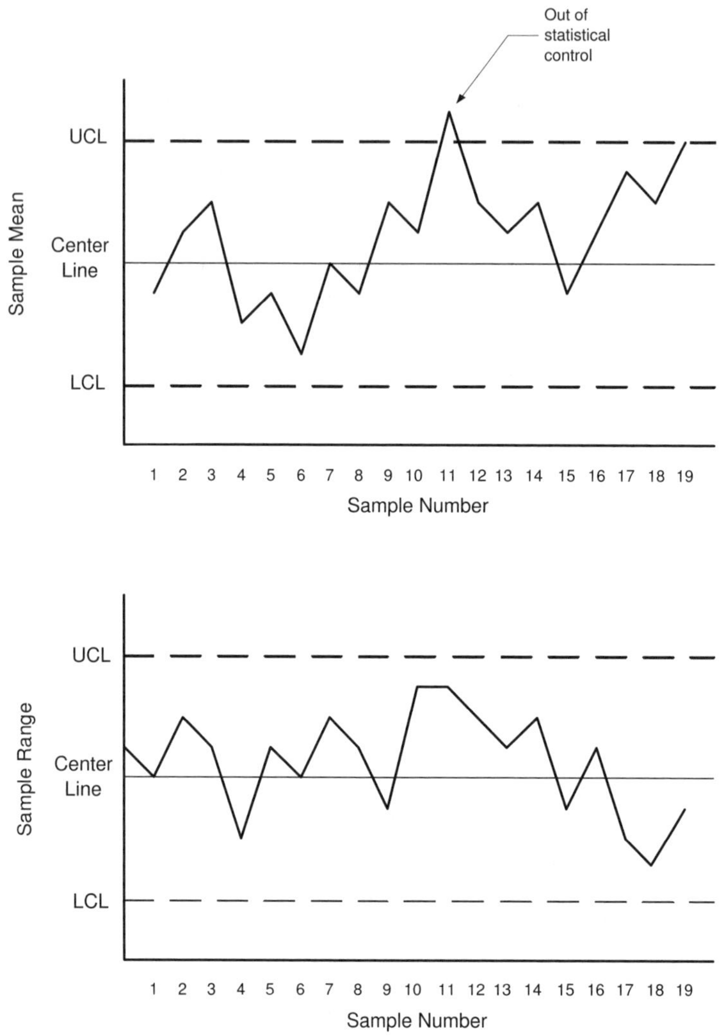

FIGURE 12.9
Control charts for mean and range.

than 3.4 defects per million occurrences of the process, but its main goal is continuous improvement.

Six sigma uses the improvement process called DMAIC (Defining, Measuring, Analyzing, Improving, and Controlling). Six sigma uses extensive statistical tools to aid analysis. The main tools used are process mapping, process capability tool, fishbone diagrams, pareto charts, hypothesis testing, failure mode effect analysis, design of experiments, and statistical process control. Six sigma is usually performed as a project, where a problem is identified and the project team then follows the DMAIC process to improve the process. Six sigma has become very popular in industry and certificate programs are available to qualify people as "black belts" indicating they have gained and demonstrated knowledge of the six sigma process and tools.

Improvement Opportunities

To do continuous process improvement, the analysis teams need to identify improvement opportunities. Some commonly encountered improvement opportunities are:

- Eliminate waiting time (a waste)
- Eliminate wasted movement or effort
- Minimize inventory
- Eliminate repair and rework
- Minimize material movement
- Minimize inspections
- Reduce variation of inputs, process, and outputs
- Reduce cycle time
- Improve machine reliability
- Improve flexibility of resources

In some instances, improving quality is equivalent to improving the other process measures. For example, In healthcare, quality includes low waiting times, so reducing waiting times improves the patient's perception of quality. However, other aspects of service quality may be at odds with other performance measures such as cost and cycle time.

12.6 Summary

This chapter started by defining common process goals of time, cost, and quality. The chapter showed how to calculate the theoretical minimum cycle-time and theoretical maximum throughput rate to bound the process performance. Actual processes have to operate within these limits. The chapter presented a model of process work as consisting of activity work, coordination work, movement work and waiting time. It showed how coordination work affects process design because the coordination work element changes as the process structure changes. As an example, the chapter showed how it affects the change from a sequential process to a parallel process.

Process capacity is an important part of process design. Process capacity was defined as the sustainable throughput rate of a process. The chapter showed how to calculate the capacity and discussed some of the considerations in designing system capacity. Equations to understand the cost of a process were presented with a discussion of how to estimate the cost per work unit outputted by the process.

Process quality is achieved by embedding quality consciousness into the organization, developing feedback loops in the process so that the workers know what type of quality is being generated, and instituting continuous quality improvement programs to constantly measure and improve process quality. The key to the feedback loop is to have a measurement of process performance and the process's output. In the analysis phase statistical methods are used to understand the process.

Review Questions

1. Describe scenarios when it would be preferable to use simulation instead of analytical models, and scenarios where it would be preferable to use analytical models instead of simulation.
2. List three common process goals. For each goal type, provide an example.
3. Describe the view of quality from the customer and the enterprise. Explain how these two views are related to each other.
4. What does "poka-yoke" mean and how can it help improve process quality?
5. Describe ways in which Taguchi methods might be used to improve the quality of MP3 music players.
6. The airline luggage delivery process starts when the airline assumes possession of your luggage and ends when you luggage is returned. Define the attributes of a quality luggage delivery process. List potential special causes that might prevent the luggage delivery process from meeting customer needs and expectations.
7. Describe the six sigma program and how it is used in enterprises.
8. Explain how control charts are used to monitor process performance.
9. Contrast special-cause variation from common-cause variation.

Exercises

1. The Casa Manufacturing Company includes a five-station production line with data as provided in the table below.
   (a) Calculate the production rate for each work station in parts/hour.
   (b) What is the theoretical maximum production-rate for the production line?
   (c) What is the theoretical minimum cycle-time for a part in this production line?

(d) If they run the production line so that the bottleneck has a utilization of 0.90, then what is the actual production rate and the actual cycle-time?

Work Station	Service Time (min)	SCV of Service	Number of Servers
1	20	0.1	2
2	15	0.1	1
3	30	0.1	2
4	25	0.2	1
5	20	0.5	2

2. La Fea Hair Salon provides a luxury service that has 4 tasks performed in sequence; each task has a service rate of 4 jobs/hour with a SCV = 0.1. Currently, the La Fea does not schedule work, clients arrive randomly. The arrival rate is 3.75 jobs/hour with a SCV = 1.0. What reduction in cycle time is possible if La Fea schedules arrivals and the effective SCV for arrivals is reduced to 0.1? (The SCV will not equal 0 because some clients will be late and contribute to variability).

3. A maintenance department receives an average of 12 work orders per day, which has a SCV of 1.0. There is a single clerk, maintenance supervisor, and three technicians per day. The process data are:

Activity	Resource Pool	Mean Service Time (min)	SCV
Create work order	Clerk	5	0.5
Assign technician	Supervisor	4	1
Update work order status	Clerk	2	1
Repair equipment	Technician	75	1
Update work order	Clerk	5	0.5
Close work order	Clerk	2	0.5

(a) Calculate the throughput rate for each resource pool.

(b) Which resource pool is the bottleneck?

(c) Calculate the minimum theoretical cycle-time.

(d) Calculate the maximum theoretical throughput-rate.

(e) Use the queueing network analyzer to calculate the actual cycle time.

(f) What is the utilization of the clerk?

4. An insurance claims process receives 5 claims per hour. The interarrival times are *iid* and exponentially distributed. The service times are provided in the table and are also exponentially distributed (you may assume M/M/1 or M/M/n queues). There is a single clerk, single supervisor, two field agents, and a single representative. For the process do the following:

(a) Calculate the throughput rate for each resource pool.

(b) Which resource pool is the bottleneck?

(c) Calculate the minimum theoretical cycle-time.

(d) Calculate the maximum theoretical throughput-rate.

(e) Use the queueing network analyzer to calculate the actual cycle time.

(f) Create a redesign of the process to improve cycle time.

Activity	Resource Pool	Mean Service Time (min)
Record claim	Clerk	5
Create claim file	Clerk	7
Assign claim	Supervisor	4
Review claim	Representative	8
Investigate claim	Field agent	25
Create claim report	Field agent	17
Evaluate loss	Representative	7
Negotiate claim	Representative	12
Settle and close claim	Representative	9

Bibliography

[1] D.M. Byrne and S. Taguchi. The taguchi approach to parametric design. *Quality Progress*, December:19–26, 1987.

[2] W.J. Hopp and M.L. Spearman. *Factory Physics.* Irwin McGraw-Hill, Boston, MA, 1996.

[3] L. Jiang and R.E. Giachetti. A queueing network model to analyze the impact of parallelization of care on patient cycle time. *Health Care Management Science*, 11(3):248–261, 2008.

[4] V. Krishnan, S.D. Eppinger, and D.E. Whitney. A model-based framework to overlap product development activities. *Management Science*, 43(4):437–451, 1997.

[5] D.C. Montgomery and G.C. Runger. *Probability and Statistics for Engineers.* John Wiley & Sons, Hoboken, NJ, 2007.

[6] A. Nuñez, R.E. Giachetti, and G. Boria. Quantifying the amount of coordination work as a function of the task uncertainty and interdependence. *Journal of Enterprise Information Management*, 22(3):361–375, 2009.

[7] K.T. Talluri and G. Van Ryzin. *The Theory and Practice of Revenue Management.* Springer, New York, NY, 2005.

[8] T. Williams, F. Eden, F. Ackermann, and A. Tait. Viscious circles of parallelism. *International Journal of Project Management*, 13(2):151–155, 1995.

Part IV

Information View

13

Information Modeling

"Everybody gets so much information all day long that they lose their common sense."
– Gertrude Stein (1874-1946), American writer.

An information model describes the data content of the system and how the data are structured. The information model is important to systems development because almost all systems contain a lot of data without which the system could not operate. Information model is performed during the analysis life-cycle phase. In this book we focus on how to model and design the information models. In implementation, the information model is almost always implemented as one or more databases in the enterprise or as a part of a computer application.

There are essentially two means to model information: either the traditional entity-relationship model or an object-oriented model. The entity-relationship model describes data in terms of entities, their attributes, and the relationships between those entities. An object-oriented model describes data as classes, their attributes, and the relationships between those classes. The majority of business software uses the entity-relationship model to represent the data structure and then implements the storage in a relational database system. Object-oriented models are usually used to model the data in a computer application.

The distinction between an information model and a data model is fuzzy. Here, we follow the logic presented by Schenck and Wilson [1] that a data model is designed with the intention of describing the structure of data in a computer system. Information modeling is done to describe the information content of a system. The information model may or may not be directly implemented in a computer system. Thus, the distinction is a matter of the intended use of the model. In enterprise design, we are concerned with understanding the information needs of the enterprise, so we will call it an information model. However, we recognize that eventually the models would be implemented in one or more systems.

After completing this chapter, you should be able to:

- Describe the three-schema architecture and the benefits it provides.
- Name the components of an information model.
- Define entities and their primary keys.
- Define entity attributes, their data types and domains.
- Define relationships between entities.
- Create an information model using the entity-relationship model.
- Migrate primary keys to foreign keys.
- Differentiate between identifying and non-identifying relationships.
- Create unary, binary, and ternary relationships.
- Explain the need for data standardization.
- Explain how the information structure conveys different model semantics.

13.1 Information Model

There are many different views of the same data in a system. Information modeling provides a way to reconcile the very different end-user views of the nature and role of data. The information model represents the structure and semantics of information within an environment or system. Information models are logical models in that they do not indicate how the information is physically stored. Thus, the analysis team can focus on capturing and representing the system information without being distracted by technical considerations. While this is the ideal, in reality the team needs to consider how the information model will be implemented because there are two main methods of information modeling and the choice is often dictated by the eventual implementation strategy used. The two main modeling approaches are the entity-relationship model and the object-oriented model.

The entity-relationship model is used to depict the data for implementation in relational databases, the most popular and widely used database technology. The strength of the relational database model is it has a mathematical foundation and standard, declarative query-language called Structured Query Language (SQL). The object-oriented model is used to depict the information for use in many object-oriented programming languages such as Java, Visual Basic, and others. The most popular object-oriented modeling language is the Unified Modeling Language (UML). The UML consists of fourteen different models, to model information you would use class diagrams. In this book, we concentrate on the entity-relationship model because it is the most widely used to model information, and even when the application will be developed with object-oriented technology these applications usually use a relational database for the persistent storage of the data.

13.1.1 Three-Schema Architecture

The ANSI-SPARC three-schema architecture[1] provides a useful means to understand the relationship between different views of the same data. The architecture separates the user's view of the data from the system's view and how it is physically stored on the hardware. The three views are:

- External Schema – Defines a view of the data as seen by a specific application or end user. There can be multiple external views, all of them are independent such that changes to one external view do not affect other external views.

- Conceptual Schema – Defines the global, logical view of the data from the perspective of the system designer. The conceptual schema is often implemented as the system database.

- Physical Schema – Defines the internal organization of the data and how it is stored on the hardware. The database management system handles the mapping from the conceptual schema (logical view) to the physical storage of the data on hardware.

Figure 13.1 illustrates the three schemas and the relationships between them. In this example, the scheduling department would have information to patients, doctors, and their appointments but not the sensitive patient medical data. Meanwhile, the medical department would have access to the medical data as well. Insurance would have a different view to support the payment and reimbursement functions. The conceptual schema would include the complete logical model of all the data. The data itself is physically stored on one or more databases.

[1] A "schema" is a model or plan.

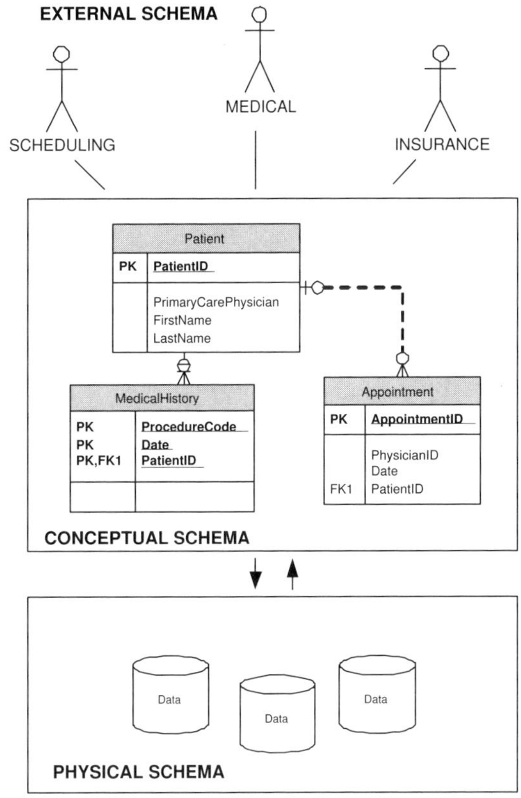

FIGURE 13.1
ANSI three-schema architecture.

The benefit of the three-schema architecture is it hides the physical storage of the data from the users. Users do not need to deal with or understand how the data are physically stored. A system designer is primarily concerned with developing the conceptual schema and then one or more external schemas for different user groups.

The information model captures the static structure of the information content, often through the identification and formalization of business rules. A *business rule* describes how the enterprise operates. These business rules are discovered during requirements gathering. Examples of business rules are:

- A cruise reservation is for one or more cabins on a cruise.
- A patient has a single primary care provider at a given time.
- A doctor can be the primary care provider for many patients.
- A purchase order contains one or more purchase order items.

Within a single industry, you would expect most enterprises to have similar information models. For example, all airlines have flights, airplanes, crew, reservations, etc. Yet, the precise definition and structure of the information model would differ from business to business because they often have different business rules. Some business rules cannot be graphically depicted in most information models, yet their identification and documentation is necessary to the design of a good system. For example, the business rule, "a procedure that costs more than $500 must be pre-approved by the insurance company" cannot be implemented in the information model, yet it is an important rule to capture and implement in the system.

13.2 Entity-Relationship Model

An entity-relationship diagram models the information content of a system as consisting of entities, their attributes, and the relationships between entities. To understand the entity-relationship model we need to understand each model component.

Entity

An *entity* is a set of real or abstract objects with common attributes or characteristics about which an enterprise needs to store data. An entity can describe a person, place, object, transaction, event, concept, or group. To be an entity, there should be multiple members of the set, and each member should be distinguishable from other members. We call each individual member an *entity instance*. To illustrate, for a university we could define an entity `Student` that describes the set of students in the university. Student is an entity because there are both multiple students (members of the set student), and each student is distinguishable from other students. An example of something that is not an entity is a potato chip. There are many potato chips, but they are indistinguishable from each other.

An entity is drawn as a rectangle. The entity name is a singular noun and goes in the top of the box. Within a single model no two entities can have the same name. Figure 13.2 shows the entity `Student` and some of its instances.

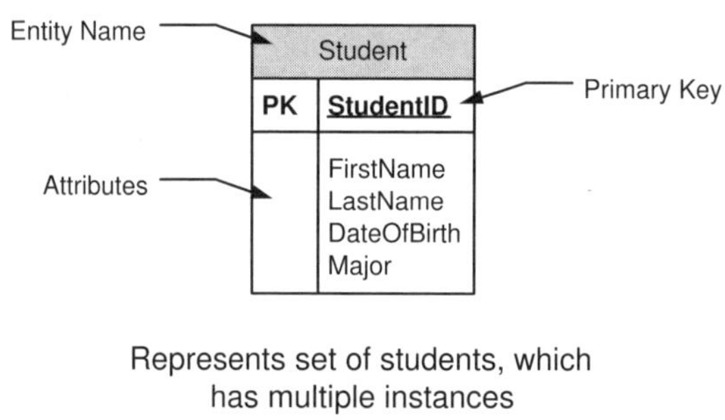

StudentID	FirstName	LastName	DateOfBirth	Major
475234	Giacomo	Grilli	05/15/1985	Systems Engineering
525342	Silvia	Brunn	01/23/1987	Business Administration
767342	Adriana	Garcia	10/03/1987	Systems Engineering
889234	Anthony	Delgado	03/30/1986	Industrial Engineering

FIGURE 13.2
Entity definition.

Attribute

Entities are described by their attributes. An *attribute* is a descriptive property or characteristic of an entity. A single entity will have many attributes. One way to think about entities and attributes is that entities are nouns and attributes are adjectives describing the noun. The attributes describing the entity are shown in the box below the entity name (see Figure 13.2).

Each attribute has a data type and a domain. The *data type* specifies the types of data values that can be stored in the attribute. Data types are: number, text, logical, currency, and date/time. The *domain*, if it exists, defines the set of allowable values for the attribute. Domains can be defined by the combination of operators including $<, >, \leq, \geq$ and also the logical operators (and, or, not). Alternately, a domain can be enumerated by listing all the valid values it can assume. Table 13.1 shows some attributes with their data types and domains. Some attributes such as `FirstName` will not have a domain because any combination of letters is allowable. `BirthDate` has a datatype of date and the domain is defined such that it can be any date prior to today's date. The attribute `State` shows the domain defined as an enumerated list of all the allowable two-letter abbreviations for states (including the District of Columbia and Puerto Rico). Notice the data type for `HomeTelephoneNumber` is text even though you might expect a number. The reason is you do not do mathematical operations on telephone numbers, so to prevent this we define its datatype as text.

TABLE 13.1
Example Attribute Data Types and Domains

Attribute	Data type	Domain
FirstName	text	-
BirthDate	date	`Date` $\leq$`Today()`
State	text	{AL, AK, AZ, AR, CA, CO, CT, DE, DC, FL, GA, HI, ID, IL, IN, IA, KS, KY, LA, ME, MD, MA, MI, MN, MS, MO, MT, NC, ND, NE, NV, NH, NJ, NY, OK, OK, OR, PA, PR, RI, SC, SD, TN, TX, UT, VT, VA, WA, WV, WI, WY}
TelephoneNumber	text	-
Sex	text	{M, F}

Primary Key

Every entity has a special attribute or subset of attributes to uniquely identify each instance of the entity. The attribute or subset of attributes that uniquely identifies an instance of the entity is called the *primary key*. Primary keys are essential to the operation of a database because they are the only attributes shared between entities. For example, an information model can have the entity `Student` to describe the students in a university. The entity describes the class of students. To identify a unique student you need to know their `StudentID`, which is the primary key for this entity. `StudentID` provides a unique identifier for each student (instance) of the `Student` entity. The *entity integrity* rule states that every entity must have a primary key, and the values in the primary key must be unique and not null. Not null means that when data are input into the entity there must be a value inputted; you cannot leave it blank, which is called null.

The primary key can be a single attribute or a group of attributes. If it is a group of attributes, we call it a compound primary key. Figure 13.3 shows some example primary keys. Oftentimes, when no suitable primary key exists naturally, an enterprise will define an

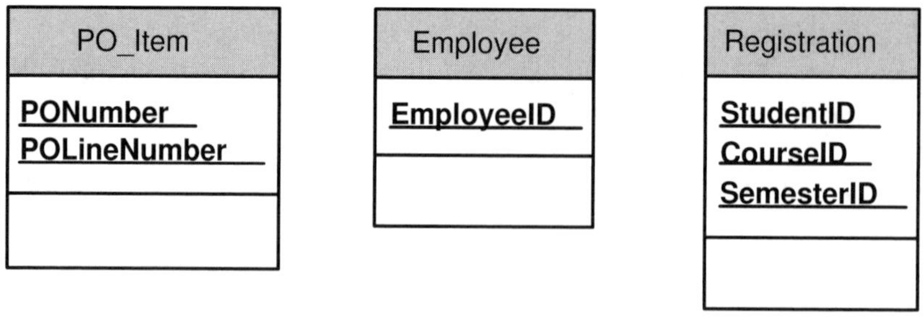

FIGURE 13.3
Example primary keys.

artificial primary key. For example, PurchaseOrderNumber, StudentID, or EmployeeID are all artificial primary keys. This is especially common, since for security and privacy reasons, most enterprises are forbidden from using government-issued numbers such as the Social Security Number in the U.S.

Relationship

A *relationship* describes the manner in which members of one entity are associated with the members of the same entity or another entity. Each end of the relationship has a cardinality that describes the number of entity members that are included in the relationship. The allowable cardinalities are shown in Figure 13.4. These cardinalities enumerate the minimum and maximum number of instances that are involved in a relationship. The minimum number in a relationship is either 0 or 1 and the maximum number is either 1 or many. The relationships also have a verb phrase to describe the relationship. The verb phrase is next to the relationship line and has a '/' to separate the left-to-right verb phrase from the right-to-left verb phrase.

In the model, the bi-directional relationship is depicted as a single line connecting the entities. The bi-directional relationship is read as if it represents two separate, single-directional relationships. Figure 13.6 shows a relationship between the Ship entity and the Cabin entity, which is read from left to right as,

One Ship has one or more Cabins

The same relationship is then read from right to left as,

One Cabin is in one Ship.

Notice when reading the relationship, you always start with a single instance of the entity (i.e., "one") and use the relationship verb phrase and cardinality to express how it is related to the other entity.

Foreign Key

The relationship between two entities is made by the migration of a primary key to the foreign key. A *foreign key* is an attribute (or attributes) in a table that refers to the primary key of another table. The *referential integrity* rule states the value of a foreign key must equal the value of the corresponding primary key it references or be completely null. In the relationship between Passenger and Reservation in Figure 13.5, the attribute PassengerID is a primary key in Passenger and a foreign key, as denoted by FK1, in Reservation. This means that any value of PassengerID in Reservation must match a value in Passenger or be null.

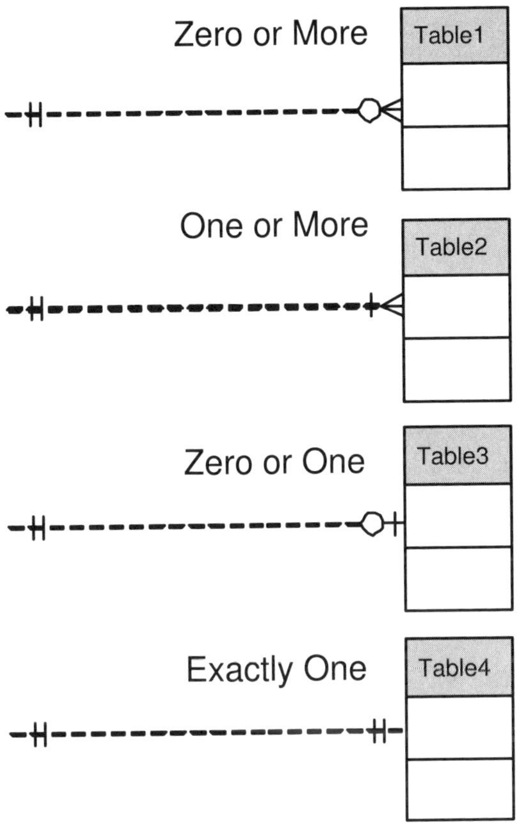

FIGURE 13.4
Relationship cardinalities.

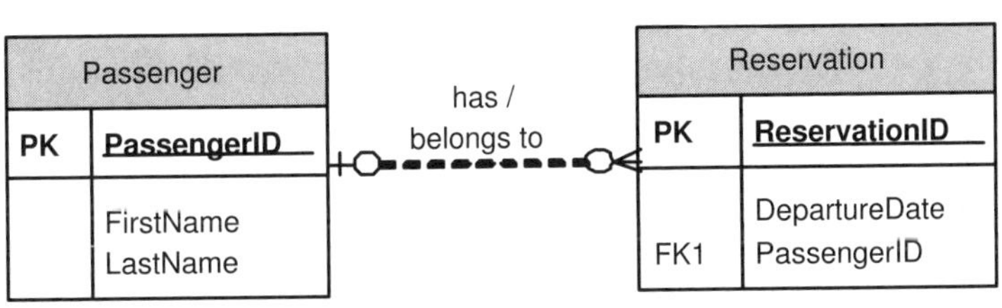

FIGURE 13.5
Foreign key definition.

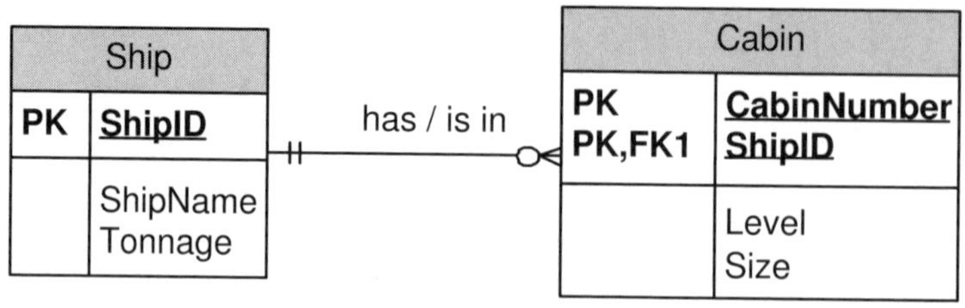

FIGURE 13.6
An identifying relationship.

Primary Key to Foreign Key Migration

In an entity-relationship model the relationships are created by the migration of primary keys to foreign keys. Rules governing the migration of primary keys to foreign keys are:

1. All primary keys must migrate from the one side of a relationship to the many side of the relationship.
2. The entire primary key must migrate across the relationship.
3. Only primary keys migrate across relationships.
4. When there is a compound primary key, each component of the primary key must migrate together.

The first rule implies that an entity-relationship model only has 1-to-many relationships, which is correct for the final entity-relationship model. During the intermediate development of the model, the team might identify many-to-many relationships but these relationships must later be "resolved" by introducing associative entities to replace the many-to-many relationships. The last rule says that when the primary key consists of two or more attributes, these attributes must stay together when they migrate across the relationship.

Non-identifying and Identifying Relationships

Relationships are either identifying relationships or non-identifying relationships depending on how the primary key migrates to become the foreign key. An *identifying relationship* is when the existence of one entity depends on another. In Figure 13.6, a `Cabin` depends on the existence of the `Ship` and is depicted as a solid line connecting the two entities. In identifying relationships, the primary key of the parent entity (on the one side of the relationship) migrates to become both a foreign key and part of the primary key of the child entity (on the many side of the relationship). In this model, the relationship is identifying because if we do not know the `ShipID` then we cannot identify a particular cabin. For instance, just knowing cabin number 703 is insufficient because the cruise line has many ships that have cabins with the number 703. So, we would not be able to identify a specific cabin. By also knowing the `ShipID` then we can identify a specific cabin.

A *non-identifying relationship* is when there is no existence dependency between the entities. In Figure 13.7 an instance of the entity `Employee` can exist without being related to an instance of the entity `Department`. In the non-identifying relationship, we can identify an employee independently of the department he or she works in. The primary key `DepartmentID` of the parent entity migrates and becomes a foreign key of the child entity, but not part of the primary key as was done in the identifying relationship. A non-identifying relationship is depicted as a dashed line.

In an entity-relationship model, all the relationships must be either a one-to-many or

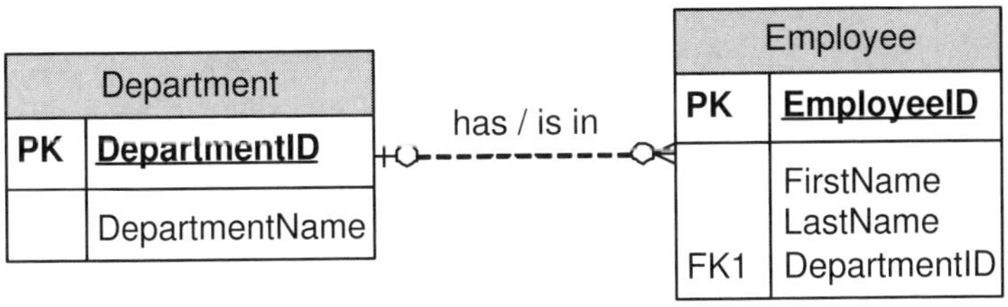

FIGURE 13.7
A non-identifying relationship.

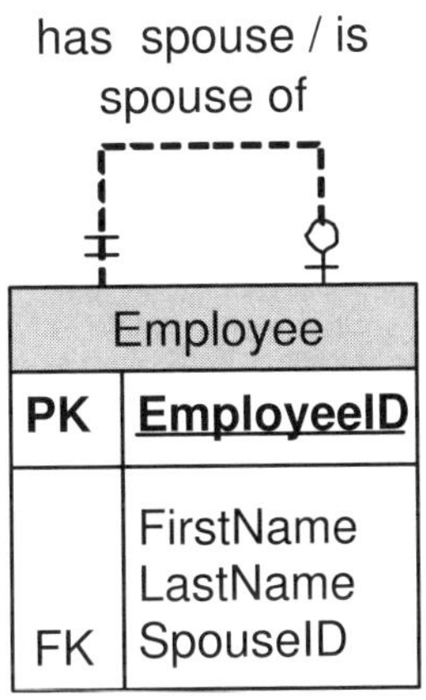

FIGURE 13.8
Recursive relationship.

one-to-one type of relationship. In every relationship, the primary key must migrate from the one side of the relationship to the many side of the relationship. On the many side of the relationship, the attribute that migrated is called the foreign key. The *foreign key* is either a single attribute or a combination of attributes whose values must either match the values of the primary key in the other entity or be null. It is via the migration of primary keys to foreign keys that information is related throughout the information model.

Relationship Degree

The degree of a relationship indicates the number of entities that participate in the relationship. A unary relationship is when the relationship involves a single entity, a binary relationship involves two entities, and a ternary relationship involves three entities.

Unary relationships are recursive relationships between instances of the same entity. These types of relationships can occur in many situations. Figure 13.8 shows a recursive relationship in which the model allows for the possibility of two employees being married. In the relationship, the primary key `EmployeeID` migrates and becomes the foreign key SpouseID of another instance of the entity because you cannot have two attributes with the same name.

The employee-to-employee relationship described above is a one-to-one relationship. It is also possible to have a many-to-many recursive relationship such as shown in Figure 13.9 for a course and its prerequisite courses. The business rules are that a `Course` can have as a prerequisite zero or more `Courses`, and a `Course` can be the prerequisite for zero or

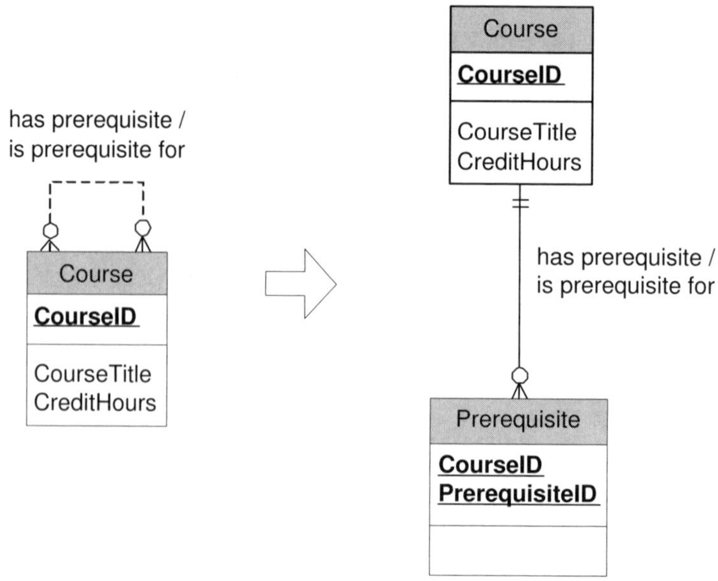

FIGURE 13.9
Many-to-many recursive relationship.

more `Courses`. The many-to-many relationship cannot be implemented as shown, instead we need to introduce another entity called `Prerequisite` and redraw the relationship.

Figure 13.10 shows a ternary relationship between `Passenger`, `Cabin`, and `Cruise` through the `Rerservation`. A ternary relationship cannot be done directly, it must go through an entity that joins all the entities. It is possible to have higher degree relationships but it is not common.

13.3 Standardization and Semantics

One of the main goals of the information model is to standardize terminology of the information content of the system. This avoids misunderstanding that can affect the entire enterprise system project and the system implementation. The information model formalizes the semantics to avoid any confusion. Semantics define the meaning of the model, compared to the syntax, which defines the grammar and rules of the model. Seemingly small modifications of an information model can have significant ramifications for describing how the enterprise operates. Figure 13.11 shows three very similar model fragments. Each model is for an online bookstore and shows the relationship between books and the books they have in inventory. The book entity describes all the information about a book and the inventory book entity describes all the information related to an actual book kept in inventory. Each model fragment shown in the figure conveys a different semantics about the enterprise's operations. In the top model, `Book` has an identifying relationship with `InventoryBook`, which means the enterprise only keeps inventory for books in their database. The second model has a non-identifying relationship. Since the foreign key, `ISBN`, can be null, then it is possible this store will have a `InventoryBook` for a `Book` that is not defined in their system. The third model also has a non-identifying relationship. But unlike the second model, it

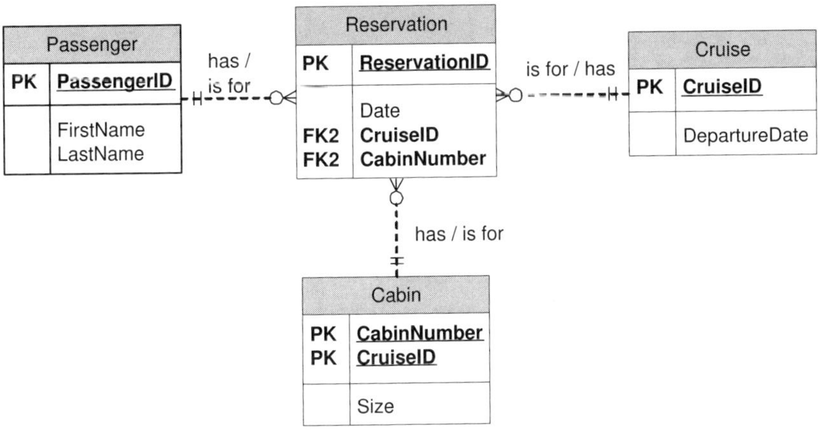

FIGURE 13.10
Ternary relationship.

excludes zero in its cardinality. In this model, if a `Book` is in the system, then it has at least one `InventoryBook`. In the second model, if a book is in the system, there may or may not be any kept in inventory. These differences describe different store operations. The second model is for a store that has the business philosophy that it will maintain extensive information on many different books that it does not have copies of, but is willing to obtain a copy if a customer expresses an interest in that book. Whereas, the third model describes a store philosophy that it only carries inventory if it has a description of that book in its system.

The following excerpt describes a more humorous account of what can happen when two parties have different concepts of what a single term means. This joke was told on March 7, 1960 by Jack Harold Parr (1918-2004), an American talk show host for *The Tonight Show*.

> An English lady, while visiting Switzerland, was looking for a room for a more extended stay, and she asked the schoolmaster if he could recommend any to her. He took her to see several rooms, and when everything was settled, the lady returned to her home to make the final preparations to move. When she arrived home, the thought suddenly occurred to her that she had not seen a "W.C." [Water Closet, a euphemism for bathroom] around the place. So she immediately wrote a note to the schoolmaster asking him if there were a "W.C." near the room. The schoolmaster was a very poor student of English, so he asked the parish priest if he could help in the matter. Together they tried to discover the meaning of the letters "W.C.," and the only solution they could come up with for the letters was for a Wayside Chapel. The schoolmaster then wrote the following note to the English lady:
>
> Dear Madam:
>
> I take great pleasure in informing you that the W.C. is situated nine miles from the room that you will occupy, in the center of a beautiful grove of pine trees surrounded by lovely grounds. It is capable of holding about 229 people and it is only open on Sunday and Thursday.
>
> As there are a great number of people who are expected during the summer months, I would suggest that you come early; although, as a rule, there is plenty of standing room. You will no doubt be glad to hear that a good number of people bring their lunch and make a day of it. While others who can afford to go by car arrive just in time.

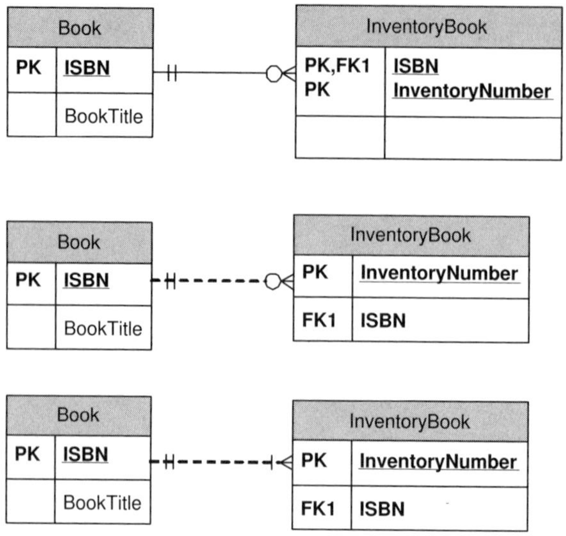

FIGURE 13.11
Model semantics.

I would especially recommend that your ladyship go on Thursday when there is a musical accompaniment. It may interest you to know that my daughter was married in the W.C. and it was there that she met her husband.

I can remember the rush there was for seats. There were ten people to a seat ordinarily occupied by one. It was wonderful to see the expression on their faces. The newest attraction is a bell donated by a wealthy resident of the district. It rings every time a person enters.

A bazaar is to be held to provide plush seats for all the people, since they feel it is a long felt need. My wife is rather delicate, so she can't attend regularly. I shall be delighted to reserve the best seat for you, if you wish, where you will be seen by everyone.

For the children, there is a special time and place so they will not disturb the elders.

Hoping to have been of service to you, I remain,

Sincerely,

The Schoolmaster.

13.4 Summary

This chapter describes information modeling. An information model represents the static structure of the information content of a system. To depict the information model the entity-relationship model is presented. An entity-relationship model has three components: entities, their attributions, and the relationships between the entities. Every entity must have a primary key that obeys the entity integrity rule. The relationships between entities must obey the referential integrity rule. Relationships are described by their cardinality and degree. Most relationships will be binary, but unary and ternary relationships are also possible. The attributes in an entity are described by their datatype and domain, if applicable.

Our description of the entity-relationship model focused on the syntax of the model symbols and how they are put together. However, the model also conveys the semantics of the system. During the course of creating an entity-relationship model the team will likely need to generate many unique codes and so forth – this is one of the goals of information model, to generate standard information definitions.

The information model is created during the analysis phase. It documents the information requirements for the system. The deliverable is the entity-relationship model. The next chapter continues with entity-relationship modeling and explains how to create the database definition model.

Review Questions

1. List and summarize each of the three schemas.
2. Describe a benefit of the three-schema architecture?
3. What is the purpose of information modeling?
4. Define an entity and list the five types of entities.
5. Differentiate between an entity and the instance of an entity.
6. What is the difference between a data type and a domain?
7. In a university database, why would the student's name not be a good primary key?
8. What does the rule that a primary key cannot be null mean, and why do we need that rule?
9. What is an appropriate data type and domain for AutoManufacturer in a data model for a car dealership?
10. What is the relationship between a foreign key and a primary key?
11. Explain the difference between an identifying relationship and a non-identifying relationship.
12. What is the purpose of introducing associative entities into a model?
13. Explain how semantics are embedded in the information model.

Exercises

1. A faculty is uniquely identified by its FacultyCode. A faculty is also described by its first name, last name, department code, building number, room number, and telephone number. A faculty can have one or more degrees. Degrees are the type (e.g., PhD, MS, BS, etc.), the major, and the year awarded. Draw the entity-relationship model to represent this information.
2. A `book` has one or more `authors`. An `author` may write one or more `books`. A `book` has one or more `chapters`. A `chapter` is part of only one `book`. Each `book`

is uniquely identified by its ISBN. An `author` is uniquely identified by his or her First Name, Middle Initial, and Last Name. A `Chapter` is uniquely identified by the chapter number and book ISBN.

(a) Draw the entity-relationship diagram for the above problem description showing the entities, relationships, and primary key to foreign key migration. If needed, create an associative entity to resolve any many-to-many relationships.

(b) Is the relationship between `book` and `chapter` an identifying relationship or non-identifying relationship?

3. A `music CD` has one or more `artists`. An `artist` may perform on one or more `music CD`s. An `artist` may sing one or more `songs`. A `music CD` has one or more `songs`. A `song` is part of one or more `music CD`s. Each `music CD` is uniquely identified by its SKU. An `artist` is uniquely identified by his or her First Name and Last Name. A `song` is uniquely identified by the song name and artist first name and last name.

   (a) Draw the entity-relationship diagram for the above problem description showing the entities, relationships, and primary key to foreign key migration. If needed, create an associative entity to resolve any many-to-many relationships.

   (b) Is the relationship between `artist` and `music CD` an identifying relationship or non-identifying relationship?

4. A soccer `team` has many `players`. A `player` may play on only one `team`. A `player` may score zero, one, or many `goals` during a `match`. A `goal` is attributed to only one `player`.

   Each team is uniquely identified by its Team ID. A player is uniquely identified by his or her first name, last name, TeamID, and jersey number. A goal is identified by who scored the goal and by the match number. A match is identified by the match number.

   (a) Draw the entity-relationship diagram for the above problem description showing the entities, relationships, and primary key to foreign key migration. If needed, create an associative entity to resolve any many-to-many relationships.

   (b) Is the relationship between `player` and `team` an identifying relationship or non-identifying relationship?

5. In a data model for a supplier, specify an appropriate data type and domain for the following attributes: SupplierCode, SupplierName, Street, City, State, ContactTelephoneNumber, and SupplierZip.

6. An aerospace manufacturer has an entity in its production information model called component. The following business rules apply:

   (a) A component can contain zero, one, or many other components.

   (b) A component can be part of zero, one, or more other components.

   (c) A component has a unique ComponentCode, a description, and tracks the number of parts in stock.

   Create the entity-relationship model for this recursive relationship.

7. A test preparation company has the following business rules. They operate many examination centers identified by a unique centerID. An examination center can offer many different exams. An exam is identified by the exam type and exam code. An exam can be offered at none, one, or many examination centers. An applicant can take zero, one, or many exams but at different times. An applicant can take the same exam type multiple times, but of course at different times. An applicant is identified by his or her social security number.

   Create the entity-relationship model for this system showing all entities, primary keys, foreign keys, and relationships. If the above information is insufficient, state reasonable assumptions you make in order to complete the model.

8. A Health Maintenance Organization (HMO) allows its clients to select one and only one doctor, called a primary care physician. The HMO has a list of doctors available. Each client is assigned a 5-digit client ID number. The client name, age, address, and company they work for must also be included in the system. Each doctor has many clients and is described by a primary care number, which is 4 digits. Doctors must also have information on their name, address, and specialty. There are about ten speciality types in the system. Clients have a medical record that lists their identifying information (client id, name, etc) and their medical history including: medicines allergic to, dates of all doctor's visits, description of each visit, medicine received on past visit).

   Create the entity-relationship model for this system showing all entities, primary keys, foreign keys, and relationships. If the above information is insufficient, state reasonable assumptions you make in order to complete the model.

9. Two tables are shown below. The first table shows the trucks and where they are assigned. The primary key is `TruckNum`. The second table shows the Stations, and the primary key is `StationCode`. For these tables is the entity integrity rule broken? Is the referential integrity rule broken? Answer yes or no and explain why.

TruckNum	**StationCode**	**TruckType**	**PurchaseDate**	**TruckVIN**
01	201	FB	Jun 13 03	ASD-12234-22-0987
02	201	HT	Aug 30 99	LKJ-97899-23-0981
03	205	FB	May 22 01	OUI-94583-33-2349
04	203	HT	Dec 1 03	JOK-98234-98-1234
05		DT	Nov 15 05	HKI-83543-88-8932
06	204	DT	Oct 30 04	KDF-23423-89-9888
07	203	FB	Oct 29 04	TEF-35939-91-8728

StationCode	**City**	**State**	**StationMgr**
201	Atlanta	GA	Richard Anderson
202	Dayton	OH	Mary Smith
203	Tampa	FL	Lucia Gomez
204	Dallas	TX	Bob Marsh
205	Modesto	CA	Brian Quinnlin

10. Tuffy Ltd makes specialty vehicles by contract. The company operates several departments, each of which builds a particular vehicle, such as a limousine, a truck, a van, or an RV.

    When a new vehicle is built, the department places an order with the purchasing department to request specific components. Tuffy's purchasing department is interested in creating a database to keep track of orders and to accelerate the process of delivering materials.

The order received by the purchasing department can contain several different items, each listed as a separate line item. An inventory is maintained so that the most frequently requested items are delivered almost immediately. When an order comes in, it is checked to determine whether the requested item is in inventory. If an item is not in inventory, it must be ordered from a supplier. Each item may have several suppliers.

Create the entity-relationship model for this system showing all entities, primary keys, foreign keys, and relationships. If the above information is insufficient, state reasonable assumptions you make in order to complete the model.

Bibliography

[1] D. Schenck and P. Wilson. *Information Modeling the EXPRESS Way.* Oxford University Press, New York, NY, 1994.

14

Information Design

"Where is the Life we have lost in living? Where is the wisdom we have lost in knowledge? Where is the knowledge we have lost in information?" – T.S. Eliot (1888 - 1965) poet.

This chapter describes a methodology to build an entity-relationship model. In this chapter we present best practices for the design of an information model with an eye towards its implementation as a relational database. The chapter introduces SQL, a query language for relational databases. The chapter explains how to use the four SQL commands necessary to create, read, update, and delete data. To deliver a quality database, it needs to be non-redundant and avoid data anomalies. This is done through normalization. The chapter describes how to bring an information model to third normal form.

After completing this chapter, you should be able to:

- Describe the methodology to build an entity-relationship model.
- Resolve many-to-many relationships in entity-relationship models.
- Write the four SQL commands: Select, Insert, Update, and Delete.
- Describe the database anomalies that can occur.
- Normalize an information model to third-normal form.
- Create a data dictionary.

14.1 Method To Build Entity-Relationship Model

The entity-relationship model can be developed in two stages with three steps in each stage. There is some iteration and parallelization in the steps, but adherence to the sequence helps ensure proper consideration of all the components of the information model. The stages with their steps are:

1. Build the Business Model.
   (a) Modeling project initiation.
   (b) Entity definition.
   (c) Relationship definition.
2. Build the Database Design Model.
   (a) Primary key definition.
   (b) Resolve many-to-many relationships.
   (c) Non-key attribute definition.

TABLE 14.1
Source Material Log

Number	Name and Description	Source	Comments
1	Purchase order – form sent to purchasing department	Paper form	
2	Telephone book	Paper book (25 pages)	
3	Organization chart	Paper document	May be out-of-date

After Stage 1, the intermediary model developed is called the Business Model. The business model is a high-level description of the information content and relationships. The final deliverable after Stage 2 is a Database Design Model. It shows all the entities, their primary key and other attributes, the relationships between the entities, and the definition of each attribute. It is the final design model prior to database implementation.

14.1.1 Information Modeling Project Initiation

The modeling team should define the purpose and scope of the information model. Model conventions should be discussed and agreed upon. For example, naming conventions (is it `StudentID`, `StudentNum`, or `StudentNumber`), spelling agreements (is it `Dept` or `Department`), or notation agreements (`Employee_ID` or `EmployeeID`). Throughout this book, we prefer the later notation of not using underscores to connect parts of the name. In all cases of notation, do not leave spaces in a name (e.g., never have `Employee ID`) because computer languages such as SQL would interpret the space as the start of a new variable.

In this phase, the modeling team needs to devise a strategy for source data collection (refer back to Chapter 8 for a detailed discussion of data gathering). Source materials include all existing forms, reports, manuals, instructions, Web pages, and other documentation that exists in the enterprise. Additionally, interviews should be conducted to determine what are the entities, relationships, and attributes that are needed in the model. The source data materials should be collected, and then they should be logged into a Source Materials Log.

The *Source Material Log* (see Table 14.1) documents and tracks all the source material collected to build the data model. Each source material is given an identification number. The source material name and description is recorded. The type of source is recorded (e.g., paper form, Web page, policy manual). Additionally, the analyst should record any comments with respect to the source material item.

For each item in the source material log, you create an entry in the source data list. The *Source Data List* (see Table 14.2) includes all relevant data items that are candidates for inclusion in the data model. Each data list item is given an identification number, a name, data properties, and is cross-referenced to the source material log.

14.1.2 Entity Definition

To identify entities you go through the source data list and find all the nouns. Some helpful questions to identify entities are:

- Can it be described?
- Does it have qualities?

TABLE 14.2
Source Data List

Number	Name	Source Material Number	Data Properties	Comments
111	Purchase order	1	14-digit alphanumeric character	Document name
112	Purchase order number	1, 22, 35	5-digit number	Tracks PO, is unique
113	Vendor number	1, 22, 15	10-digit alphanumeric character	Unique identity of vendor

- Are there several instances of it?
- Can one instance be separated or identified from another instance?
- Does it refer to or describe something else? For this last question, if the answer is yes, then the item is probably an attribute rather than an entity.

TABLE 14.3
Entity Identification

Entity Type	Description
Person	An entity can describe a person in the system such as student, employee, customer, guest, patient, doctor, spouse, child, or pet.
Place	An entity can describe any geographical location in the system such as a classroom, building, state, sales zone, store, or country.
Event	An entity can describe any event in the system such as arrival, trip, departure, award, or meeting.
Object	An entity can describe a physical real object in the system such as part, inventory, movie, product, tool, or machine.
Transaction	An entity can describe a transaction the system processes such as a reservation, sale, contract, purchase order, deposit, or withdrawal.
Concept	An entity can describe a concept in the system that does not have a physical manifestation such as a class, semester, procedure, account, block of time, or work assignment.
Group	An entity can describe a group of people such as a department, team, division, supplier, or company.

In analyzing the source data list, you will often find many nouns that are instances of an entity. Recall an entity describes a class of similar objects, so the analyst must be able to identify when multiple instances must be synthesized into a single entity (see Figure 13.2 for illustration).

The entity pool (see example shown in Table 14.4) is used to document the candidate entities that were identified. The entity pool should be exhaustive, listing every possible

TABLE 14.4
Entity Pool

Entity #	Entity Name	Entity Definition	Source	Comments
E1	Purchase Order	A document authorizing the purchase of goods	2	
E2	Carrier	A vendor who provides delivery service	2, 3	
E4	Department	The organization unit that employees are assigned to	2, 3, 5	
E5	Invoice	A document that requests payment for services from a client	3	

entity. Later, it can be pared down by combining terms into a single entity or removing redundant entity definitions.

When analyzing the entity pool, the following rules need to apply:

- Each entity must have a unique name, and the same meaning must always apply to the same name. Furthermore, the same meaning cannot apply to different names unless the names are aliases.
- An entity must have a primary key, which is one or more attributes that uniquely identify the entity. This is called the *entity integrity rule.*

14.1.3 Relationship Definition

Once all the entities are identified, the next phase is to define the relationships. The ultimate goal is to have only parent-child relationships in the model. To define relationships between entities, we suggest the following two steps. First, determine whether two entities are related. To help do this, the modeling team can create an Entity-Relationship Matrix, which is a binary matrix simply showing whether two entities have a relationship between them. Once you establish the existence of a relationship, the relationship must be specified with the cardinality and a relationship phrase. Remember the relationships are bi-directional, so to define the cardinality first start with one entity and say, "one instance of this entity is related to how many instances of this other entity." Then ask the same question but in the other direction.

The completion of this phase results in the business model (see example business model in Figure 14.1). The business model describes the entities in the model and the relationships between those entities. The business model is used to go back to the stakeholders to validate and verify the model before the details of the information model are added. The stakeholders can determine whether anything is missing from the model, they can clarify any relationships that are not well defined, they can identify any relationships that might be missing. Once the business model is verified and validated, then the details can be added in the next phases.

14.1.4 Primary Key Definition

In this stage, the primary key is defined for each entity. A good primary key uniquely identifies every instance; it has a value that does not change over the life of the instance it

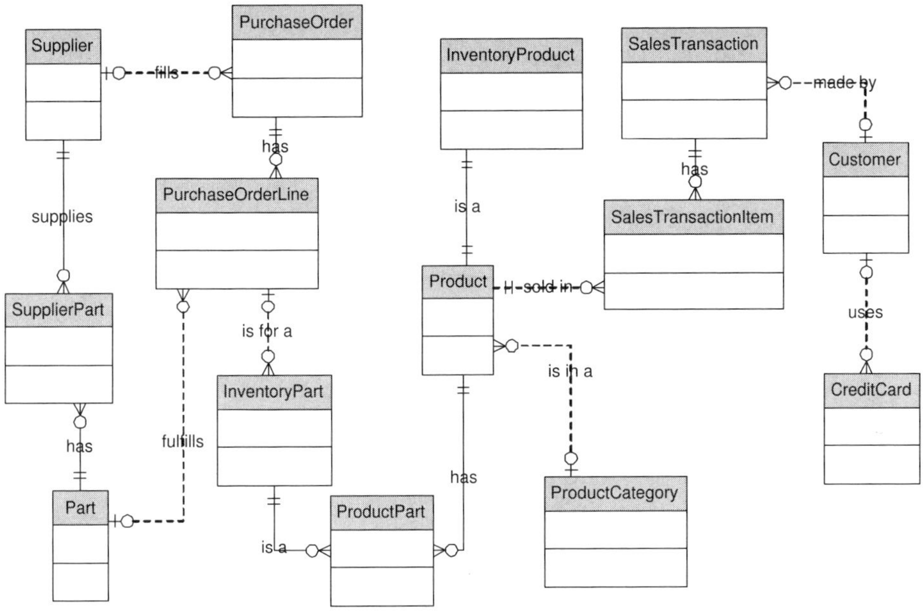

FIGURE 14.1
Business model.

represents; it is preferably a single key; and lastly, it has a data type of either integer or short, fixed-width alphanumeric characters. Of these criteria, the foremost criterion for a good primary key is that it uniquely identifies every entity instance.

Sometimes it is possible to identify naturally occurring attributes that have the properties of a primary key. For example, the ISBN number on books might be an excellent candidate for a book store information model. Other times, it may be necessary for the information modeling team to create a code or identification number that can serve as the primary key.

One practice is for the system to generate an internally, unique primary key for each instance. Some large ERP systems will do this for the data in their systems. The benefit of this approach is it guarantees a unique primary key, etc. For example, a university using PeopleSoft's ERP package will create a 7-digit `StudentID` that is unique. This number is used throughout the university to identify the student for course registration, grading, library, parking, etc. A disadvantage is that it creates yet another number for the students to remember since the `StudentID` is then used to access all university systems.

Some attributes although they are unique make a poor primary key choice. For example, in the U.S. the social security number (SSN) is a unique ID issued by the federal government. Several problems are associated with its use as a primary key. Many people view it as private and do not want it used for such purposes in a database since its wide spread availability could lead to fraud. Another problem is what to do with somebody who does not have an SSN, such as a non-resident alien or just a tourist visiting the U.S.? The modeling team needs to discuss these issues while developing the information model.

14.1.5 Primary Key to Foreign Key Migration

Primary keys must migrate through all specific relationships from the parent to the child entity (see Figure 14.2). The entire primary key must migrate; you never split up a compound

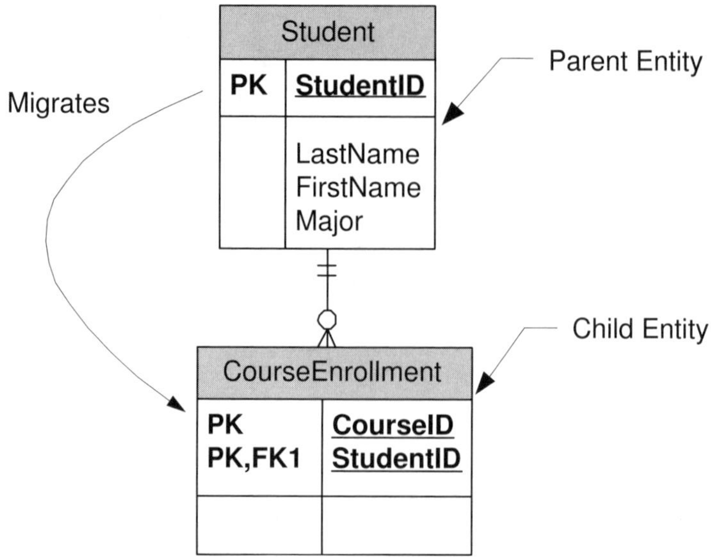

FIGURE 14.2
Primary key to foreign key migration.

primary key and migrate only part of it. Only the primary key migrates in a relationship; this is done so as to minimize redundancy in the database model. Consequently, non-key attributes will appear in only a single entity in the data model. The repercussions of this design requirement are significant. Consider a university database that has the attribute `Address` in the `Student` entity. What would happen if the `Address` attribute was located in multiple entities? Then when a student moves, the university would have to change the address in multiple locations in the database. While there are triggers and other means to automate this, it is far easier if the `Address` attribute only appears in a single location. Moreover, if the university does not have integrated database systems, then students might be in the position of having to change their address with each office (registrar, academic department, parking and traffic) that they deal with.

14.1.6 Resolve Many-to-Many Relationships

It is possible that some many-to-many relationships were defined in the business model. The final information model does not allow these non-specific relationships. To resolve a many-to-many relationship we usually add an associative entity such that the associate entity replaces the many-to-many relationship between the entities. Figure 14.3 shows a many-to-many relationship between `Cruise` and `Crew` such that a single crew is assigned to zero or more cruises and a single cruise can have one or more crew. To resolve the many-to-many relationship, we add the associative entity, `CrewAssignment`, and create two one-to-many relationships between the original tables and the new entity. Notice the original entities will have the cardinality "1" on them and the many side of the relationships will be on the new associative entity. The primary keys from both entities migrate and become part of the primary key for the associative entity.

If the data model has redundant relationships defined as more than a single path between any two entities, then usually, the direct relationship is the redundant relationship and can be removed from the model (see example in Figure 14.4).

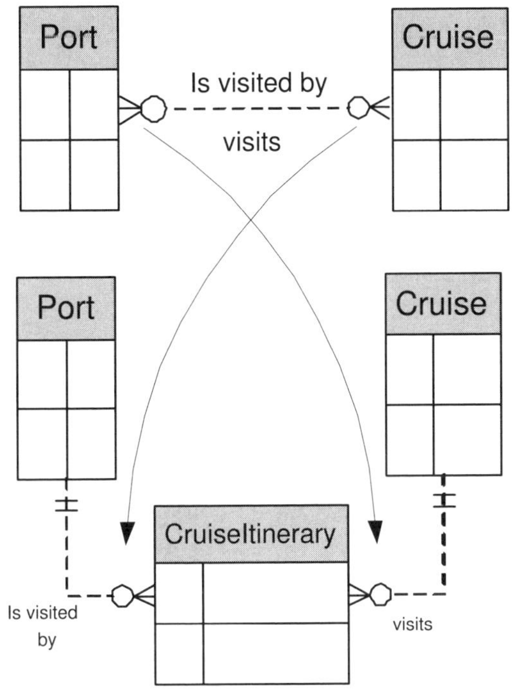

FIGURE 14.3
Resolving a many-to-many relationship.

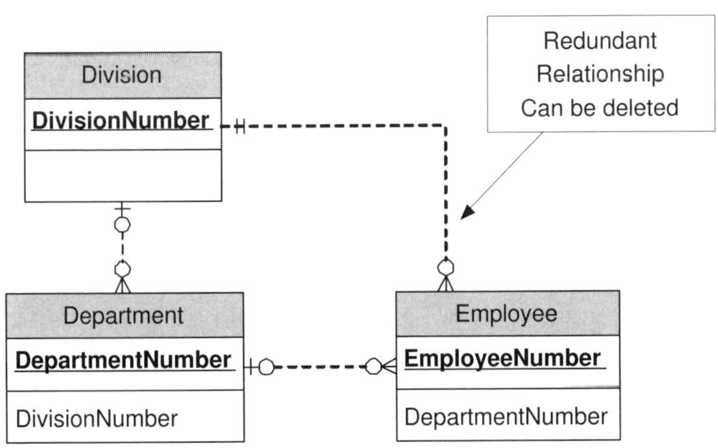

FIGURE 14.4
Redundant relationship.

14.1.7 Database Rules

Two database rules help maintain a good database design. The *entity integrity* rule says that all primary key entries are unique, and no part of the primary key can be null. Entity integrity guarantees that each instance of an entity can be uniquely identified and ensures that foreign key values can properly reference primary key values.

The *referential integrity* rule says that the foreign key's values are required to match those of the primary key or be completely null. Referential integrity ensures that the data in a child entity can be matched up with the data in the parent entity. One implication of referential integrity (when the information model is implemented) is that deleting an instance of an entity that has a matching foreign key is blocked. These two rules are important to the implementation and operation of a database.

14.1.8 Non-Key Attribute Definition

The focus of the information design is identifying the entities, their relationships, and the primary keys. Once this crucial aspect is completed, then you add all the non-key attributes. Every data element that the enterprise needs to include in the information model should be described by one or more attributes. The data attributes are assigned to the entity that they most belong to. The information model should not have redundant non-key attributes, each piece of data should appear in only one entity in the entire model. For each attribute, you need to define the data type and domain. This is documented in the data dictionary.

14.2 Relational Database

The entity-relationship model can be implemented in a relational database management system (DBMS). The DBMS is a large record storage system that maintains all the data in a system. The DBMS allows a user to store, read, update, and delete data in the system without having to know how the data are physically stored or organized. This is because the DBMS handles all the mappings between the conceptual schema and the physical schema. The user can work solely with the logical structure of the data, which is much easier.

In the entity-relationship model we have the terms entity, attribute, and relationship. These names change in the database.

Information Model Term	Database Term
Entity	Table
Attribute	Field
Entity instance	Record

Each entity in the entity-relationship model becomes a table in the DBMS. The attributes become fields in the DBMS. A row in the table is called an entity instance in the model but is called a record in the DBMS. A properly designed entity-relationship model has almost a one-to-one correspondence with its database implementation. This makes it easy to quickly build a database for the model.

14.2.1 SQL

The Structured Query Language (SQL), pronounced "sequel," is a declarative language for accessing data in a database (see Figure 14.5). SQL has a data definition language and a data

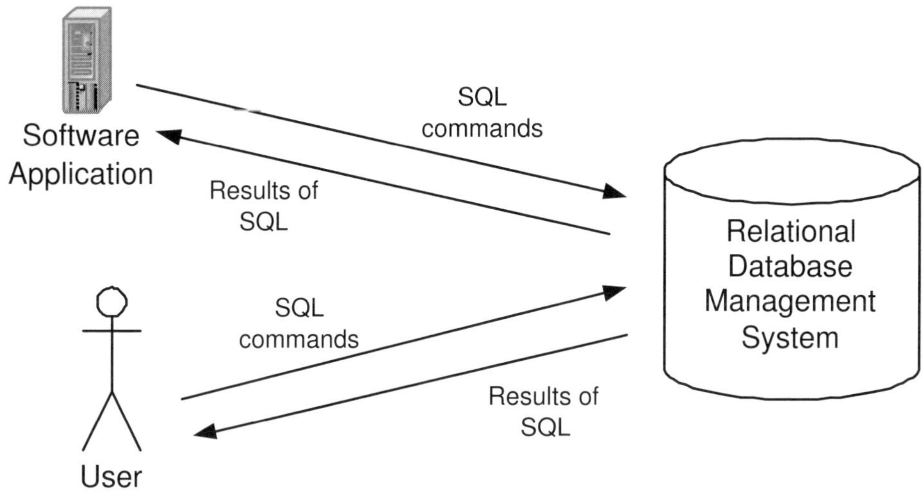

FIGURE 14.5
Relational database management system.

manipulation language. Here we concentrate on only a small, but important, subset of the language that allows us to create, read, update, and delete data in the system (commonly abbreviate CRUD). In Chapter 10 it was explained that CRUD describes the four main operations that are done on a database. Here we show how CRUD is implemented in SQL using the SQL `Insert`, `Select`, `Update`, and `Delete` commands. SQL is not case sensitive and ignores extra spaces, but good practice says you structure the SQL commands to make them easy for people to read.

SQL Insert

SQL uses the `Insert` command to enter new data into a table. The `Insert` command's syntax is as follows:

```
INSERT INTO TableName
([AttributeName1], [AttributeName2], ...)
VALUES (Value1, Value2, ...);
```

In the SQL command uppercase text denotes SQL commands, and *slanted text* denotes where the user enters the specified value. The `[` and `]` surround what is an optional value to enter. In the `Insert` command, if you enter all the attribute values in the table, then you do not need to specify the attribute names. Otherwise, if you enter only partial data for the record, you need to specify which attributes the values correspond to. In the `Insert` command, there is a one-to-one correspondence between the attributes and values.

To illustrate, suppose we are entering a new shore excursion for the Bahamas. The SQL `Insert` is:

```
INSERT INTO ShoreExcursion
(ExcursionCode, ExcursionName, Duration, PortCode)
VALUES ('E502', 'Swim with Sharks', 4, 'BAH');
```

The text data and date data are entered with apostrophes delimiting them, and numerical data are entered without apostrophes. Also, not all the attributes were inserted, for instance, we did not insert values for the `Cost`, `WheelChairAccessible`, or `ActivityLevel` attributes. Note the attribute names and the values are entered in the same order. In every Insert command the primary key is the only attribute that cannot be left null. All SQL commands end with a semicolon.

SQL Select

To read data from the database SQL uses the `Select` command. The `Select` command's syntax is as follows:

```
SELECT [DISTINCT] AttributeName1, AttributeName2, ...
FROM TableName(s)
[WHERE] Predicate
[ORDER BY] Attribute ASC or DESC;
```

In the `SELECT` clause, the optional `DISTINCT` command removes any duplicates from the returned data. Each attribute that is desired in the output is listed by name. Alternately, the symbol * is a wildcard that will return all the attributes in the table. The `FROM` clause is used to list the table or tables where the data are to be retrieved from. The optional `WHERE` clause is used to specify a predicate to filter the records. A predicate is any statement that has one of two truth values: true or false. A predicate can be built using any of the following operators: =, >, <, AND, OR, NOT. The optional `ORDER BY` clause orders the output data in either ascending (ASC) or descending (DESC) order in the attribute specified. If you leave out the `ORDER BY` clause, then the output is ordered as it is found in the table, which is unordered.

The SQL `Select` command is very powerful and versatile. To simply read all the data in a table you could write:

```
SELECT *
FROM Cruise;
```

This command use the wildcard '*' without any `WHERE` predicate to return all the data in the table.

Suppose you wish to find a shore excursion for when the cruise stops in the Bahamas, where the duration is less than 3 hours, and the cost is less than $50. The command to obtain this data is:

```
SELECT ExcursionName, Description, Cost, Duration
FROM ShoreExcursion
WHERE Duration <= 3 AND Cost <= 50 AND PortCode = 'BAH';
```

Suppose the `ShoreExcursion` table has the data as shown in Table 14.5. Then the above `Select` command will return the following data:

ExcursionName	Description	Cost	Duration
Island Tour	A jeep tour of the island	$45	3

Of the four records in the table, only one record satisfies all the predicates in the WHERE clause. The second record is too expensive, the third is too expensive and too long, and the fourth is in the Virgin Islands and not the Bahamas.

If the data we want are in more than a single table, then we can join the tables together through the shared primary keys they have as a result of the primary key to foreign key migration. This is done in the WHERE clause of the Select command. Also, note there is no

TABLE 14.5
ShoreExcursion Table

Excursion Code	Excursion Name	Cost	Duration	Description	Wheel Chair Accessible	Activity Level	PortCode
E501	Island Tour	$45	3	A jeep tour of the island	Y	1	BAH
E502	Jet Skiing	$60	2	Jet skiing in the bay	Y	2	BAH
E504	Scuba Diving	$100	5	Dive off wreck	N	3	BAH
E610	Para-sailing	$50	3	Fly 100 ft over beach	N	2	VIR

requirement to output the primary key with the results. In fact, it is often the case that the user is disinterested in the primary key because it is a code or number with little meaning to the user.

```
SELECT Cruise.DepartureDate
FROM Cruise, Ship
WHERE Ship.ShipID = Cruise.ShipID;
```

In the command the "dot-notation" is used to distinguish which table the attributes are found in. The dot-notation is `TableName.AttributeName`. The command returns only the departure date from the cruise table. It does this when the value of the `ShipID` foreign key in the `Cruise` table matches the value of the `ShipID` primary key in the `Ship` table.

There is no limit to the number of tables that can be joined together as long as there are primary keys that can be matched to foreign keys.

SQL Update

If a data record already exists in the table, then to change any of the attributes the SQL Update command is used. The syntax is as follows:

```
UPDATE TableName
SET AttributeName = Expression [,attribute name = Expression, ...]
[WHERE Predicate];
```

To illustrate, we can update all the prices of the shore excursions in the Bahamas by $10 with the Update command:

```
UPDATE ShoreExcursion
SET Cost = Cost + 10
WHERE PortCode = 'BAH';
```

In this example, SQL recursively takes the current value of `Cost` and adds 10 to it to arrive at the updated cost.

SQL Delete

To delete an existing record or records from the database, the SQL `Delete` command is used. The syntax is as follows:

```
DELETE
FROM tablename
[WHERE predicate ]
```

As an example, the following command deletes all the shore excursions from the Bahamas.

```
DELETE
FROM ShoreExcursion
WHERE PortCode = 'BAH';
```

SQL Summary

To summarize, SQL is a query language that allows users to interact with databases. It contains both a data definition language and a data manipulation language. Here we concentrated on four commands in the data manipulation language that allow users to Create, Read, Update, and Delete (abbreviated CRUD) data. These actions on a database are performed with the SQL commands: `Insert`, `Select`, `Update`, and `Delete`.

TABLE 14.6
Anomalies Explained

Anomaly Type	Description	Example
Insert	A record cannot be inserted into a table unless information is first inserted into another table or phantom data are created.	In the cruise table, if we insert a new ship, we need to create a phantom cruise because it is part of the primary key.
Update	A record cannot be updated without changing data in many different places. This can be quite unwieldy.	In the cruise table, if we update the ship's name, then we need to update it in multiple rows.
Delete	A record cannot be deleted without deleting a record about a related entity.	In the cruise table, if we delete a cruise we also delete the ship.

14.3 Normalization

Normalization is the process of recognizing undesirable properties in the information model and converting them into a more desirable form. Normalization does two things: it removes data redundancy by having any piece of data appear in only one entity, and it ensures data dependencies make sense. By doing this, normalization leads to a data structure that will avoid insertion, update, and deletion anomalies (see Table 14.6 for an explanation of the possible anomalies).

Normalization occurs through a series of stages, called normal forms, such that each successive stage builds on the previous stage. To be in second normal form, a data model must first be in first normal form. Normalization is applied to the entities in the data model; all entities must adhere to the normal form to be declared in that normal form.

First Normal Form

A data model is in first normal form (1NF) when all the attributes are defined such that they can contain only single values. Alternately, we say first normal form requires no repeating groups, where a repeating group is an attribute that can contain multiple values or multiple occurrences of the same attribute. A common example of violating first normal form is shown in Figure 14.6 where the `Passenger` entity has an attribute called `Children`. A passenger can have zero, one, or many children, so it is possible the attribute can contain multiple values as occurs for the second record (see Table 14.7) – thus violating first normal form. To fix the data structure so that it obeys first normal form, we remove the attribute causing the problem, and create another entity named `Child`. The new `Child` entity has a relationship to the `Passenger` entity. The new structure, shown in Figure 14.7, allows for one passenger to have any number of child without violating first normal form.

Table 14.8 shows the cruise table for a cruise line. The `Cruise` entity is defined to show the ports visited during each cruise. Even though each attribute contains only single values, the names of the attributes `Port`, `Departure`, and `Arrival` repeat. This is a problematic structure because to accommodate a cruise with more than three ports in its itinerary, we need to add more attributes, but then this means that shorter cruises will have many blanks.

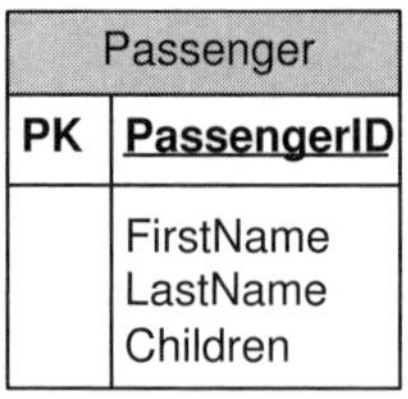

FIGURE 14.6
Example violation of first normal form.

TABLE 14.7
Passenger Table That Violates First Normal Form

PassengerID	FirstName	LastName	Children
52344-1	Bob	McGuire	-
52345-1	Yuly	Sanchez	Carlos, Gabriela
52345-2	Carlos	Sanchez	-

To fix this violation of first normal form, the repeating group consisting of the attributes `Port`, `Departure`, and `Arrival` are removed from the entity. These attributes are used to create a second entity, called itinerary. A one-to-many relationship is created between Cruise and Itinerary, the two primary keys migrate over the relationship to give the resulting data structure shown in Figure 14.8.

Once all the entities in the model are in first normal form, then the model is said to be in first normal form.

Second Normal Form

To be in second normal form (2NF), the model must be in first normal form and all non-key attributes must depend on the entire primary key. In other words, there are no partial dependencies in which a non-key attribute depends on only part of the primary key. It is only possible to violate second normal form in entities with compound primary keys (i.e., two or more attributes form the primary key).

Returning to the cruise example in Figure 14.8, we check whether each non-key at-

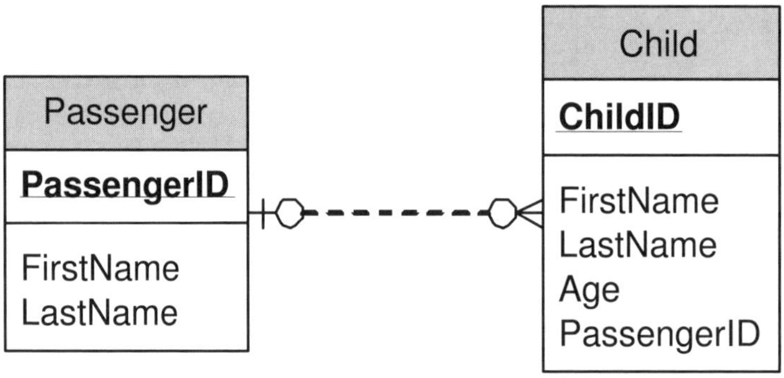

FIGURE 14.7
First normal form.

TABLE 14.8
Cruise Table

Ship	Name	Cruise	Departure	Port	Arrival1	Port1	Departure1	Arrival1	Port2	Departure2
25	Sea Wind	2008-05MIA	3/10/08	MIA	3/11/08	BAH	3/12/08	3/13/08	NQX	3/14/08
18	Coral Sky	2008-06FLL	3/10/08	FLL	3/12/08	KST	3/13/08	3/13/08	VXT	3/14/08
25	Sea Wind	2008-06MIA	3/14/08	MIA	3/15/08	BAH	3/16/08	3/17/08	NQX	3/18/08
12	Blue Fish	2008-03NQX	3/12/08	NQX	3/14/08	CAN	3/15/08	3/15/08	COZ	3/16/08
05	Starfish	2008-05MIA	3/15/08	MIA	3/16/08	NQX	3/17/08	3/19/08	COZ	3/20/08

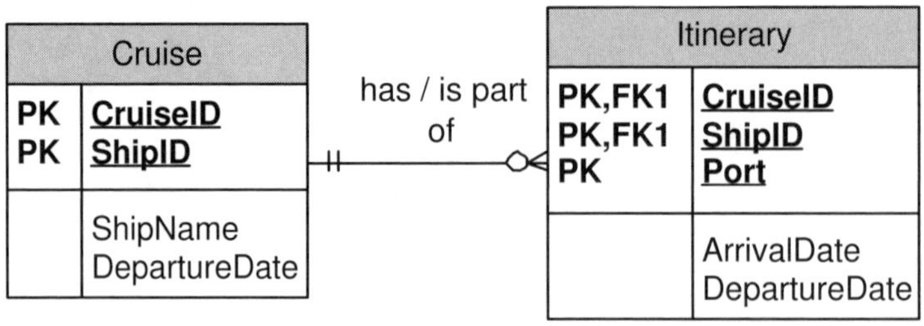

FIGURE 14.8
First normal form - cruise example.

tribute depends on the full primary key. The attribute `ShipName` only depends on the `ShipID` because if we know the `ShipID` then we can determine the `ShipName`. The attribute `DepartureDate` depends only on the `CruiseID` because if we know the `CruiseID` then we can determine the `DepartureDate`. Consequently, both `ShipName` and `DepartureDate` are only dependent on part of the primary key. Therefore, the `Cruise` entity violates second normal form.

In the `Itinerary` entity to know the `ArrivalDate` or the `DepartureDate` requires knowing the `CruiseID` and the `Port`, but not the `ShipID`. Thus, this entity also violates second normal form because `ArrivalDate` and `DepartureDate` are partially dependent on the primary key.

The problem with the structure is the `Cruise` entity should only contain information about the cruise and not the ship. To correct the problem we create another entity called `Ship`. All the attributes that had partial dependencies are removed from Cruise (in this case the `ShipName`) and put in the new `Ship` entity. We create a non-identifying relationship between `Ship` and `Cruise`, and therefore make `ShipID` a non-key attribute in `Cruise`. Doing this also requires us to remove `ShipID` from `Itinerary`. Figure 14.9 shows the new structure that is in second normal form.

Third Normal Form

A data model is in third normal form (3NF) when it is in second normal form and there are no transitive dependencies in the entities. A transitive entity is when one non-key attribute depends on another non-key attribute. In the cruise example, the `Cruise` and `Itinerary` entities have no transitive dependencies, but the `Ship` entity does have a transitive dependency (see Figure 14.10). The `Age` attribute depends on the `ConstructionDate` which depends on the primary key. To put the model in third normal form we only need to remove the `Age` attribute. It is not needed because the age of the ship can be calculated at run time. Oftentimes, attributes whose values are calculated based on other attributes violate third normal form such as in this case.

14.4 Data Dictionary

The information model relies on mutually agreed-upon set of data elements with clearly defined names and definition. To define, document, and standardize the data elements, we use a data dictionary. A *data dictionary* is a centralized repository of information about

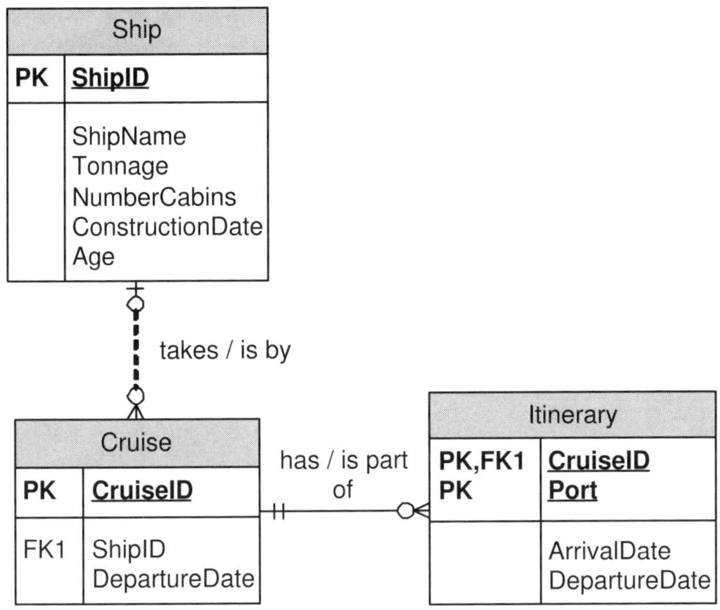

FIGURE 14.9
Second normal form.

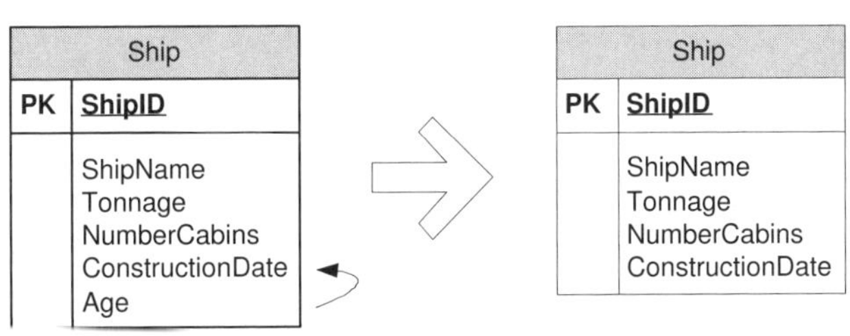

FIGURE 14.10
Resolution of third normal form.

data such as data type, meaning, relationships to other data, origin, usage, and format of each attribute. Table 14.9 shows a partial data dictionary.

14.4.1 Data Design

The information model design should be such that users do not need to type excessive amounts of data. For example, suppose the `Student` entity has attributes for the student's address including the state he or she lives in. The `State` attribute has the data type text. If the designer does not specify a domain, then a user of the system would be free to type in whatever he or she wants. This can lead to problems during operation of the system because different users may input different values for the same state. Some possible values might be "FL," "Florida," "FLA," and even misspellings such as "Florda." Now, when somebody wants to query the system to, for example, find all the students who live in Florida, then he or she needs to try to write the query to cover all possible values for the state attribute and even still may miss some students (such as the ones who misspelled Florida). We can state two general design principles from this example:

1. Minimize the amount of user-typed data in the system.
2. Use codes whenever possible.

14.5 Deliverables

The final output of information modeling is a database definition model. It includes the:

- Final entity-relationship model, called a Database Definition Model.
- The Source Material Log.
- The Source Material List.
- The entity pool.
- The attribute pool.
- The Data Dictionary.

The final database definition model shows all the entities, their primary keys, their non-key attributes, and the relationships between them. A syntactically correct entity-relationship model only has relationships of one-to-many type, and all primary keys (the entire primary key) migrate from the one side of the relationship to become foreign keys on the many sides of the relationship. Figure 14.11 shows the final entity-relationship model for the cruise line.

The source material log, material list, entity pool, and attribute pools are retained because they provide traceability from the entity-relationship model to the source data, and they provide design rationale so that somebody could try to reconstruct the decisions that went into the entity-relationship model.

TABLE 14.9
Data Dictionary

Field Name	Description	Reason for Data	Data Type	Field Size	Format	Validation and Other Rules	Required?	Indexed?	Application Found in
Patient_ID	Unique number for patient (PK)	To track patients	Text	6	n/a	n/a	Y	Y	all
FirstName	The patient's first name	-	Text	25	n/a	n/a	Y	N	all
LastName	The patient's last name	-	Text	30	n/a	n/a	Y	Y	all
State	The state the patient resides in	-	Text	2	n/a	{AL, AK, AR, CA .... WY}	Y	N	all
Phone	The patient's contact phone	-	Text	10	(999)999-9999	n/a	Y	N	all

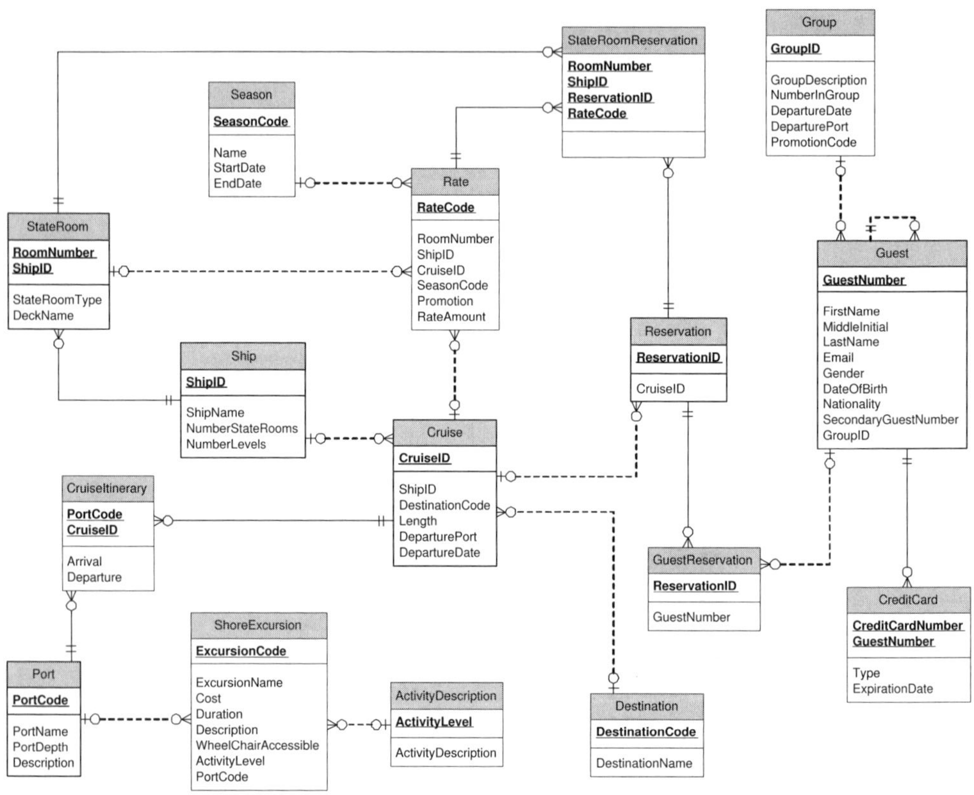

FIGURE 14.11
Database definition model.

14.6 Summary

This chapter presented a two-stage, six-step methodology to create an entity-relationship model. The entity-relationship model provides the information perspective of the enterprise architecture. The final model created, called a database definition model, describes the information content of the system, including all of its objects and the relationships between those objects. In addition a data dictionary defines all the attributes, their data types, and domains. Together these documents are sufficient to construct a database.

The chapter also introduced SQL and the four SQL commands that implement CRUD. These queries allow a person to do the four transactions that are done to databases. Normalization of the database removes structures that can lead to data anomalies during database operation. The chapter showed how to bring an information model to third normal form.

Review Questions

1. What are the seven types of entities?
2. Describe the correspondence between the entity-relationship model and its implementation in a database management system.
3. What are the four SQL commands required to implement CRUD?
4. Why do we normalize a database?
5. Describe the three types of data anomalies that can occur.
6. Describe the first three normal forms.
7. What should be put in the data dictionary?

Exercises

1. In the diagram below is a many-to-many relationship showing the primary keys. Resolve the relationship with an associative entity and show the primary key to foreign key migration.

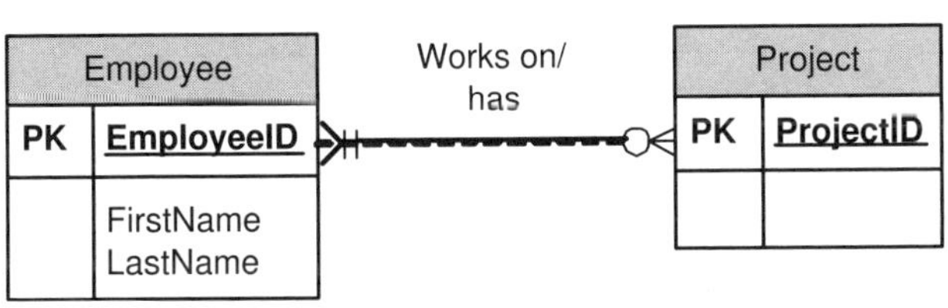

2. In the diagram below is a many-to-many relationship showing the primary keys. Resolve the relationship with an associative entity and show the primary key to foreign key migration.

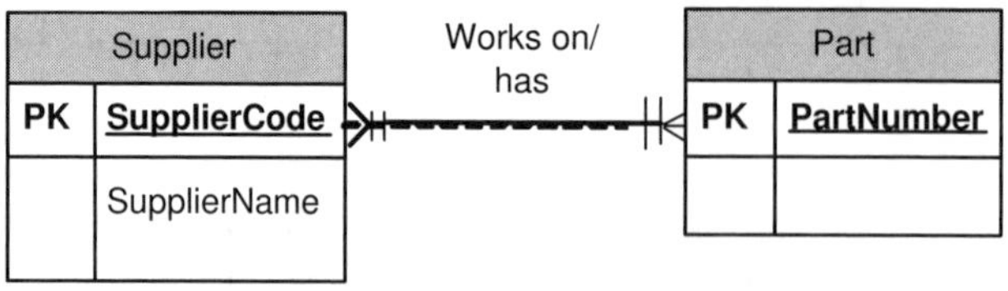

3. To keep track of office furniture, computers, printers, and other office equipment the ACME company uses the table structure shown here:

ItemID	ItemName	RmNum	BldgCode	BldgName	BldgMgr
2315-a	HP DeskJet 89	325	EC	Wilson Hall	Smith
3422-b	HP Toner	325	EC	Wilson Hall	Smith
7812-a	Canon Sx	123	PC	Primera Casa	Schultz

Explain which, if any, normal form the table violates and draw the entity-relationship diagram for 3NF.

4. For the information model in Figure 14.11 write the SQL for the following queries:

   (a) Create a new guest with the following data: Guest number is 542, first name is Kevin, last name is Reagan, Email is kreagan@xmail.com, gender is "M" and date of birth is May 31, 1970.

   (b) Write a query to obtain all the CruiseIDs for cruises that depart from the port of Miami (port code is MIA).

   (c) Write a query to change the Wheel Chair Accessible from No to Yes for the shore excursion code BAH-203.

   (d) Write a query to list all the Cruises that have a destination code PAC and depart after May 1, 2009 and before June 1, 2009.

   (e) Write a query to obtain the room number and ship ID for the guest Sandra Morales with guest ID of G-98237. (Hint: you must join several tables.)

5. The three tables: Student, Course, and CourseEnrollment are shown below. Answer the following questions based on the data in the tables.

StudentID	LastName	Major
105	Dion	Industrial Engineering
106	Houston	Chemical Engineering
108	Esteban	Computer Science
110	Felliciano	Information Systems

CourseID	CourseTitle	InstructorID
ISM4100	Information Systems Design	T55
ENT5250	Enterprise Engineering	T55
SYS7001	Requirements Engineering	T40
SYS7003	Systems Design Method	T33

CourseID	**StudentID**
ISM4100	105
ISM4100	108
ISM4100	110
ENT5250	105
ENT5250	106
SYS7001	108
SYS7003	105

(a) Write the data returned by the query:

SELECT CourseID, CourseTitle
FROM CourseEnrollment, Course
WHERE StudentID = 105 AND Course.CourseID = CourseEnrollment.CourseID;

(b) Write a query to obtain the student last names of all the students who are enrolled in a course with instructorID of T55.

(c) Write a query to enroll student ID 106 in the course SYS7003.

(d) Write a query to remove student ID 108 from the course ISM4100.

(e) Write a query to change the student Houston's major to Systems Engineering.

6. For the database modeled by Figure 14.11, do the following:

   (a) Write a query to obtain the credit card number and expiration date for the guest Jonathon Rose.

   (b) Write a query to obtain all the cruises that depart from Miami.

   (c) Write a query to insert yourself as a guest. Use your data.

   (d) Write a query to change SeasonCode "SumEnd09" to start on August 10 and to end on September 3.

   (e) Write a query to obtain all the cruises for the ship "Beauty of the Seas."

   (f) Write a query to delete a passenger with guest number G089234.

   (g) Write a query to increase the cost of all shore excursions by $5.

7. In Coconut Grove, FL a boat charter company wants to keep track of its charter operations. The company has twenty boats. For each boat they need to keep track of the boat manufacturer, boat length, fuel capacity, maximum number of people, range in miles, purchase date, and minimum draft. Each boat has an outboard engine with manufacturer, HP, shaft size, and hours. They want to keep track of their clients' names, address, and other contact information. The boats rental rate depends on the boat and the season (summer, fall, winter, and spring). They need to be able to make reservations for boats so that they know what is available and what is booked. They also want to keep track of the maintenance on the boats so that they can schedule maintenance when boats are not in use. Maintenance involves routine procedures such as bottom painting, changing the engine oil, tune-up, and so forth.

   Create an entity-relationship model for the boat charter company. Show all entities, attributes, and relationships. All relationships should be 1-to-many type of relationships.

8. A distributor of health foods and nutrition operates a small warehouse in Wichita, KS. They want to keep track of their inventory and customer orders. Their customers are retail stores in the region. For each customer they keep track of

the store, its location, and the points of contact in the store (usually a store manager, assistant manager, or district manager). The orders they receive are for products and the order includes the request date, quantity, and price. For each product in inventory they need to know the quantity on hand, cost, price, weight, description, expiration date and the location in the warehouse it is stored. Warehouse location is a matter of aisle, shelf, and level. It is coded as 5-2-3 for aisle five, shelf two, and level 3. The system should support the following activities: receiving customer orders, picking the orders from the warehouse, and answering queries about inventory status.

Create an entity-relationship model for the health food distributor. Show all entities, attributes, and relationships. All relationships should be one-to-many type of relationships.

9. Given the table shown for a small car rental company in Old San Juan, Puerto Rico, make the modifications necessary (you may need to add new tables) to bring it to third normal form.

CarNum	CarModel	CarRent	CarSeats	CarMileage
9432R	Ford Taurus	$49	5	16,580
9234A	Ford Taurus	$49	5	12,288
9874F	Honda Civic	$29	5	13,623
8796C	Ford Focus	$30	4	10,008
6783A	Ford Focus	$30	4	17,823
8798X	Honda Odyssey	$69	7	12,982
8734X	Toyota Sienna	$69	7	13,122

10. Equipment maintenance within a manufacturing facility is a vital activity that can impact capacity, runtime, and safety. Maintenance activities vary in scope, skills, and time required to accomplish the tasks. To manage the equipment in a facility so that it operates efficiently and effectively most facilities have a maintenance department that does both preventive and unplanned maintenance.

The maintenance work order process is triggered by one of two events: scheduled preventive maintenance or a failure. The process for both is explained below:

For scheduled maintenance events, the database triggers the work order. Essentially, each piece of equipment in the database has an associated preventive maintenance schedule. When the scheduled maintenance date approaches, then the system triggers the work order. For failures, the maintenance department clerk is notified by telephone from a factory worker (hereafter called the "originator") which piece of equipment failed. The clerk creates the maintenance work order. The clerk takes the information from the originator including the point of contact for the failure, equipment number that failed, date it failed, and a description of the failure. The system assigns a maintenance work order number, which is a unique ID for tracking, to the work order and the clerk tells the originator this number. The maintenance work order is forwarded to the maintenance supervisor who assigns a technician to the work order. Assignment is based on the technician's knowledge, skills, and schedule. For example, each technician is qualified only on certain pieces of equipment, or some only do mechanical problems, while others do electrical problems. Once assigned, the work order status is updated to show which technician is responsible for the work order and the estimated time of repair is sent to the originator. At the start of every shift, the technicians receive their schedule of work orders. The technician troubleshoots the equipment problem. Then the technician repairs the equipment. Once the equipment is repaired, the technician updates the work order to show what repair was done. The clerk then closes out the maintenance work order.

For the maintenance department described above:

(a) Create an entity pool.

(b) Create an entity-relationship model for the health food distributor. Show all entities, attributes, and relationships. All relationships should be one-to-many type of relationships.

Part V

Organization View

15

Organization Design

"Remember upon the conduct of each depends the fate of all." – Alexander the Great (356 BC-323 BC), Greek conqueror of Persia and the Near East.

In designing the enterprise, the human component is often the most important to the success of the enterprise, yet in many projects, especially projects with significant amounts of technology the organizational view is under-served. Frequently, the organizational design is treated indirectly as users of the information systems or resources within the process. These views treat people as if they are equivalent to mechanical resources [4, 23], which disregards the significant body of knowledge on the design of organizations.

This chapter addresses the organizational viewpoint of the enterprise architecture. A brief review of organizational theory is followed by a discussion of each of the main organizational design factors. Emphasis is placed on how to structure an organization, design the reward system, design the job, and motivate people. After completing this chapter, you should be able to:

- List the five main decisions in organizational design.
- Describe what organizational structure is and the factors that are considered in determining an appropriate organizational structure.
- Diagram organizational structure in an organization chart.
- Describe organizational decision making.
- Explain organizational culture and how it affects organization design and performance.
- List the factors that are considered in job design.
- Describe how the reward structure works and explain its importance to the organization.

15.1 Organizational Design

Organizations have been a subject of study for a long time. As a result, the literature contains a range of theories concerning how organizations should be designed. The more significant organizational theories that have emerged are: the classical theory, the human relations theory, the organization decision-making theory, and contingency theory. The later theories did not replace the earlier theories, but instead build on them by refining the earlier concepts, rejecting some ideas, and introducing new ideas. So modern organizational theory is influenced by all of these earlier theories.

The core principle of the classical theory is striking a balance between the division of labor into separate organizational units in order to increase productivity and the coordination required to integrate the organizational units. It was taken up by the Scientific Management movement as was described in Chapter 1. Classical theorists also treated man as rational, meaning a person has the ability to gather information, evaluate all the alternatives, and make a decision based on the facts.

The human relations theory of organizations was strong influenced by psychology and rejected the classical notion of a rational man. Chapter 1 reviewed this school of thought and described the famous Hawthorne studies as an example. Human relations management stresses how people are motivated. Maslow's hierarchy of needs provides a classification that is useful in designing organizations, jobs, and reward systems.

March and Simon [15] disagree with the structuring of decision problems in the field of operations research that searches for a unique optimal solution, and argue that a manager's decisions are sub-optimal because of bounded rationality, which means a person is limited by how much information he or she can process. They introduce the term of a satisficing solution instead of an optimal solution. March and Simon stress the decision-making tasks within the organization and the information flows required to support it. Galbraith [3] takes an information-processing view of the organization and argues the organization is structured so as to most efficiently and effectively process information.

Classical ideas on the division of labor concepts are not incorrect, but one of the major premises of Scientific Management is no longer thought valid; this is the idea that there is a single best way to organize work. Most modern theories accept some variation of contingency theory, which says the best organizational structure is contingent on the organization's environment. Contingency theory recognizes that organizations are open systems so the interaction with the environment is a significant design variable. So the organizational design problem is to find the best fit between organizational structure and the external environment [13].

These theories identify important organizational design variables, explain how they are related and affect organizational performance, and provide advice on how to best design an organization. In this text, we follow information-processing view of the organization and use contingency theory to understand how various organizational factors and the organization's environment fit together.

Based on these theories, we divide the organization design problem into five aspects:

1. Design the organization structure – The organization structure describes how the organization is divided into organizational units and how they are related to each other. Organizational structure is more than the ubiquitous organization chart; it has many dimensions describing various characteristics of the organization including both formal reporting relationships and informal communication relationships.
2. Design the reward system – The reward system describes how individuals in the organization are rewarded in terms of perks, financial rewards, promotions, and prestige.
3. Design the people – Design the people, refers not to actually designing the individuals in the organization, but to designing the training program and the organizational culture that dictates how they interact.
4. Design the job – Job design involves specifying the set of tasks that constitute a job.
5. Design the decision processes – To design the decision processes involves specifying how decisions are made, how people are managed, and how the work is coordinated. It is process design, but design of the managerial processes not the work processes.

15.2 Organization Structure

The organization structure describes how the organization is divided into organizational units and the relationships between those units [16]. An *organizational unit* is a distinct, identifiable grouping of people. There can be multiple criteria for grouping; it can be by function, by common output, or by geography. The purpose of grouping is that the people in the organizational unit share common resources and work toward common measures of performance. Grouping establishes common supervision of the unit and allows us to deal with people at a higher abstraction level than as individuals.

Grouping occurs at multiple levels to form an organization hierarchy. For example, a company may have divisions, which are divided into departments, which are further divided into workgroups. In universities, the university is organized into colleges, which have schools, which have departments.

An organizational unit will have relationships with other organizational units. The relationship most commonly represented is the authority relationship that depicts how one organizational unit reports to another organizational unit; in other words, the relationship between a supervisor and his or her subordinates. When an organizational unit or employee reports to a supervisor, then this implies the supervisor has: budget authority, the authority to review and reward the employees work, the power to direct the tasks performed by the employee, and the power to intervene in the employee's work.

The potential basis for grouping is:

1. Functional
   (a) Knowledge and skill – People are grouped according to their knowledge and skills. Examples are hospitals that have departments based on medical specialties such as Pediatrics, Cardiology, etc.
   (b) Work function – People are grouped according to the business function in the organization. This is very common in industry. For example, purchasing department, accounting department, marketing department, and so forth.
2. Divisional
   (a) Product – People are grouped by the output they produce. A manufacturing might have a product structure such that each division is for a different product (output) they produce. For example, an engine manufacturer.
   (b) Customer – People are grouped so as to provide complete service to a single customer type. For example, a company has separate groups for consumer customers, business customers, and government customers.
   (c) Geographical – People are grouped according to the geographical area they serve. For example, Northeast Region, Southeast Region, Latin American division, North American division, European division, and Asian division.
3. Matrix – Simultaneously both functional and divisional.

Figure 15.1 shows the four more common organizational structures. These structures represent different decisions on how to form organizational units. The *functional structure* groups organizational units according to the work they do, their skills, or expertise. A common functional structure in industry defines organizational units such as Marketing, Human Resources, Accounting, Engineering, Manufacturing, and so on. The benefits of a functional structure is that each unit groups people with similar skills and knowledge that facilitates developing greater expertise in each functional area. Also, a functional structure

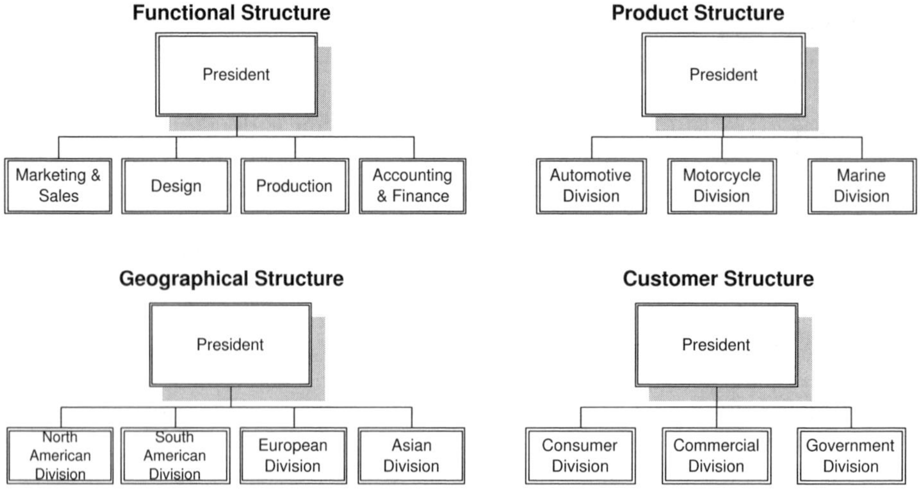

FIGURE 15.1
Four common structures.

benefits from pooling, as described in Section 12.3.1, because the resources are centrally located and available to all other units as needed. However, the functional structure has been highly criticized because most business processes cross the functional boundaries. For example, the process new product development crosses boundaries between Marketing and Design, and then Design and Production. Information flow across organizational boundaries is less efficient than the flow within an organizational unit. Also, it is more difficult to coordinate activities between organizational units and to align the goals of each unit with the overall organization's goals. These difficulties give rise to integration problems that result in lack of customer response, long cycle-times, and lower product quality among other problems.

The *product*, *geographical*, and *customer structures* are all variations of the *divisional structure*. The divisional structures tend to avoid the integration problem because all the functions are collected under one group. These structures tend to benefit from closer contact with the customer, but at the cost of duplicating or having surplus functional resources in each division. For example, in the product structure, organizational units are defined according to the product output. So an engine manufacturer may have an automotive division, motorcycle division, and marine division for each of the three types of engines they produce. Each division will have an engineering design group, an inventory group, and a production group. This introduces an inefficiency because, as was demonstrated in Section 12.3.1, it is more efficient to pool or collect resources together rather than distribute them as is done in the divisional structure. In the divisional structure, the decision is made to forego the efficiencies of pooling because it is easier to manage the information flow between departments, the coordination of activities, and alignment of subunit goals when all the units are in the same division. So organizations that use a divisional structure trade off the efficiency of pooling like resources together for what is believed a more effective structure for coordination and integration of activities.

Beyond the more common hierarchical structures, there are matrix and hybrid structures. The *matrix structure* has units report to both functional managers and project managers. In this way, it combines aspects of the functional and product structures. A typical scenario is where an employee is assigned to a functional group (e.g., marketing) and works on one or more projects where the employee reports to a project manager. The problem of reporting to multiple managers is often cited with the matrix structure. A network structure organizes the work around team delivery, but the teams frequently work in combination with other teams to deliver the product to the customer. A *hybrid structure* combines attributes of any of the above structure. For example, a hybrid structure could have a product structure for the overall enterprise and within each division a functional structure.

The above discussion of the various organizational structures reveals that how the organizational units are specified depends largely on a compromise between specialization and coordination [13]. Specialization is the keeping of similar people together, which leads to greater expertise, knowledge, productivity, and efficiency. But the more specialized a group becomes, then the greater are the potential coordination problems between the groups. Coordination within a unit is almost always easier than between units. The observation that coordination is the primary factor driving grouping decisions calls for examining coordination in greater depth.

Coordination is managing the activities, resources, and organizational units so that they work together effectively and efficiently. The need for coordination arises because we divide the organization, and consequently also the work, into smaller organizational units. The organizational units are interdependent on each other for information, inputs, outputs, direction, or resources. So, if we organize functionally, then the design department must coordinate with the manufacturing department because they are interdependent – the output of design is the input to manufacturing.

Interdependence is either an information, control, or resource relationship such that one organizational unit depends on another organizational unit.

Coordination is the management of the interdependence.

Much of a manager's work is the coordination of activities of the employees under their authority. The manager directs the employees' work, relays information, and ensures all the employees contribute harmoniously to the unit's output. In some settings such as a production line, the coordination is achieved through technology and process design. Coordination can also be achieved via workflow management systems such as used in insurance claims processing.

Thompson [22] conducted seminal work on coordination, and defined three types of interdependence in increasing strength and difficulty to coordinate as pooled, sequential, and reciprocal. Pooled interdependence is the weakest kind and is based on shared resources. Sequential interdependence is when the outcome of one task is the input of the next task. Reciprocal interdependence is when tasks need to be performed in parallel because the output of one becomes the output of another and vice versa. These dependencies are described in greater detail in Section 16.4.4.

Thompson describes the basis for grouping shown in Figure 15.2. The three interdependence types as well as teams are classified as various types of workflow interdependence. Ideally, groups are formed that minimize inter-group interdependence. This would lessen the need for coordination between groups. A system's engineering perspective would describe the grouping problem as seeking high cohesion within a group and low coupling, or interdependence, between groups. Achieving this design would reduce the need for coordination between units, which improves the performance of the organization.

Other than workflow reasons for grouping, there is process interaction, economies of scale, and social reasons. Process interaction is when people need to work together such as

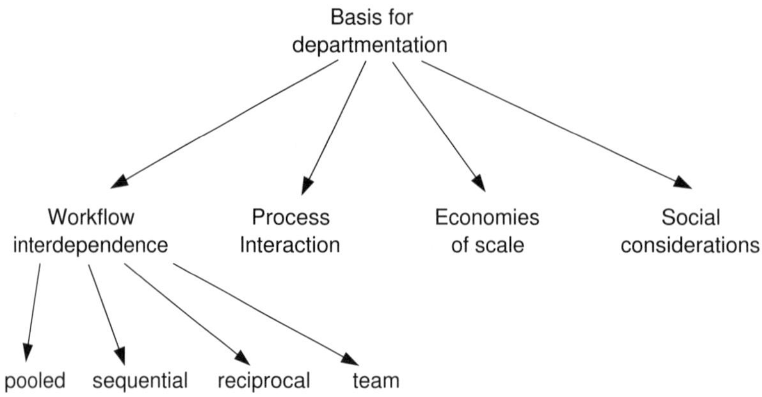

FIGURE 15.2
Grouping.

lawyers who consult each other to take advantage of specialized knowledge. Grouping can also be done to achieve economies of scale. For example, having a separate maintenance unit for every department would poorly utilize the human resources, so having a single maintenance unit is better. Lastly, there may be social considerations that argue for one type of group formation over another type.

An additional factor influencing the grouping of organizational units is the limits to span of control. The span of control describes the number of subordinates reporting to a manager and limits how large a single-person lead group can grow. Eventually, as a group grows beyond a certain limit it makes sense to split the group into subgroups, each with its own leader. Management science recommends the ideal span of control is between four and six employees in order to maximize performance [18]. Of course, the actual number depends on many contextual variables. The nature of the organization, type of work supervised, abilities and style of the manager, and the amount of coordination required for the work tasks. Consequently, it is possible for a single supervisor to oversee as many as twenty employees on an assembly line. The assembly line work is repetitious, there is low diversity in work tasks, and little coordination is needed. When task diversity is high, then optimal span of control is narrower. A faculty can only effectively supervise about four to six PhD students because their work is highly specialized, diverse, and knowledge-intensive.

The organization structure is diagrammed in an organization chart. The Organization Chart is a diagram that shows the division of the organization into units and the authority relationship between those units (see example in Figure 15.3) . Each organizational unit is labeled with the title of that unit. The organizational chart is drawn vertically such that the organizational unit at the top has authority over organizational units below it.

Organization charts are frequently and highly criticized because it is said they do not represent the real power relationships in the organization, they do not account for the nature of the people in the roles, and they miss all the informal relationships in the organization. Yet, an organization chart is the one model that almost every enterprise has. So, it is a model that many enterprises find useful regardless of all its shortcomings. An organization chart should be approached as how the enterprise wants to be, rather than how it actually works. Lateral relationships due to social groups, information flows, and teaming relationships are generally not shown in organization charts.

There is no single best structure for all organizations; the structure is contingent on the

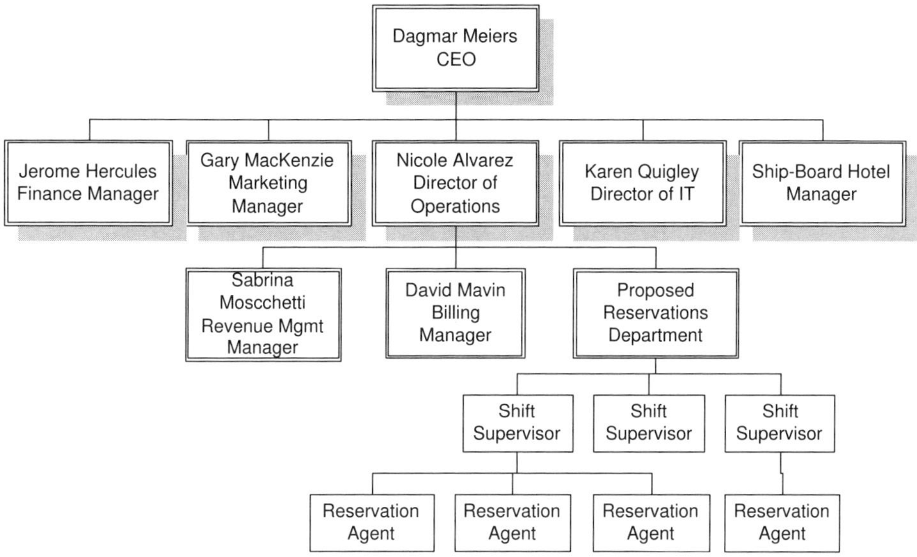

FIGURE 15.3
Partial organization chart of cruise line.

environment, the size of the organization, and the enterprise strategy. Table 15.1 compares the three dominant structures.

15.3 Decision Processes

An important function of the organization is to solve enterprise problems by making decisions. A decision is the selection of a course of action from among alternatives. Decision making is the process of making decisions and includes the activities to identify problems, understand the problem, generate alternatives, and select the best alternative – i.e., make a decision. Decision making is at the heart of management responsibilities. Organizational performance, even survival, is often tied to the quality of the decisions the organization makes [2].

Decisions are classified according to their scope as either strategic, tactical, or operational. Chapter 7 discussed strategic decisions for setting the direction of the entire enterprise. This section is more concerned with establishing the organizational structure for making tactical and operational decisions. These decisions are sub-classified into either routine decisions or non-routine decisions. *Routine decisions* are recurring, well-defined decisions that an organization must make. Examples of routine decisions include store clerks deciding whether to refund a purchase, clerks deciding whether to reimburse a travel expense, or a car rental agent deciding whether to offer a customer an upgrade. Routine decisions can be programmed and even frequently automated. *Non-routine decisions* are novel and poorly defined. Examples of non-routine decisions include how a company deals with a product release that causes safety problems, or how does a company deal with the sudden realization that it is the victim of fraud.

Designing the organizational decision process involves:

TABLE 15.1
Comparison of Structures

	Functional	**Divisional**	**Matrix**
Division of labor	By skills	By outputs	By skills and outputs
Decision making	By functional managers	By division managers	Division managers and coordination responsibility by functional managers
Importance of informal structure	Low	Modest	Considerable
Resource efficiency	High	Poor	Moderate
Time efficiency	Poor	Good	Moderate
Responsiveness	Poor	Moderate	Good
Adaptability	Poor	Good	Moderate
Accountability	Good	Excellent	Poor
Environment best suited for	Stable, low uncertainty	Heterogeneous, moderate to high uncertainty	Complex with multiple customer streams
Strengths	In-depth skill training, economies of scale within functions	Strong focus on customers, increases coordination in divisions	Coordination within and between product lines; flexible sharing of individuals
Weaknesses	Poor horizontal coordination between functions, creates narrow perspectives	Poor coordination across product lines, loss of economies of scale, fosters rivalry among divisions	Conflicts arise between functional and divisional units; dual authority can cause problems in decision making

1. Designate who in the organization has the authority to make what types of decisions.
2. Develop systems and processes to obtain the information required for the person to make the decision.
3. Establish policies governing how the decisions are to be made.

Who makes decisions?

Where in the organizational hierarchy decisions are made is described by the *centralization* of the organization. Highly centralized organizations tend to concentrate decision making at the top of the hierarchy with little or no input from lower-level employees. Decentralized organizations delegate more decisions to be made at lower levels.

Centralization consolidates power and decision making at the top of the organizational hierarchy, which means that decisions are more uniform, consistent with organizational goals, and quick to react to organization-level issues because only a few individuals participate in decision making. Centralization works well when the work is simple. It also works well if the organization is small and/or if the workforce is highly uniform because it separates the decision making from the execution of the task. However, there seems to be many more disadvantages of centralization and the trend has been towards greater decentralization. First, centralization is limited by the size of the organization because as the organization gets larger, then upper management will become overwhelmed with the information coming from below and will not be able to make the decisions fast enough. Risk is also high due to the concentration of power – if the decision makers become incapacitated then the organization suffers. Other disadvantages are that these decision makers are often further away from the problem situation, thus they may not make as good decisions for local issues, and also many employees will feel less motivated because their ideas about running the organization are not valued.

Decentralization delegates decision-making authority to lower-level employees. If the work is complex, then decentralized systems tend to perform better than centralized organizations on most measures of performance because the decisions are made close to the problems or work where they occur. Local problems and minor issues are resolved more quickly because the manager on the scene is able to make the decision.

Decentralization is accomplished by delegating authority, the formal power associated with a management position, to the right level in the organization. When delegation is done correctly, a person has the authority he or she needs to execute decisions that fall within his or her responsibility. Then this person can be held accountable for the quality of the decisions made. Problems arise when people have responsibility but lack authority or vice versa. Related to delegation is empowerment.

Traditionally, many organizations tended towards greater centralization. A reason many organizations were centralized has to do with the available workforce and technology. When workers were less educated, had less skills, and less training, then this favors a centralized organization. Also, decentralization was frequently difficult to implement due to the lack of technology to share information among decentralized decision makers. Nowadays, both these factors have changed. Many countries have highly educated workforces, many workers have expectations that they will have some decision-making role, and enterprises now have the information infrastructure that makes decentralization possible.

Information for decisions

Information is a key input to the decision-making process. To make a decision, the manager first needs to know a problem exists, then identify alternative courses of action, evaluate the courses of action, and select the best course of action.[1] Each phase of the decision-making process relies on information. The quality and appropriateness of the information provided to decision makers determines to some extent the quality of the decisions made. The enterprise design team needs to design an information infrastructure that identifies the right information and delivers it to the right person at the right time. Too much information is as much a problem as too little information. People become overwhelmed with too much information because it competes for their time, and a situation of information overload can be detrimental to the decision-making process.

Centralization versus Decentralization

Centralization

- Top-down control and vision leads to consistent decisions aligned with organizational goals.
- Uniformity of decisions and lack of dissent.
- Able to react quickly to major issues because decisions can be made fast and implemented.
- Slow to react to local problems or minor issues.

Decentralization

- Bottom-up decision making is more democratic and participatory.
- Can be slow to adapt to major issues and changes because of the diffused responsibility.
- More flexible because lower-level units can make decisions that address problems unique to them.
- Fast to react to local problems and minor issues.

The design of the information infrastructure involves information technology. Two important technologies are decision support systems and data warehouses. *Decision Support Systems* (DSSs) are interactive computer-based applications intended to help decision makers utilize data and models to identify and solve problems and make decisions. A decision support system generally contains one or more data sources, a decision model, and a user interface. The data sources utilized by DSS are frequently from large transactional databases called On-Line Transactional Processing (OLTP). OLTP is performed by large enterprise-wide systems such as MRP, MRPII, and ERP, which automate the transactions of the organization. The immense amounts of data contained in these systems provide an opportunity for organizations to make better decisions. However, the structure, format, and content of the OLTP databases do not facilitate decision analysis. For this reason, recent trends have been toward developing On-Line Analytical Process (OLAP) systems, often called data warehouses. These systems use different schemas to organize the information, frequently in an aggregate format to better support decision making. Data warehouses are described in Section 16.6.

[1] Of course, this describes an ideal situation.

Decision policies

Policies are the rules that guide how individual decisions are made. Policies differ for routine decisions and non-routine decisions. For routine decisions, policies tend to be didactic, specifying rules, procedures, and documentation to guide the decision. For example, many retail stores have check cashing policies that state the conditions under which a check can be accepted from a customer. Policies for non-routine decisions usually specify broad, flexible processes for making decisions. An organization might have policies to bring issues to a bi-weekly department meeting where the issue is discussed and decisions are made. Oftentimes, the policies are informal and embedded in the organizational culture. In this situation, managers learn from experience how to proceed to make decisions, and each manager has a lot of leeway in their own personal approach.

Some enterprises have many policies for every little problem (e.g., the U.S. Department of Defense has specifications for the construction of urinals). *Formalization* refers to the degree that rules, procedures, and written documentation prescribe the rights and duties of employees. Highly formal organizations have lots of rules, procedures and written documentation in an attempt to control most aspects of employee decision making. Formal organizations leave little room for employee discretion. Informal organizations have a looser structure, allowing for greater amount of employee judgment and discretion in decision making. Part of designing the decision processes is specifying policies that dictate how decisions are made in the organization.

15.4 Job Design

Job design is the specification of the content of work that an individual or group undertakes and the methods they use to do their work. The work itself, and the requirements for the work are specified as part of process design. Recall, that process design specifies a network of activities to produce a product or service. Job design determines how to group the activities into a defined job that is assigned to an individual or organizational unit. Consequently, job design links the organization design to the process design. Job design and process design are interdependent activities because when designing the process the designer specifies workers for each activity who in turn place constraints on what the activity involves due to the skills, knowledge, and capabilities of those workers. The goal of job design is to specify the work content of a job to satisfy the process design requirements while simultaneously achieving an effective fit between the person, job, and organization.

Job design has two dimensions: breadth – the number of tasks in each job, and depth – the degree of control the person has over the work. Job design involves answering the following questions:

1. What tasks should be put together in the same job?
2. What skills and training are needed for the job?
3. What decisions should the employee doing the job be allowed to make?
4. What job characteristics will motivate employees to be productive, do high quality work, contribute toward enterprise goals, and be motivated to continuously improve individual skills and knowledge as well as organizational capabilities and knowledge?

The traditional approach of job design hails from the early work conducted by Taylor and others of the Scientific Management movement (see Section 1.4.1). In this approach, the primary goal of job design is productivity, and the human worker is seen as little more than a machine that can also think. A cornerstone of the approach is division of labor, which explains how workers become very proficient when jobs are narrowly defined. Jiang and Giachetti [9] developed a model that incorporated productivity gains from specialization into the queueing model to analyze these benefits. Indeed, in many instances the gains from division of labor are impressive, validating one of the basic premises of Scientific Management. Jobs designed according to scientific management tend to be narrow in breadth and low in depth. This is because one of the tenets of scientific management is the separation of doing the work from managing the work.

The behavioral scientists argue that people react better to self-actualization and autonomy. Here the approach is to design jobs so they are more intrinsically rewarding, which leads to greater employee satisfaction and greater job performance. In this approach, job design is not a straightforward technical task but is grounded in the social context, which changes with time and culture. Jobs designed according to this criteria will have both greater breadth and depth. Two job design strategies used are job enlargement and job enrichment. *Job enlargement* designs a job to have breadth by adding more tasks to a single job. The belief is that greater breadth leads to a more interesting and thus motivating job. When job breadth is achieved by cross-training employees to perform multiple tasks, then a benefit realized is greater labor flexibility. If a manager sees a process bottleneck, the manager can shift workers from other tasks to the bottleneck when they are cross-trained for multiple tasks. *Job enrichment* designs a job so as to allow workers the freedom and authority to make decisions concerning their own work. Again, the idea is that when people have greater control over their work they will be more motivated and perform better. Essentially, both job enlargement and job enrichment run counter to Taylorist principles: job enlargement is the opposite of specialization and job enrichment is the opposite of separating the performance of the work from the planning of the work.

Job rotation is when the organization intentionally moves employees from job to job to reduce boredom by having them perform a variety of tasks. The idea is that greater task variety better fulfills human needs and therefore the employees will be better motivated and performance will improve. Job rotation is mostly used in routine jobs such as found in factories, warehouses, or retail. Job rotation can also be used to develop employees skills and increase workforce flexibility. For example, Johnson & Johnson will rotate new engineering hires through three different jobs in their first year-and-a-half of employment. Here the reason is not to reduce boredom, but to expose new hires to multiple parts of the business with the intention that they develop a system-wide or process perspective of the company's operations.

In a study of U.S. firms, Osterman [17] found that 56.4% of respondents utilized job rotation in production environments. Job rotation is relatively easy to institute in an enterprise. One caveat in establishing a job rotation policy is that if the jobs change from one routine job to another routine job with similar tasks then it is unlikely to derive many benefits for the workforce.

The team approach replaces individuals with well-defined jobs to individuals being part of a team, and the team is assigned a task. However, this does not dissolve team members from specific roles, just that their roles are not tied directly to a process activity. Teams are becoming common in healthcare, new product development, manufacturing, and other industries as well. Cohen and Bailey [1] identified four types of teams: work teams, parallel teams, project teams, and management teams. Work teams are responsible for producing goods or providing services, and they work on an ongoing basis. Parallel teams are constituted by members from different units of the organization with the purpose of performing a function or task, usually problem solving or improvement, not assigned to any of the regular units of the organization. Project teams are formed on a temporary basis to produced a one-time output such as a new product, service or project. Management teams are created to laterally integrate independent units of the same organization, by coordinating their respective functions according to a single strategic direction.

The benefit of teams is due to the synergistic effect of multiple people working collaboratively to solve a problem or complete a task. Teams are an alternative job design strategy for tasks that cannot be completed effectively by a single individual or by the aggregated independent efforts of a group of individuals.

15.4.1 Illustration of Job Design Approaches

To illustrate traditional job design, we present an example of the process redesign in a dental insurance company. The company located in suburban Massachusetts converted from a paper-based to a computer-based office environment in 1981 [26]. The office achieved 30 to 40% productivity increases in the first year of implementation and 105% productivity increase by the end of the second year. Moving to a computer-based environment standardized and automated the work, which greatly reduced the job enrichment. In general, most clerks were unhappy with the work changes because they felt they essentially became just another "cog in the machine" because the work no longer required knowledge about claims processing. Actually, the IT consultants recommended a further reduction in job span to increase productivity even more, but this was rejected by management because they felt further removal of variety would make it difficult for them to attract "... reasonably capable people" ([26] p. 134).

What this example illustrates is that the gains from specialization (narrow, routine jobs) combined with automation greatly increased productivity so that even with reduced worker satisfaction the organization realized tremendous productivity improvements. What is not said is the long-term effects of the policy. A meta-analysis[2] with data from over 50,000 workers, found that job satisfaction was modestly correlated with job performance, $r = 0.30$ (where r ranges from 0 to 1, with higher values indicating higher correlation). Given the complexity of the relationship between job satisfaction and other factors such as differences in worker personality, other job characteristics, organizational norms, and so forth, social scientist consider this a positive correlation. Other studies have found similar levels of correlation between job satisfaction and job performance [8, 21, 25]. These studies do not conflict with the dental insurance example recounted above because here we are talking about the correlation with job performance of which productivity is just a single component. Job performance includes quality, reliability, customer service, and other components besides productivity.

Some of the improvements in performance might be due to other changes that happened concurrently with the job enlargement. Kelly [11] analyzed a number of cases of job enlargement, and found increases in productivity of the order of 20%. He also says the productivity increases may not necessarily be due to an increase in job satisfaction, but it may be due to such factors as removing delays that occurred when workers had to wait for each other to pass on materials; by improving methods of work, such as using both hands; and better through better designed work stations. So, it is possible the job enlargement improved productivity because it simultaneously reduced the coordination work when the job was enlarged, and not because the workers are more satisfied.

What this discussion of job design highlights is that research on human work is fraught with difficulty because of the many factors that influence human behavior and performance. Isolating job satisfaction (even just defining job satisfaction) is challenging, and then linking

[2] A meta-analysis is when the researchers statistically analyze many previously published studies.

it to job performance in light of all the other factors influencing performance makes it more vexing. So, even though the correlation is weak to moderate, the research seems to show a positive correlation between job satisfaction and job performance. What must be kept in mind is the magnitude of improvement can be much smaller than alternate job designs based on specialization and scientific management principles. The job designer needs to trade off competing concerns to arrive at a satisfactory job design that meets several goals concurrently.[3]

Another element of job design is the ergonomics of the job. Ergonomics is the study of the physical demands work places on a person's body and how to best design the work to mitigate any negative consequences. Ergonomics is especially important in jobs requiring physically repetitive work (e.g., meat-packing industry, automotive assembly), jobs that use the same muscles repetitively (e.g., secretaries typing and the incident of carpel tunnel syndrome), and also in office jobs for good posture to avoid back problems. The project team should have an ergonomist consultant to review possible job designs to identify potential problem situations and make recommendations for improvements with respect to the ergonomics of the job.

The output of job design is a job title and job description. The *job description* sets expectations for performance so that the employee understands what is required of the job, how performance is rewarded, and what the employee's relationship to the organization will be. The job description should be broad enough to allow flexibility in the employee working at various activities because a very specific job description can impede process redesign and innovation. A job description includes:

- Job title.
- Job summary.
- Job tasks, responsibilities, and authorities.
- Job qualifications needed to carry out the work: education, training, certifications, skills, experience, and knowledge required.
- Job reporting relationships such as the employee's supervisor and any positions that might report to the employee.
- Possible career progression, criteria and timing of performance reviews.
- Total renumeration package including salary, wage, bonus, and other benefits.
- Work location and any travel required.

15.5 Reward System

The reward system or incentive system is a formal mechanism in the organization for recognizing employee achievement of the organization's goals. The reward system defines expectations for individual performance and specifies what rewards employees can expect when they meet or exceed expectations. The performance side of the system defines measures of performance for each individual, how it will be evaluated, and how the organization will

[3]Throughout this discussion, we ignore any moral or social obligation the enterprise has to providing meaningful work for people.

provide feedback to the individual. The reward side defines the types of rewards including bonuses, salary increases, promotions, or other awards. The reward system plays an important role in motivating employees to perform for the good of the organization.

Reward systems can be a vital aspect of any organization. They can actively engage and renew the overall sense of community and mission of an organization. A properly administered system of rewards can provide incentive for quality workmanship and staff performance. Likewise, a poorly administered reward system can lead to low morale, unproductive performance, and even lead to a high percentage of staff turnover. A reward system is successful when the staff interprets its policies as even-handed, consistent, and relevant [24].

The reward system defines the relationship between the organization and the individual member by specifying what the organization expects of the individual and what the individual can in return expect from the organization. It specifies what the expected contributions from individuals to the organization and the behavior norms they must conform to. When individuals meet or exceed these expectations, then the reward system specifies what they can expect to receive as a result of their performance. In this way, a reward system is a powerful means to influence an organization's culture.

One design issue in a reward system is the extent you reward the individual versus the group. Overemphasis on rewarding the individual might lead to problems of employees learning that political efforts are rewarded more than performance, it might work against teamwork, and it does not contribute to concern for organizational performance [19]. When designing a reward system, the design team should also determine how the effectiveness of the reward system can be measured. This way the enterprise can evaluate the effectiveness of the reward system after it has been implemented for a reasonable period of time.

To design a reward system, each job needs to have a goal or work standard to establish expectations for the performance of that job. When the employee meets or exceeds those expectations, then an incentive is offered. Incentives can be monetary reward (a raise or bonus), a promotion, time off, or recognition by peers (e.g., employee of the month plaque).

To specify effective goals for a job requires the company to consider two aspects. First, the individual employee goals need to be defined so they align employee performance with the overall organizational goals. Second, the employee goals need to be defined such that performance against the goals is within the control of the employee. In other words, the employee's performance should not be contingent on events outside of their control.

There are countless cases in the literature where both these guidelines are violated. A common problem is goal misalignment between the individual, organizational unit, and the organization that leads to sub-optimization [14]. Sub-optimization occurs when different subunits each attempts to reach a goal that is optimal for that unit, but that may not be optimum for the organization as a whole. Kerr [12] called this, "the folly of rewarding B while hoping for A." Misalignment occurs frequently because management rewards what is easy to measure rather than reflective of what performance is desired, rewards the individual when group effort is desired, or rewards a lower-level unit for performance without considering how attainment of the performance goal will affect high-level units.

A common scenario is when employees are rewarded based on productivity that is easy to measure without regard to other performance measures. The result is the employees are motivated to produce output, sometimes without regard to quality or without a ready customer to purchase the output. For example, call centers might measure an agent's performance based on the number of calls they can handle per hour, but this motivates the agent to quickly end a call. As a result, the agent might not provide as high a quality of service, or the customer may have had other questions, but he or she did not think to ask the question before the agent ended the call. There is common tension between efficiency of the operation (handling many calls per hour) and the effectiveness of the service (quality of call). In this simple case, the call center agents should be measured against a few performance criteria to better match the goals of the call center, even if measuring the service quality is difficult to measure.

People respond to incentive systems

A construction company in South Florida instituted a policy that all its employees would get a $50 bonus at the end of the month if there were no safety incidents. The purpose of the incentive was to reduce the company's insurance premiums by reducing the number of claims. On the last day of one month, the company's vice president and son of the owner noticed one of the workers was limping and had a bloody shoe. The worker was injured when something fell on his foot, but was not going to report it because he didn't want to jeopardize his bonus or that of his co-workers.

If the employee is measured against goals that are outside of his or her control, it can be demoralizing. Deming did an exercise to demonstrate the effects of rewarding/punishing employees for what was statistical noise. He had a bucket with mostly white marbles and a few red marbles. The red marbles were "defects." Students were given a paddle to reach in and collect marbles. If the student collected all white marbles, then he or she was rewarded. If the student got red marbles, he or she were demoted, and if it happened again the student was fired. What marbles a student collected was completely random and outside of the control of the students. The exercise demonstrated the vagaries of measuring performance to something outside of the employee's control, in this case to what was a random event. A student who was rewarded for collecting only white marbles would on his next turn most likely collect a red marble and be demoted for "poor performance." A student who collected a red marble but then subsequently collected all white would be congratulated on his or her efforts for good performance – unfortunately, many reward systems are equally arbitrary.

In summary, the reward system needs to let individuals know what is expected of them. The employees need to have the skills, abilities, resources, and support to be able to fulfill those expectations. Next, the employees need to know that, when they do accomplish those things, they will be rewarded in ways that are personally valued and meaningful to them. To accomplish this, every job needs to have a goal or work standard associated with it; the goals need to be aligned to organizational goals; the goals need to be within the employees ability to control performance; and management needs to monitor the performance of employees against the goals, and provide periodic feedback to the employee.

15.5.1 Compensation

Employee compensation is closely related to the reward system. Compensation is the total remuneration paid to employees for the work they perform. Compensation includes the wages and benefits the employees receive (paid healthcare being the largest non-wage benefit). Three dominant compensation systems are:

1. Salaried workers. Salaried workers are paid a set salary for a period of time such as biweekly regardless of how many hours they work. Many professionals such as engineers, doctors, and managers are salaried employees.
2. Hourly workers. Hourly workers are paid a wage per hour they work. The company must keep track of the hours worked on time cards so the employees are compensated correctly.
3. Commissioned workers. Workers on a commission are paid for performance. Many

salespeople are on commissions. They earn a base wage (usually very low) and are paid a percentage of each sale they make. In the service industry (waiters, taxi drivers, hotel bellboys), it is common that tips make up the bulk of the compensation for these employees.

The wages paid to employees are set by market forces in most economies, although many of these countries have minimum wages that must be paid to employees. If the employees are unionized, then the wages are negotiated as part of a collective bargaining agreement. In this case, all employees in the same job title will receive the same hourly wage. Otherwise, without a union, employees will have to negotiate the wages with the company.

The setting of wages is determined largely by market forces, whether the wage is negotiated by a union on the employee's behalf or the employee negotiates alone. What this means, is that the wage is set according to supply and demand. If the supply of workers with the skills needed for a job is larger than the demand for those workers, then this will keep the prevailing wages low. If the job requires workers of a rare skill or knowledge, then the company will be pressured to keep wages high to attract those employees. These forces can be seen at work in almost every industry. In healthcare, the demand for skilled, experienced nurses is greater than the supply of qualified nurses in many communities. Consequently, hospitals and other business that employ nurses must offer higher wages and are known to even offer sign-on bonuses to new nurses. However, wages are not the only way to attract employees. Non-financial compensation such as daycare for dependents, flexible hours, time off, job security, and work conditions can also serve as important inducement (or discouragement) to prospective employees.

Non-financial compensation to attract employees

Nationwide in the U.S., there seems to be a persistent shortage of nurses. Due to cultural and historical reasons, most nurses are female. Nursing is a regulated field, and depending on qualifications, a person can either be a Registered Nurse (RN) or a Licensed Practitioner Nurse (LPN). The healthcare industry requires nursing staff around the clock, all days of the week. In order to attract and retain nurses, many companies allow for great flexibility in their work schedules. Nurses may work 3 12-hour days, 4 10-hour days, or a more traditional schedule of 5 8-hour days each week.

What this means is that a business needs to know the prevailing wage for each job category. Depending on other factors such as health benefits, location, etc. the business needs to offer wages comparable to the prevailing wage. If the business offers wages that are seen as significantly lower than what employees can get elsewhere, then the good employees are likely to either not take jobs at the company, or they will be the first to leave once they find a better job elsewhere.[4] It seems that paying much more than the prevailing wage does not necessarily obtain better performance for the company. This goes back to Maslow's hierarchy of needs in which, beyond a certain level, more pay does not translate into greater performance.

Merit pay or pay for performance tries to link the compensation to the employee's performance. When done right, the relationship is positive: i.e., the employee's performance improves due to the incentive of earning more pay [5]. A good merit pay system carefully considers the measurement of performance, the establishment of pay increases, and the linking of pay to performance. The merit pay system may be for individual performance or

[4]One university is suspected of paying its faculty less than the average salary but not more so than 10% less because they find that few faculty will switch jobs for a raise less than 10% – the switching cost.

group performance. As an example of merit pay tied to group performance, a small construction company offers $50 to each employee at the end of an "accident-free month." They do this so as to lower their workmen's compensation insurance premiums. This example, shows effective alignment of employee performance with the organization's goal. In some industries, end-of-year bonuses are very common (e.g., the financial industry). The bonuses are merit pay, and are used as incentives to motivate employees to dedicate time and effort to the company.

15.6 The People

Many leaders have said how the enterprise is nothing without its people. An enterprise cannot "design" its people but it can decide who to hire (and who to fire), how to train and develop people, and to lead people. These tasks fall under the domain of the human resources department in the enterprise. The task of the enterprise designer is to establish the organizational infrastructure and policies to support human resources.

The organization needs to be able to find, attract, and hire new employees. To do this, the enterprise defines what jobs it needs to fill, who they report to, and the compensation for the job. The organization needs to determine what it looks for in employees and determine how to find people that fulfill those needs. Some organizations look for people with desired personality traits that fit well with their organizational culture.

Two aspects of hiring people into an organization involve the questions of how many people and their level of knowledge, skills, and specialization. These decisions are co-dependent with each other, other enterprise design decisions, and environmental influences on the organization. The number of people hired depends on the needs of the enterprise and the level of technology automation employed. It is also influenced by the labor market and the availability of qualified applicants and the prevailing market wages. For example, an airline that operates in the U.S. and in Central America can be observed to have far more ground crew in the Central American airports than in the U.S. airports. The reason is labor is much cheaper in Central America, so they are able to hire more employees to ensure better service. In the U.S., they try to rely more on technology and more efficient work processes. The design of the work and availability of knowledgeable, skilled employees are also co-dependent. Generally, more complex work will require more knowledge and skills. If employees with the requisite knowledge and skills are unavailable, then organizations can attempt to redesign the work to require less knowledge or they can institute programs to hire people and develop those skills. Consequently, job design and who to hire need to be decided together.

Training is necessary for new hires – for existing employees to develop new skills, for the introduction of new processes or systems, and for promoting employees to new positions. Actually, because so many enterprises operate in highly dynamic environments, training is necessary for survival itself. Too many companies short-change training, yet it is important to the success of the organization's operations. This is especially true when rolling out new systems or processes. Training orients new hires to the policies and procedures of the organization. If they do not have the skills required, then skills training can provide them these skills. Determining the type of training, amount of training, and how to deliver the training is an important organizational design decision.

Cross-training is when employees are trained in several different jobs with the aim to increase organizational flexibility. If an employee can do several jobs, then the employee is more valuable to the organization, and the organization has greater flexibility in meeting variations in demand. There is a limit to the extent of cross-training; if a person is trained in too many tasks, then because of the lack of regular use in conjunction with cognitive limits the person will not be very productive in all those tasks.

The managers in the organization are the leaders of that organization. The organization needs to understand what they want in a leader, and then develop programs to find people with the traits, skills, and knowledge to fill the leadership role, and to develop those leaders.

In summary, an organization needs to make the following decisions concerning the people in the organization:

1. How many people to have in the organization.
2. What knowledge, skills, and expertise these people should have.
3. What training to provide.
4. What type of leaders to have for the organization.

These decisions are dependent on other enterprise design decision and also influence how other decisions are made.

15.7 Organization Culture

It is not the intention to provide an in-depth exploration of culture, a subject that is discussed at great detail in works by others (see for example [7, 20]). Rather, in the context of organizational design, we review how culture affects the other organizational design factors and describe how it can be indoctrinated into the organization. We refrain from saying design the culture, since design implies an ability to directly control the outcome through specifications. For organization culture, it is more of shaping the culture by first determining what organizational values and beliefs you wish to promote and then through example, communication, and visible behavior you institute and continually reinforce the culture.

Organizational culture is "the set of values, beliefs, and norms that are shared by people and groups in an organization and that control the way they interact with each other and with stakeholders outside the organization [6]." *Values* are expressions of what the organization believes. Values cannot be evaluated as either true or false, they are simply values held by the people in the organization such as "the customer is always right" or "quality is job one." Values are expressed in mission statements, vision statements, corporate policies, and the core values statement. *Beliefs* embody particular views about how the world works. Beliefs form the world-view of the person; to the degree they are shared, they are then called organizational beliefs. Beliefs describe a cause-effect relationship and they are therefore open to debate. *Norms* describe the behavioral manifestations of values and beliefs. They are the set of expectations about how people will conduct themselves in ways that are consistent with the organization's core values and beliefs.

Schein [20] presents a model of organizational culture. At the first and most cursory level of Schein's model is organizational attributes that can be seen, felt and heard by the uninitiated observer. The observable portion of organizational culture is the behavior patterns of the people and the outward manifestations of the culture: dress codes, how people address each other, layout of the work space, and perks provided to executives are examples. Included are the facilities, offices, furnishings, visible awards and recognition, the way that members dress, and how each person visibly interacts with the others and with organizational outsiders. All may be visible indicators of culture but they are very difficult to interpret. The values and beliefs are more difficult or even impossible to observe.

Company slogans, mission statements, and other operational creeds express the values. To dig deeper calls for interviews and questionnaires of people in the organization. It is these values that form the foundation for the observable behavior.

Schein [20] also includes the tacit assumptions of the organization which are the unseen elements of culture that are not cognitively identified in everyday interactions between organizational members. Sometimes, these are the elements of culture which are often taboo to discuss inside the organization. Many of these "unspoken rules" exist without the conscious knowledge of the membership. Those with sufficient experience to understand this deepest level of organizational culture usually become acclimatized to its attributes over time, thus reinforcing the invisibility of their existence. Surveys and causal interviews with organizational members cannot draw out these attributes – rather much more in-depth means is required to first identify and then understand organizational culture at this level. Notably, culture at this level is the underlying and driving element often missed by organizational behaviorists.

Walmart in Germany

In 2005, Walmart introduced their ethics manual in Germany. It was not well accepted. German employees interpreted a caution against supervisor-employee relationships as as a puritanical, over-reaching ban on office romance. They interpreted a call to report improper behavior of co-workers as an invitation to rat them out. This and other organizational norms that work in the U.S. do not translate well to Germany. When Walmart offered services such as bagging customers' groceries they were surprised at the negative reaction. It seems that Germans do not want strangers handling their groceries. What the story reveals is that a highly, successful company in the U.S., in this case Walmart, needs to understand and incorporate local cultural values into their foreign operations or face embarrassing blunders as described or even worse. – Ewing, J., Walmart: Struggling in Germany, *BusinessWeek*, April 11, 2005.

An enterprise's culture will reflect the society's culture of which the enterprise is a part. This brings up questions about international enterprises that are located in many different societies. Hofstede and Hofstede [7] demonstrate that there are national and regional cultural groupings that affect the behavior of organizations. Their research was based on observed differences they found in IBM international around the globe. They identify five dimensions of culture in their study of national influences:

- Power distance – refers to the degree to which a society expects there to be differences in the levels of power.
- Uncertainty avoidance – refers to the extent to which a society accepts uncertainty and risk.
- Individualism vs. collectivism – refers to the extent to which people are expected to stand up for themselves, or alternatively act predominantly as members of the group or organization.
- Masculinity vs. femininity – refers to the value placed on traditionally male or female values. Male values for example include competitiveness, assertiveness, ambition, and the accumulation of wealth and material possessions. Female values emphasize relationships and quality of life.
- Long- vs. short-term orientation – refers to a society's "time horizon," or the importance attached to the future versus the past and present. In long-term oriented societies, thrift

and perseverance are valued more; in short-term oriented societies, respect for tradition and reciprocation of gifts and favors are valued more.

What this research suggests is that an enterprise that operates internationally cannot expect to have the same exact organizational structure, processes, and work environment exactly the same in every country or region of the world. See the box entitled, "Walmart in Germany," for an example of cultural misunderstandings.

15.7.1 Culture and Organization Design

Culture is important to organization design because it affects how people work together. It influences the day-to-day activities of people: how they communicate; how they form work relationships; what behavior is acceptable or not; and how power and status is allocated in the organization. Culture also influences how the organization adapts to the external environment.

Culture strength varies – some organizations have very strong cultures and others have weak cultures. IBM is an example of a company that had and has a strong culture. IBM employees would talk about, "the IBM way" to describe how things would be done in IBM. Strong cultures can be a hindrance to the change necessary to respond to external threats or they can help the organization adapt to the new challenges. The box entitled, "Cultural Change at IBM," summarizes how culture was a major issue in redesigning IBM so that it could succeed in the computer solutions market.

There are a few ideas concerning culture that can be used proactively by enterprise designers to attain their goals.

- Culture should also be aligned with the enterprise strategy.
- Continuous Process Improvement (CPI), kaizen, and other ideas need to be embedded into the organization's culture.
- To accomplish system implementation or change then the culture must change.

Culture influences our thinking in ways we do not consciously think about.

Konosuke Matsushita of the Matsushita Electric Industrial Company made the following statement:[5]

> We [Japanese] are going to win and the industrial west is going to lose; there's nothing much you can do about it because the reasons for failure are within yourselves. Your firms are built on the Taylor model; even worse so are your heads. With your bosses doing the thinking while the workers wield the screwdrivers, you're convinced deep down that this is the right way to run a business. For you, the essence of management is getting the ideas out of the heads of the bosses and into the hands of labor.

He then went on to explain how the Japanese were beyond Taylorist ideas and used the full potential of their employees to deal with the complexity inherent in modern-day enterprises. What Matsushita is arguing is that the western countries are following an enterprise design paradigm that no longer fits the environment. A paradigm is a shared set of assumptions, concepts, values, and practices that constitutes a way of viewing reality. A paradigm is unconscious in that we think this way without pondering why. So Matsushita suggests that

[5]Reading this makes one think about the joke defining progress. Asked how to define progress, the person answers it took the British 500 years to become arrogant, the Americans 100 years, and the Japanese only 50 years.

the Japanese do not adhere to the western paradigm and the way they view the world will be more successful for business competition.

One aspect of organizational culture that enterprise needs to establish is the perception and delivery of quality. Largely influenced by the perceived quality gaps between U.S. manufactured and Japanese products during the 1970s through 1980s, the idea of Total Quality Management (TQM) took root in many organizations. TQM is an organization-wide emphasis on quality that involves all employees striving to adhere to quality principles and continuous improvement. TQM requires a change in cultural attitudes and norms so that the enterprise can provide superior products and services. The cultural changes involve getting all employees to recognize quality and the lack thereof, of developing an attitude of doing things right the first time, and of developing an attitude towards waste and continuous efforts to reduce waste. To attain a TQM culture, it helps if the organization has an open and cooperative culture that encourages employees to assume responsibility for the quality of their work, to take pride in their work, and to be responsible for customer satisfaction. This is accomplished by including employees in the development of enterprise visions, strategies, and plans. The reward system can encourage this shift in attitude by rewarding quality work, introducing incentives for defect-free work, etc. The point here is that the success of many initiatives such as TQM rely on an acceptable organizational culture. The enterprise can shape this culture by showing the way itself, training, structuring the work processes, establishing appropriate reward system, and incentive plans. For example, quality guru Juran [10] saw employees as the source of a company's competitive advantage, because only those involved in a process can improve it. That capability is derived from long tenure in the workplace and from the repetitive performance of the jobs in that workplace.

Many large enterprise system projects fail to achieve their promised benefits not for technological shortcomings but because the implementation of the system into the organization went poorly. Change management is the technique to prepare the organization for and manage the transition to a new system or way of working. People seem to naturally prefer the familiar to the new. As a result, the initial reaction to a new system is to resist it. Change management works to make the new system familiar before it is launched and thus facilitate its full adoption and success. Change management includes informing stakeholders of the impending changes, training stakeholders to use the new system, and monitoring the implementation phase. In introducing new systems, organization culture plays a large role. If the organization culture is built around trust, then employees are less likely to feel threatened in their job, and the project may be more successful in its implementation.

15.8 Summary

This chapter describes organizational design as touching on six main decisions of designing the organization structure, decision processes, reward system, people, job, and culture. Contingency theory says there is no single best organization design, instead the best design of the organization is the match between the decisions in these areas, other enterprise design decisions, and the enterprise's environment. The organization designer needs to consider these many factors simultaneously to find a good fit.

The design of the decision processes cannot be made in isolation of the organizational structure decisions. The structural organizational design affects the decision processes, especially the structural aspects of formalization and centralization. Additionally, the decision process design must be designed in conjunction with the information view because information is a primary input to decision making.

Job design is important to realize the full potential of the organization's employees. It is also an important link between the organizational design and the process design. Modern job design approaches include job enlargement, job enrichment, and job rotation.

The reward system establishes how performance is rewarded in the organization. It is important to align the rewards with the organizational goals so that employees are motivated to do what is important to the organization. The reward system is also pivotal in shaping the organization's culture, which affects the success of many new projects.

Cultural Change at IBM

IBM was, and remains, notable for having a strong culture. In the 1990s, IBMers would all dress in white dress shirts – an indication of a strong corporate culture. However, it was IBM's strong culture that hindered its reaction to market changes. IBM had a cumbersome management system that demanded everything be done "the IBM way." This included reaching consensus through meetings. If there was disagreement, then further meetings were held. This culture became problematic as the computer industry shifted rapidly from mainframes (where IBM made its money) to personal computers, yet IBM could not adapt fast enough. Another outward sign of IBM's culture and deep-held values was its life-long employment commitment. Again this impeded the company's efforts in reorganizing to meet the external market challenges.

It was not until a new CEO, Louis Gerstner, an outsider not immersed in IBM's culture, came into IBM and changed the culture that IBM was able to adapt. Gerstner, in his book *Who Says Elephants Can't Dance*, likens his arrival at IBM to stepping through a time warp and arriving back in the 50s. A massive, difficult, and painful reengineering feat was required to get the insular IBM to focus on bringing value to the customer in the marketplace. Ultimately, though, this led to the "new" IBM. Gerstner writes, "Until I came to IBM, I probably would have told you that culture was just one among several important elements in any organization's makeup and success – along with vision, strategy, marketing, financials, and the like. I came to see, in my time at IBM, that culture isn't just one aspect of the game; it is the game. In the end, an organization is nothing more than the collective capacity of its people to create value." Gerstner credits his realization that culture was the most important aspect of reshaping IBM to compete in the computer industry as a provider of customer solutions. – Loomis, C.J., Dinosaurs? *Fortune*, May 3, 1993, 36-42 and Louis V. Gerstner, *Who Says Elephants Can't Dance*, Harper Business, New York, NY, 2002.

Review Questions

1. Describe the environmental and organizational conditions under which centralization will perform better than decentralization.
2. Describe the environmental and organizational conditions under which decentralization will perform better than centralization.
3. Contrast and compare job enlargement and job enrichment.
4. List five common organizational structures.
5. Why does increasing job specialization lead to the need for more coordination in an organization?
6. What is the relationship between interdependence and coordination?

7. What is meant by the term span of control?
8. Describe the decision-making process.
9. What is the difference between routine decisions and non-routine decisions? Provide an example of each.
10. Describe the two main approaches to job design.
11. Explain the role a reward system plays in an organization.
12. List and briefly describe three common compensation plans.
13. Explain how the enterprise's environment affects decisions on the hiring of people into an organization.

Exercises

1. List some examples of when an individual's goals are misaligned with the organization's goals. Think of examples from government, university education, business, and others.
2. Select an organization that you are familiar with. Describe the culture of the organization. Provide examples that illustrate aspects of that organizational culture.
3. Use the Web to search for organization charts of some large organizations. Find one that is structured functionally and one that is structured divisionally. Describe the environment each organization operates in and explain why it may have chosen the organizational structure it did.
4. Research how compensation is done for different types of jobs. For example, how are most engineers compensated? How are most car salesman compensated? How are traders in the financial industry compensated? Explain the differences for the financial reward systems used by each company and how it affects employee motivation.
5. Writing mortgages is a complex process that involves completing a mortgage application, checking the credit history of the applicant, getting an appraisal of the property, and determining the loan amount, terms, and conditions. To design the job for this process, one option is to have a single loan officer do all the tasks. Another option is to break it up into smaller tasks such that one person only does getting appraisals, etc. Compare the options by discussing the advantages and disadvantages of each with respect to job design as discussed in this chapter.

Bibliography

[1] S.G. Cohen and D.E. Bailey. What makes teams work: Group effectiveness research from the shop floor to the executive suite. *Journal of Management*, 23(3):239–290, 1997.

[2] R.L. Daft. *Organization Theory and Design.* Prentice Hall, Englewood Cliffs, NJ, 1998.

[3] J. Galbraith. *Organization Design.* Addison-Wesley Publishing Company, London, 1978.

[4] R.E. Giachetti, A. Kusiack, K. Toh, and M. Zelm. *Enterprise Engineering and Integration: Building International Consensus: International Conference on Enterprise Integration and Modeling Technology*, chapter A human factors taxonomy and human modeling in the enterprise, pages 75–81. Springer-Verlag, Torino, Italy, 1997.

[5] R.L. Heneman and J.M. Werner. *Merit Pay: Linking Pay to Performance in a Changing World.* IAP, Charlotte, NC, 2004.

[6] C.W.L. Hill and G.R. Jones. *Strategic Management.* Houghton Mifflin, Boston, MA, 2001.

[7] G.H. Hofstede and G.J. Hofstede. *Cultures and Organizations: Software of the Mind Revised and Expanded 2nd Edition.* McGraw-Hill, New York, NY, 2005.

[8] J. Humphrey and J. Nahrgang. Integrating motivational, social, and contextual work design features: A meta-analytic summary and theoretical extension of the work design literature. *Journal of Applied Psychology*, 92:1332–1356, 2007.

[9] L. Jiang and R. Giachetti. The optimal division of business processes into subtasks with specialization and coordination. Technical report, Florida International University, 2010.

[10] J.M. Juran. *Juran on Planning for Quality.* The Free Press, 1988.

[11] J.E. Kelly. *Scientific Management, Job Redesign, and Work Performance.* Academic Press, London, 1982.

[12] S. Kerr. On the folly of rewarding B while hoping for A. *Academy of Management Journal*, 18(4):769–783, 1975.

[13] P.R. Lawrence and J.W. Lorsch. *Organization and Environment.* Harvard Business School Press, Boston, MA, 1967.

[14] H. Levinson. Management by whose objective? *Harvard Business Review*, Jan 1:107–116, 2003.

[15] J.G. March and H.A. Simon. *Organizations.* John Wiley & Sons, Upper Saddle River, NJ, 1958.

[16] H. Mintzberg. *Structure in Fives: Designing Effective Organizations.* Prentice Hall, Englewood Cliffs, NJ, 1992.

[17] P. Osterman. Work reorganizatoin in an era of restructuring: trends in diffusion and effects on employee welfare. *Industrial and Labor Relations Review*, 53(2):180–198, 2000.

[18] C. Perrow. *Organizational Analysis: A Sociological View.* Wadsworth Publishing Company, Belmont, CA, 1970.

[19] J. Pfeffer. *The Human Equation: Building Profits by Putting People First.* Harvard Business School Press, Boston, MA, 1998.

[20] E.H. Schein. *Organizational Culture and Leadership.* Jossey-Bass, San Francisco, CA, 1996.

[21] D. Schleicher, J. Watt, and G. Greguras. Reexamining the job satisfaction-performance relationship: The complexity of attitudes. *Journal of Applied Psychology*, 89:167–177, 2004.

[22] J.D. Thompson. *Organizations in Action: Social Science Bases of Administrative Theory.* McGraw-Hill, New York, NY, 1967.

[23] T.J. Williams. *Handbook of Life Cycle Engineering*, chapter The Purdue enterprise reference architecture (PERA), pages 289–330. Springer, New York, NY, 1998.

[24] T.B. Wilson. *Innovative Reward Systems for the Changing Workplace.* McGraw-Hill, New York, NY, 2004.

[25] T. Wright and R. Cropanzano. The moderating role of employee positive well-being on the relation between job satisfaction and job performance. *Journal of Occupational Health Psychology*, 12:93–104, 2007.

[26] S. Zuboff. *In the Age of the Smart Machine.* Basic Books, New York, NY, 1988.

Part VI

View Integration

16

Enterprise Technology

"Any sufficiently advanced technology is indistinguishable from magic."– Arthur C. Clarke (1917-2008), British science-fiction author.

Integration across the enterprise is one of the most challenging and significant issues facing an organization. The decomposition of the enterprise into subsystems and views is a matter of necessity to deal with the complexity and size of enterprise systems. There are several ways to decompose the enterprise as was discussed in Chapter 15 on the organizational structure; one of the more common decompositions is functional. The benefits of the functional decomposition are derived from the differentiation of each subsystem to develop unique knowledge, information, and systems to optimally address local problems. However, the decomposition must be countered with appropriate levels of integration so that the decisions and actions of each subsystem contribute effectively and efficiently to the overall enterprise goals. The need for integration arises due to the interconnectedness between the subsystems. It is the integration of the subsystems that gives a system its superiority over a collection of elements that are not integrated. Subsystems exchange materials, people, and information. Enterprise integration ensures that all these exchanges happen efficiently and effectively. To do this, it is imperative that throughout the design process efforts are made to ensure enterprise integration. A thorough understanding of the available enterprise technologies and how they work is needed to specify a suitable technology infrastructure to support enterprise integration.

This chapter describes the enterprise integration challenges, presents a framework to classify the types of enterprise integration, and reviews the technology to achieve enterprise integration. After completing this chapter, you should be able to:

- Describe why enterprise integration is important.
- List the challenges to enterprise integration.
- Define and describe the five enterprise integration types.
- Compare and contrast the main approaches to system architecture.
- Describe how client/server computing works.
- Describe the technologies for data integration.
- Describe the technologies for application integration.
- Categorize the types of standardization.
- Discuss the trends in enterprise integration technologies.

16.1 Introduction

Integration has many different definitions depending on the context it is used in. Even within a single domain there may be multiple meanings of the term integration [6]. Many authors take a narrow focus and use the term integration to describe the interconnection and interoperability of the software applications in an enterprise [4]. Others define integration in the broadest sense to mean the entire enterprise working together as a single entity. In this book, we use integration to describe all the ways that parts of the enterprise work together. To avoid the misunderstandings that ensue from multiple meanings of integration, Section 16.4 presents a framework that describes different types of enterprise integration.

> *Enterprise Integration* makes the operation of the enterprise seem as if it is a single entity working towards achieving a known goal.

The goal of integration is to improve the efficiency and effectiveness of the enterprise by having all the enterprise subsystems work together harmoniously. The integration goal includes:

- Improving the quality and timeliness of information, and providing information on demand and where it is needed, regardless of the source system.
- Coordinating decisions made by separate organizational units so that all the units work together towards fulfilling the overall enterprise goals, and thus avoid local optimization.
- Management of the actions among people in the enterprise and synchronization of business processes to efficiently and effectively provide quality products and services.

Enterprise integration strives to achieve the benefits of integration while maintaining the autonomy of the subsystems. Separate subsystems will provide a more flexible and agile system. Moreover, autonomous subsystems are more reliable if the overall system can continue even after a subsystem failure. From a technical perspective, the subsystems are loosely coupled with location transparency, meaning that you need not know the physical location of the system in order to interface to it. In summary, the goal of enterprise integration is to streamline the interconnections between the subsystems of the enterprise. By improving the connections the enterprise can give the outward appearance of functioning like a single, cohesive whole.

16.2 Integration Challenges

Integration challenges are encountered in the three enterprise views: process, information, and organization. Process integration challenges occur when processes are not viewed as end-to-end business processes serving a customer. Without this view, the process may become fragmented and poorly integrated. Integrated processes are characterized by smooth, problem-free operations with all inputs being ready at the right time and at the right place. Poorly integrated processes are evident because workstations must wait for resources to be available, or inputs to arrive, or decisions to be made, etc. The outward sign of poor integration is waiting, unnecessary work, and poor process performance.

Figure 16.1 shows part of an order fulfillment process. If the process is well integrated then the arrival of the physical supplies and the order will be synchronized such that one does not wait for the other. On the other hand, a poorly integrated system would be beset with high levels of unwanted inventory, long waiting times to fulfill orders, and excessive communications between the customers asking where is their order and the order entry clerk. The repurcussions of poor process integration are:

- Excessive process waiting time.
- Low resource efficiency.
- Excessive need for information exchange and coordination.
- Frequent need for expediting orders because the process's coordination and information flow is not well designed.
- Quality problems because the process is not designed to mistake-proof the activities and because proper statistical quality control and other performance improvement is not in place.

Information integration challenges occur when the data are located in many different locations creating problems in using the data. As was described in Chapter 5, it is not uncommon for systems to evolve without an overall plan – in other words, an enterprise architecture. As a result, many enterprises face heterogeneity of the hardware, software, applications, and data in those applications. Heterogeneity creates problems in moving data from location to another and from understanding the data from another location. The Software Engineering Institute (SEI) at Carnegie Mellon University (CMU) suggests that almost 70% of code in corporate software systems is dedicated to moving data from system to system [10]. The repercussions of poor information integration are:

- The paradox that data are everywhere, yet it is not readily accessible or easy to use.
- Data are controlled by many different organizational units without consistent policies for access, updating, security, etc.
- It is expensive to develop, maintain, and change information systems, leading to slow deployment of new systems.

Organizational challenges occur when the decisions and actions of organizational units are not coordinated, or the goals are not aligned and units may actually make decisions that are detrimental from an enterprise perspective. Organizational culture, how it is decomposed, the reward structure, and how decisions are made all play a role in the integration of the organization. Repercussions of poor organizational integration are:

- Uncoordinated decisions made by different organizational units.
- A functional viewpoint of one's own organizational unit and its work, instead of a horizontal view of the process serving an end-customer.
- Reward systems that are not aligned with the organization's goals.

The integration challenges are overcome by enterprise design. By recognizing the integration challenges the enterprise engineering project teams can identify existing problems and avoid potential problems before they occur.

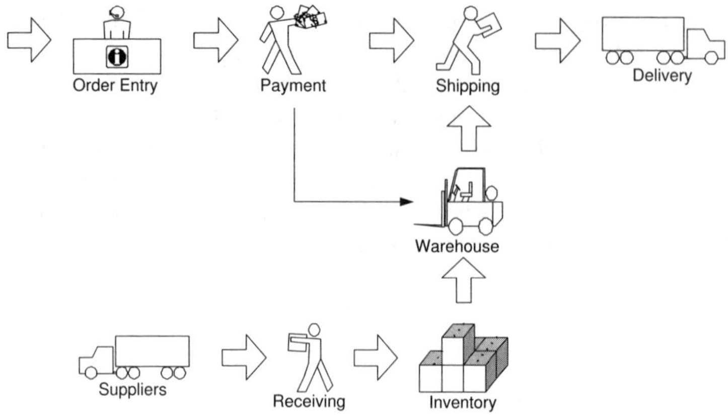

FIGURE 16.1
Process integration.

Enterprise Integration Challenges

Technical Challenges

- Stand-alone designs common to legacy systems
- Internalized data models that were not meant to be shared with other applications
- Lack of system interfaces to access data
- Heterogeneous technologies for platforms, networks, applications, programming languages, and databases
- Multiple standards
- Multiple distribution paradigms such as CORBA, COM, EJB
- Proprietary systems
- Semantic differences in the interpretation of data
- Security that limits access to data and/or systems

Management Challenges

- Working across cultures, companies, or other organizational boundaries
- Overcoming the differences associated with cultures, companies, and disciplines
- Collaborating and competing with other organizations depending on the market
- Multiple stakeholders
- Maintaining autonomy
- Data ownership
- Protection of intellectual property
- Lack of sufficient expertise for integration
- Loss of control when integrating systems or sharing data

16.3 System Architecture

The system architecture describes how the information technology components are organized into an overall system. The system architecture is one component of the enterprise architecture. In this section, we describe the system architecture patterns that are most common in industry. This section is but a summarization of the main concepts – system architecture is a tremendous topic in itself, worthy of independent study.

Most all system architectures are based on the *client-server* computing paradigm. A client is any system component that requests a service from another component. A server is any system component that receives, acts on, and replies to client requests. The terms "client" and "server" are not absolute, in one communications exchange a component can be the client and in another communications exchange it can be a server. The applicable term depends on the role the component plays in the communication.

Figure 16.2 illustrates client/server communication. The request and reply messages are formatted sets of commands and data. The communication is *synchronous* if both client and server must attend to the communication in a coordinated manner. In synchronous communication the client and server are tightly coupled and must run simultaneously for message delivery to occur. The client is blocked until the reply is received. If the network fails or something else happens, it impacts the processing of the client and server. A telephone conversation is a typical example of a synchronous communication.

The communication is *asynchronous* if the client and server send messages without being bound by time or caring whether the receiver of their message is available. After a client sends an asynchronous request, it can continue with other processing and not wait for the server's reply. In asynchronous communication the client and server are loosely coupled – they can continue running regardless of the state of the other application. The client can continue processing without waiting for the server's reply. Email is an asynchronous communication because you send email and it goes to a inbox, where it waits until the receiver acts on the email and sends a reply. Meanwhile, you can do other things while waiting for the reply.

A server is a software system, not to be confused with what computer-hardware manufacturers describe certain computers as being servers. What the manufacturers are saying is that the computer is designed specifically to work as a server because it has greater processing power than an ordinary computer. A server can be:

- File server – Receives and fulfills requests for files.
- Data server – Receives and fulfills requests for data records.
- Web server – Receives and fulfills requests for Web pages.
- Email server – receives and fulfills requests for email.
- Print server – Receives and fulfills requests to print documents, pictures, and photos.
- Application server – Receives and fulfills requests for processing.

To understand a distributed system, system architects describe the system in terms of the layers. The main layers and their functions are:

- Presentation layer – Provides all functionality for the user interface to handle system inputs and system outputs.
- Application layer – Provides the logic to support all the business processes and rules.

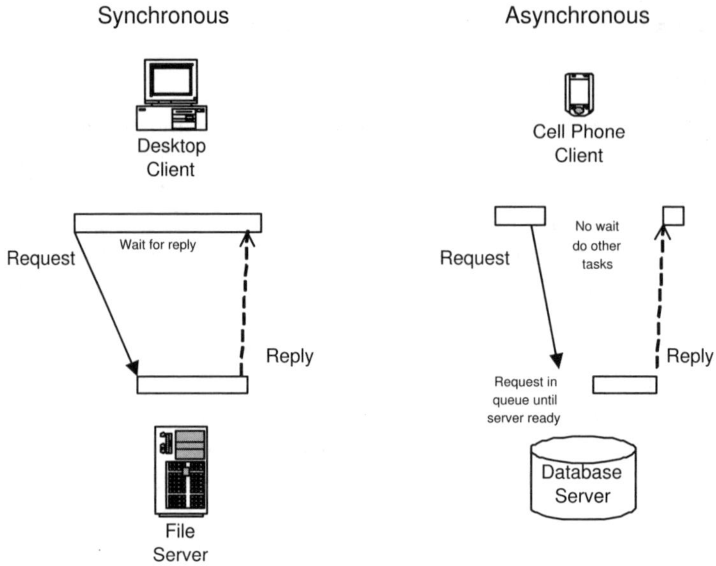

FIGURE 16.2
Client-server communication.

- Data layer – Provides the stored data and handles the data manipulation as performed by a DBMS.
- Middleware layer – Provides the services to connect different applications together so they achieve interoperability.

The decomposition of the information system into different layers facilitates understanding of how the system works and provides loose coupling to make system design and maintenance easier. Also, division into layers makes the system more scalable since it is easier to add more components to a particular layer as needed. For example, you easily add more computers to the presentation layer to serve more users.

Traditional System Architecture

The *traditional system*, frequently known as a *legacy system*, is a software application that was designed to provide certain business functionality independently of other applications. Traditional systems are monolithic, mainframe computer-based applications with the data and application logic tightly coupled, and their own interface for accessing the applications functions and data. All the processing and data storage was done by the legacy application, so the users only needed a thin client to access the system (see Figure 16.3). A typical thin-client was a CRT terminal, frequently offering the single screen color of green, with a keyboard and no CPU (central processing unit) because there was no need to do any local processing. The centralized mainframes were able to support hundreds or thousands of users concurrently, making this an affordable means to provide computing power in an enterprise prior to the advent and affordability of personal computers. As a result, these applications dominated the business software market prior to the widespread use of personal and mobile computing, and legacy applications are commonly still operating today.

The integration challenges associated with legacy applications are first and foremost due to the fact that these applications were designed as stand-alone systems. The application

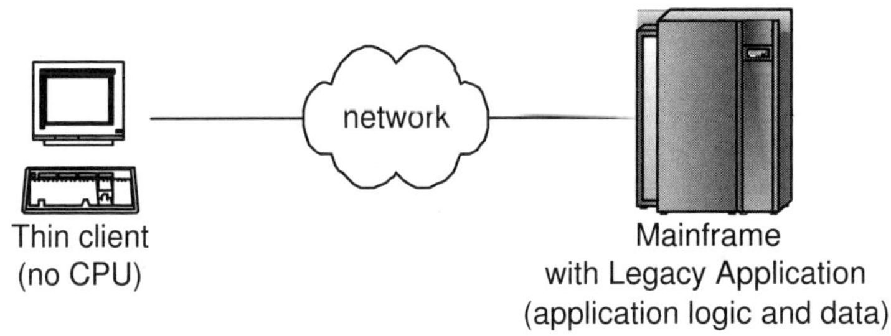

FIGURE 16.3
Legacy application.

designers did not consider whether other applications would be able to access the business logic or data in the system. These legacy applications all stored their own data, often in a proprietary format that made it virtually unusable to others. These aspects make it very difficult, but not impossible to integrate with other systems. These designs also make users reliant on a particular application vendor for upgrades, support, and maintenance. The traditional system architecture is rigid and difficult to maintain. As the enterprise inevitably changes, it becomes difficult and expensive to modify legacy systems to meet newly emerging requirements.

Distributed System Architecture

A *distributed system* architecture is one in which the system components are physically distributed to multiple locations that are connected by network technology. In the distributed system the application logic, data, and user interface logic is distributed across multiple computers in the network. It is a decentralized approach, in contrast to the centralized approach taken by legacy systems. Several architectural properties are needed for distributed systems to be successful. The architecture needs to provide physical transparency, meaning that a user does not need to know where a system component is located in order to use that component. It needs to provide reliable, safe connections, so that users are unaware of remote computer failures and unaware of other users' interaction with the system.

The decomposition into layers yields many possible n-tiered architectures. Figure 16.4 shows a simple, common three-tiered architecture. In Figure 16.4 the sequence of activities is:

1. The user interacts with the system interface and requests a service.
2. The network directs the request to the appropriate server.
3. The application server processes the request. It generates SQL statements to retrieve any needed data from the data layer.
4. The data layer returns to the application layer any data requested via SQL.
5. The application layer finishes processing the request and returns the output to the user.
6. The network directs the output to the user who requested it.
7. The presentation layer takes the output and presents it to the user.

Because of its commonality we also present an n-tiered architecture used where the network is the Internet. Figure 16.5 shows the additional tier of the Web server, which is part of the presentation layer. The Web server receives the user requests, sends application

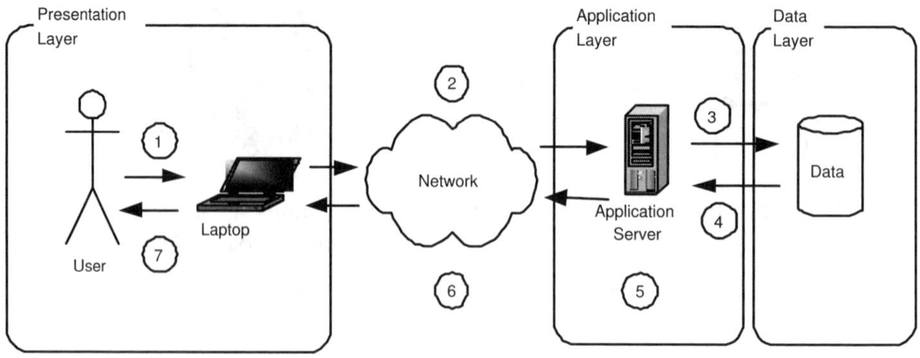

FIGURE 16.4
Three-tier architecture.

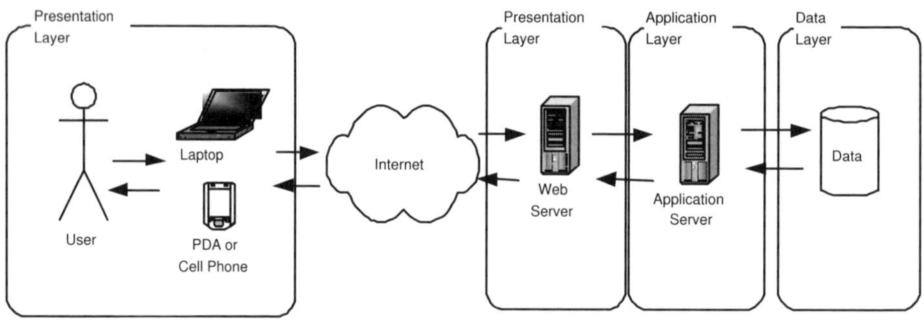

FIGURE 16.5
n-Tier architecture.

requests to the application layer, and then when it receives the response from the application layer, it processes the HTML code to produce the output and sends it back to the user. Many variations on the basic n-tier architecture are possible. If the system involves an ERP system, then this would become another layer in the system.

The n-tier architecture is tightly coupled because one component in a layer cannot function without the other components. Communication in the distributed architecture is synchronous and the original request is usually initiated by a person who is intolerant for long waits. So, while the system is decomposed into tiers, the tiers remain tightly coupled, each tier fulfills part of the overall function, and overall system operations fail if one tier is not functioning.

The middleware layer is what connects the tiers and how different components in the same tier can communicate. The next sections discuss the various types of middleware that are used. Here, the two main integration approaches are summarized. First, the connections between applications can be point-to-point type integration. In point-to-point integration, middleware is used to integrate applications in pairs. This leads to a proliferation of paths as the number of applications increases and it is very difficult to manage. The alternative to point-to-point is many-to-many such that middleware can connect many applications to many applications. The trend is towards many-to-many middleware because it provides a lower-cost, more scalable solution to application integration.

Service-Oriented Architecture

A *service-oriented architecture* (SOA) is designed to provide a service platform consisting of many service modules that can be used in combination with other services to fulfill business needs. The idea is that a business could flexibly combine and recombine services into business processes as their process needs change. In SOA, units of business functionality tied to the underlying IT are made available as components that can integrate with other services and support reuse throughout the enterprise. SOA addresses the IT problem that systems are inflexible and as the business inevitably changes in response to its environment, then the IT systems can no longer adequately support the business, yet changing them is expensive. To lower the cost of IT and support the constant evolution and change needed, a more flexible approach is to design IT architectures that provide reusable services that can be combined together to provide any number of business services. SOA proposes to do this by providing services that can be integrated into larger applications as needed. SOA is implemented using Web services, which are discussed in Section 16.7.1.

Enterprise Resource Planning (ERP)

Enterprise Resource Planning (ERP) systems are large business applications consisting of many modules designed to handle all the transactions for pre-defined business processes. ERP modules include: human resources, accounting, finance, materials management, order fulfillment, and support for many other business processes. ERP applications are single-vendor, integrated solutions that are designed around the business processes. Contrast this to legacy applications that were designed around functions and not the end-to-end business processes. In their attempt to cover the software needs for the entire enterprise, ERP packages are large and complex.

The leading vendors of large ERP packages are SAP and Oracle (including PeopleSoft and JD Edwards). For small to medium-sized companies there is a far greater number of choices in ERP from vendors including Microsoft, Sage, and many others. As a single software application with a single data repository, in theory many of the integration challenges associated with heterogeneity do not exist. However, in practice, few companies implement all the modules from a single ERP vendor. So while an ERP package eliminates some integration challenges, it is not a cure-all. Moreover, ERP projects tend to be extremely complex with high reporting of over-budget and behind-schedule projects.

16.4 Enterprise Integration Levels

Giachetti [5] presents a framework (shown in Figure 16.6) to describe the different types of integration in an enterprise. The enterprise is decomposed into five levels to reveal the different integration types. The five levels are: infrastructure level, information level, application level, process level, and organization level. Enterprise integration is defined as the achievement of all the integration types together.[1]

[1]Kosanke et al. [8] summarize the progression of enterprise integration from system integration of computer networks, application integration of business applications, business integration of process networks, and the final goal of enterprise integration through organizational networks.

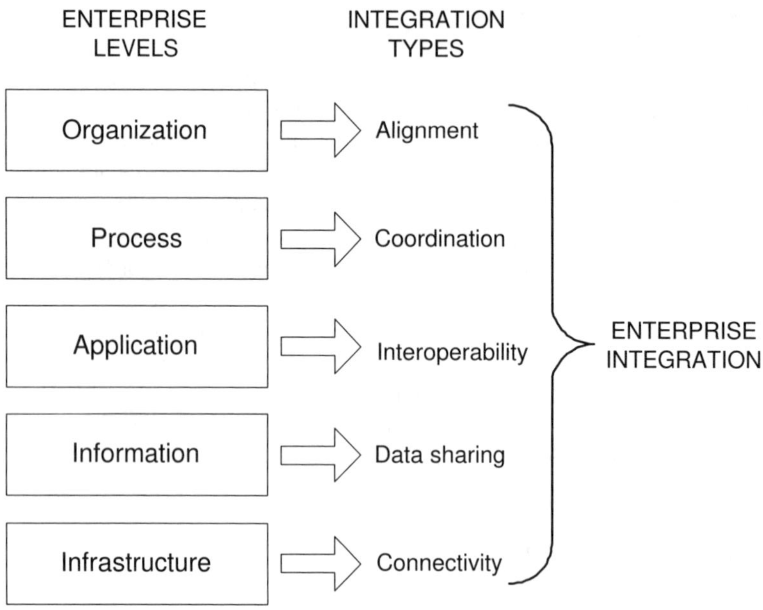

FIGURE 16.6
Enterprise integration levels.

16.4.1 Infrastructure Level

The lowest level is the infrastructure level. At this level the integration issue is the physical heterogeneity of the hardware, machines, devices, and their operating systems. The integration goal at the infrastructure level is connectivity, defined as the linkages between devices. Connectivity simply ensures that data and/or messages can be sent from one device to another device, not whether the data sent can be interpreted by the receiving system.

In practice, the integration of information and communication hardware via networks both wired and wireless has been very successful. Connectivity is provided for mostly by adhering to standards for the physical network devices and protocols defining how the communication is performed. The Internet is the foremost example because it provides a global and open network for integrating various computer hardware platforms through conformance to network protocols [2].

16.4.2 Information Level

The integration goal at the information level is to provide the ability for the enterprise applications and users to freely and easily utilize and share data. Data sharing occurs when across an enterprise people use common data definitions.

The greatest problem facing information integration is the heterogeneity of the information sources. Information sharing must address the schema diversity problems described by Batini *et al.* [1]:

1. Different perspectives or names for the same information object.

   (a) Data-name conflicts – One company may call a document an `Engineering Change Order` while another company calls the same document an `Engineering Change Notice`. Both documents represent the same concepts.

TABLE 16.1
Data Definition Mismatch

Attribute	Data type	Constraint
fecha	date	not null
fecha	number	not null
ANI	number(10)	
ANI	varchar(10)	
ANI	varchar(25)	not null

(b) Data-value conflicts – The attribute name might be the same, but different databases use different values. For example, for the attribute `State` one database might use the full name as in `Florida` and another database might use two-letter abbreviations such as FL.

(c) Data precision conflicts – Suppose the U.S. Navy and the British Navy both have a data attribute Sea Bottom, and the U.S. defines seven different types of sea bottoms, while the British define five different types of sea bottom. How do you map this?

(d) Data-type conflicts – The attribute names might be the same, but the data types are different. The data[2] in Table 16.1 provide a snapshot of some simple but troublesome mismatches identified in the data models of a Latin American telecommunications company. These mismatches are relatively easy to handle during integration because we can simply map them into the correct fields when transferring data.

2. Equivalence among information definitions. There is more than one means of modeling the information content of an enterprise. Given two information models of the same enterprise by different designers and you will likely have two different models. For example, `Male` and `Female` could be represented as separate entities with a relation to the entity `Person`. Another model might make `Sex` an attribute of `Person`. These two models are semantically equivalent but modeled differently.

3. Relationships between the domains. When two or more information models are integrated, there may be relationships between objects in one model and objects in the other model.

4. Weak semantics. Oftentimes, the semantics are informally defined or not defined at all, leading to difficulty in determining the precise meanings of the objects in the model.

16.4.3 Application Level

The applications are software systems that provide a service. Each application, programming language, or system tends to use locally defined data and message formats that lead to heterogeneity. For example, some applications might define a message using 8 bytes while another application uses 16 bytes. To overcome this problem, the integration goal is *interoperability* defined as the ability of one software application to access/use data generated by another software application or the ability to invoke services of another software application [7]. To integrate the applications, all the interfaces between software systems need to be identified, analyzed, and designed. Section 17.5 describes a methodology to do this.

[2] "fecha" is Spanish for date, "ANI" is Automatic Number Identification, and "CDCliente" is Client Code.

16.4.4 Process Level

The business processes in the enterprise will almost surely cross functional boundaries or even enterprise boundaries. Moreover, the inputs and outputs of one process will be the outputs and inputs of another process. For example, the product development process crosses many organizational boundaries; it starts in the marketing function, continues to the engineering function, and then the manufacturing function. The inputs and outputs to product development will be from other processes in the enterprise. As a result, a complex set of interdependencies emerge between the organizational units due to the process flow. The process integration problem is how to coordinate the business process by managing the interdependencies between process activities.

Interdependence is the degree to which the actions and outcomes of one organizational unit are controlled by or contingent upon the actions of another organizational unit. In this definition, interdependence is viewed as occurring between tasks, which create the interdependencies between the organizational units that are responsible for executing the tasks. If interdependence is high, then the time, cost, and effort necessary to coordinate the process will be high. Here, *coordination* is defined as managing the dependencies between entities [3].

Coordination is about first understanding the interdependencies between tasks and then identifying appropriate coordination mechanisms to manage those interdependencies. Figure 16.7 shows five interdependencies:

- Pooled resources – a resource (machine, person, information) is used by two or more activities, and the resource can only be used by one activity at a time. This is an indirect interdependency because it is through the shared resource. The coordination problem is how to schedule the use of the resource between the process activities. *Example*: A conference room is a shared resource used by all the departments in the building. To coordinate the use of the shared resource, many companies will have a schedule for those wishing to use the conference to block out time.

- Controlled sequential – the decision made in an activity controls the execution of another activity. The coordination problem is how to convey the decision to the dependent activity and to make sure the decision does not conflict with other inputs affecting the dependent activity. *Example*: Marketing's decision to have a sale on an item will create a surge in demand for the product made by manufacturing. To coordinate the activities, this information should be given ahead of time to manufacturing so that they can build inventory.

- Sequential – the output (material or information) of one activity becomes the input of another activity. The coordination problem is that the output is usable by the receiving activity, that the output is transported to the receiving activity efficiently, and that the activities are synchronized to avoid inefficiencies of the receiving activity waiting for the predecessor activity. *Example*: A university registration activity generates the invoice that is input to the cashier's office. To coordinate the activities, the university might use electronic exchange of data so that the invoice is automatically generated and sent to the cashier's office.

- Event conditional – the activity's performance is conditional on an event, either external or internally generated. The coordination problem is that the activity can sense the event and react accordingly. *Example*: A call center receives calls (the event) from customers. To coordinate the call center activities, they need to have sufficient staffing, they might monitor call volume to predict peak and non-peak hours to schedule maintenance or administrative work during non-peak hours.

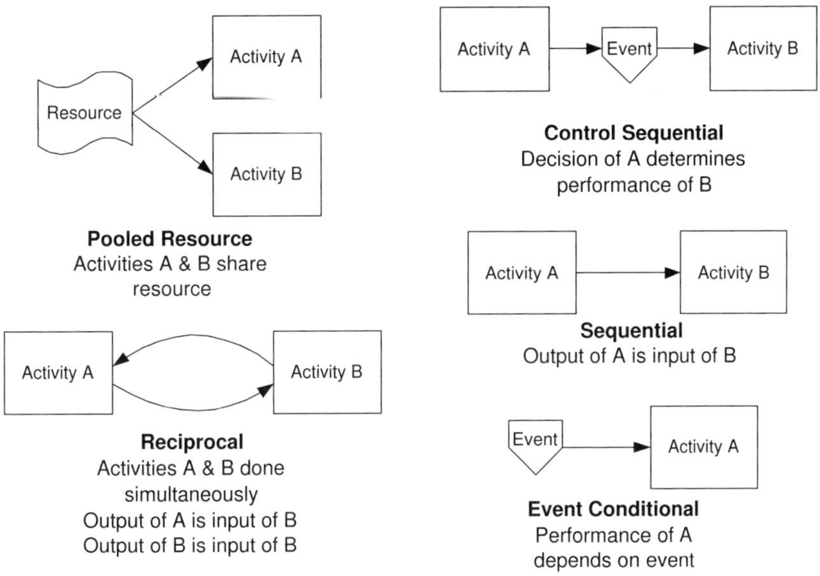

FIGURE 16.7
Interdependence types.

- Reciprocal – two or more activities are executed in parallel, and the activities exchange information. The coordination problem is how to exchange the information and minimize the number of iterations. *Example*: In the design of a car, the engine, radiator, body, etc. are designed simultaneously by separate design sub-teams. Each sub-team must constantly exchange information because all the car parts must match up (e.g., the radiator has to be sized correctly to cool the engine and it must fit in the space allocated). To coordinate these reciprocal interdependencies, a company might have weekly design meetings, co-locate the design teams to encourage greater communication, and schedule deadlines to firm up design decisions.

To coordinate the processes, the first step is to identify the interdependencies and then to determine coordination mechanisms to manage those dependencies. An enterprise or an individual can only do so many things well. They are constrained by resources or if not, at least they are constrained by insufficient time (the most valuable resource). Consequently it makes sense to identify core business processes, and to keep the number small. It is not possible to model every process in an enterprise at once.

16.4.5 Organization Level

An enterprise is decomposed into separate organizational units, each of which has a tendency to develop strong cohesion within the group. While beneficial, the strong cohesion sometimes works against the overall performance of the entire enterprise. People may act to help themselves or their immediate units regardless of whether their acts benefit the enterprise. The integration type of organizational alignment addresses this concern. Organizational alignment seeks to have all the units working together toward the enterprise mission and goals.

Organizational alignment starts by creating an organization-wide understanding of the enterprise's mission, vision, core values, and strategy. The goals and objectives must align with the strategy. Creating the mission, vision, and strategy and decomposing it into goals and objectives is part of the enterprise design problem. The next step is to make sure that it is communicated widely throughout the organization. Communication of the enterprise's strategy is necessary but not sufficient to ensure organizational alignment. The enterprise leadership must go further and develop a culture and organizational systems to support alignment.

The organizational systems that support alignment are within the domain of human resources. These systems include: the performance monitoring system, the reward system, training, leadership development, recruitment and retention, recognition and career development. All these systems can reinforce the strategic vision of the enterprise.

Foremost is the performance measurement system and the reward system. Recall in Section 2.1.6 that to control a system we need to measure the system performance and make adjustments to system levers to bring it into control. In this case, we wish to control the actions of the organizational units so that they all contribute to the enterprise-level goals. The enterprise designers need to determine what to measure at each level of the organization and put systems in place to generate those measures so that the subsystems can be controlled. It is the managers who decide given the controlled parameter what actions to take.

The reward system is especially important for getting the people in the organization aligned with enterprise goals. As was argued in the organizational design chapters, people respond to the reward structure. The reward system must provide incentives for behavior that helps the enterprise and disincentives for behavior counter to the enterprise's goals. The reward system reflects the values of the organization. People understand why and how their actions create results, both positive and negative, in the organization.

In addition to designing the control system and reward system to accomplish organizational alignment, the enterprise must provide infrastructure support. The enterprise needs to have plans for the recruitment and retention of employees that have the skills, knowledge, and personalities in support of organizational goals. The enterprise should provide training so that employees understand how their work contributes to the organization and how to be most effective in doing that job. The enterprise should have intentional career and leadership development programs so that employees can grow within the company and are motivated to work hard for the organization's success. Collectively, all of these activities develop an organizational environment that is supportive of alignment.

16.5 Integration and Standardization

One means to achieve integration is through standards. A *standard* is a formal specification to establish the technical requirements for the operation of a system. When the standard is intended for communications, then it is called a protocol. A *protocol* is a standard for data representation, data transmission, authentication, and error detection for sending and receiving information over a communications channel. The idea is that if all systems used the same standards, then this removes many of the obstacles to integration. When a system conforms to a standard this means the system follows the specification. Ideally, conformance to a standard suggests the system will work together seamlessly with other systems conforming to the standard.

Consequently, enterprise integration is replete with standards. Standards exist for programming languages, communication protocols, interfaces, data storage, hardware, and so forth. These standards greatly aid integration efforts. Standards can be: coercive (e.g., Euro conversion); collaborative (e.g., HTML, CORBA, STEP); *de facto* or competitive (e.g., MS

Windows). In the domain of enterprise integration, there are various standardization organizations that generate standards: the Institute of Electrical and Electronics Engineers (IEEE) and the International Organization for Standardization (ISO) are two such examples.

Standardization is most successful for integration of the lower levels of the enterprise integration framework. At the infrastructure level, communication protocols allow us to send and receive information without regard for how it is sent. The Internet uses the following set of protocols: TCP/IP for transmission of information packets and addressing and HTTP for Web pages. At the information level there are a fewer standards, mostly de facto standards by the dominant application vendors. For databases, SQL is a universal standard, although slight variations exist between vendors. At the application level, there are standards for interoperability but there are many choices and the standards in one domain will not work in another domain. For example, CORBA is a standard for the interoperability of software objects; Java RMI is a standard for inter-operation when using the Java programming language; and Microsoft has the .Net for Windows-based programs. Within a standards group, interoperability is possible but between groups it is more difficult.

There are many industry-specific standards. For example, in manufacturing the STEP standard is promoted so that disparate computer aided design and manufacturing systems can share product information adhering to ISO Standard 100303. The advent of XML is being investigated and applied by various industry sectors to create industry-specific definitions for the transfer of data within a supply chain.

Standards, while important are not a solution in itself. There are many competing standards, they take long to develop (e.g., it took more than 10 years to develop the Standard for The Exchange of Product (STEP)), and when generated through a standardization process the standards are a compromise solution, never satisfying every possible need. Finally, it should be noted that there are no universal standards. So, while standards are an important part of specifying the technical integration of the enterprise's subsystems they cannot solve the integration problem by themselves.

16.6 Data Integration Technologies

Data integration technologies either create a single, unified model of the data by merging databases together, or provide the tools and techniques to move data between systems. The advantage of focusing on data integration is it avoids making any changes to the applications or business processes. Since many applications have a layered architecture that separates data from application, this is often a feasible and low-cost (relatively) means of integration. Data integration is very common and has been practiced for some time now.

Successful integration between business entities requires sharing of appropriate data. Shared information is important for operating business processes, decision making, and coordination between different organizational units whether inter- or intra-enterprise. To share data across the enterprise, the options include:

1. Point-to-point integration.
2. A single, centralized database for the entire organization.
3. A federated data, which is a collection of cooperating but autonomous databases.
4. Data warehouse.

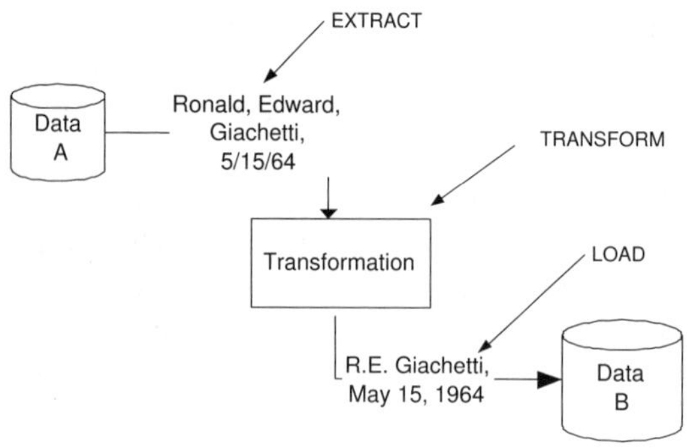

FIGURE 16.8
ETL.

Point-to-Point Integration

Point-to-point integration is connecting two databases together by defining data translators between them. Clearly, as the number of databases grows this method can get out-of-hand, but it is widely practiced because it is a straightforward approach to integrate two systems. Figure 16.8 shows the extract, transform, and load (ETL) of data from one database to another database. Enterprise-wide this can create problems because it creates redundant data across many systems.

Data Middleware

Instead of defining translators specific to a point-to-point solution, an approach is to define data-level middleware to create interfaces between the database and all other applications. A common data middleware component is ODBC. ODBC is Open Database Connectivity, a standard or open application programming interface (API) for accessing a database. By using ODBC statements in a program, you can access files in a number of different databases, including Access, dBase, DB2, Excel, and Text. It allows programs to use SQL requests that will access databases without having to know the proprietary interfaces to the databases. ODBC handles the SQL request, and using a vendor-specific driver converts it into a request the individual database system understands (see Figure 16.9).

Database gateways are APIs that allow access to many different databases (see Figure 16.10). Database gateways act interface translators that move data, SQL commands and applications from one type of database to another. The client application requesting the data, the gateway, and the database server can reside on the same platform or different ones. This allows gateways to connect clients and servers running on dissimilar networks.

Database gateways have the following functions:

1. Accept statements specified by a well-defined grammar (usually SQL) from a client application.
2. Translate the statements to a specific database format.
3. Send the statements to be executed against the database.
4. Translate the results back into a well-defined format.
5. Return the data and status information to the client.

The price of the accessibility is speed, because it is necessary to translate the client request and the server results.

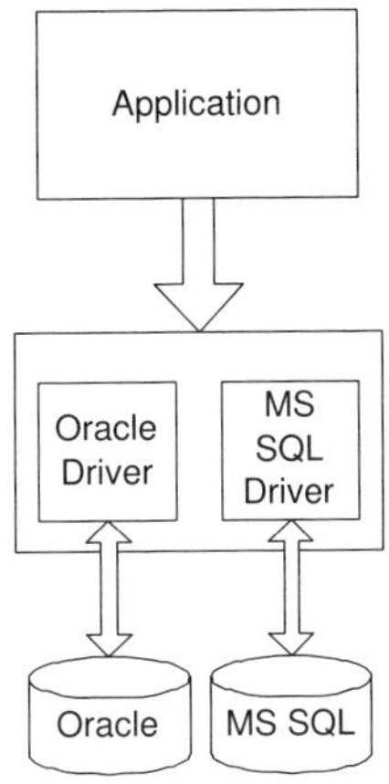

FIGURE 16.9
ODBC.

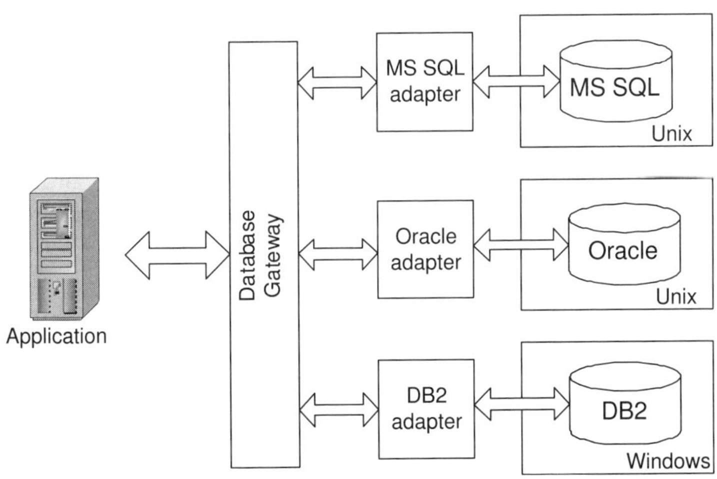

FIGURE 16.10
Database gateway connects to three different databases with different operating systems.

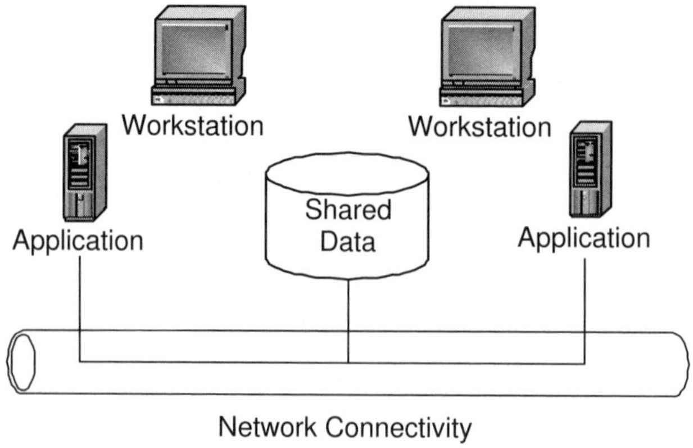

FIGURE 16.11
Shared database.

Centralized Database

A single database shared by the entire enterprise eliminates data integration problems. Every application writes and reads from the shared database (see Figure 16.11). ERP systems use a single database. The single database approach is rarely achieved in an enterprise. It is very difficult to reconcile all the disparate needs throughout the enterprise and it may even be impossible without making compromises that reduce the usefulness of the data. Centralization of a database within smaller organizational units such as a marketing database that holds all the marketing data are often achievable. However, even then, many employees will find that they cannot perform desired functions or analysis with centralized database because it is either too slow, inflexible, or structured incompatibly with their needs. These employees will then often copy the data from the centralized database into a locally controlled database or spreadsheet to do their work.

Federated Database

A *federated database* is a collection of heterogeneous, component databases over which a global view is created so that applications can treat the separate databases as a single database [9]. The system provides data transparency, which shields the users from the need to know where the sources reside, what hardware and software they run on, how they are organized, and how the data are accessed. Each component database maintains its autonomy, making federation a somewhat easier task than merging multiple databases to create a single database. The benefit of a federated database is the users see only a single logical data model, while the databases can be different types, geographically dispersed, and use different models. A federated database is achieved via data mediation, which provides a formalized model and set of services for managing data heterogeneity. In simplest terms, data mediation converts data from one format to another format. Using data mediation, the original data sources are left untouched. No effort is made to arrive at a global data model.

Data Warehouse

A *data warehouse* collects data from one or more operational databases, integrates the data, and makes it available for querying, reporting, and analysis. The term data ware-

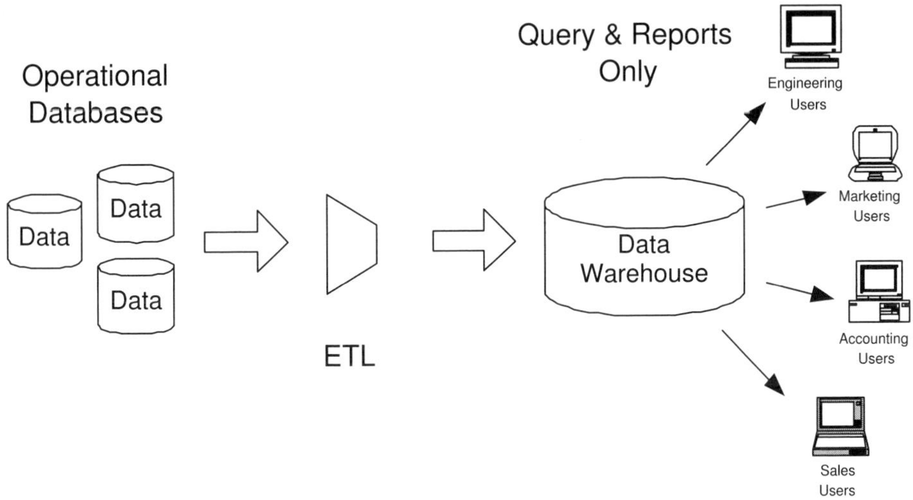

FIGURE 16.12
Data warehouse.

house usually refers to a database used for on-line analytical processing (OLAP) versus operational databases that are for on-line transactional processing (OLTP). The user of an OLAP only queries the database to obtain information that is generated and formatted in reports, graphs, and tables. They do not add new data or change existing data into the data warehouse. An OLTP system supports the business transactions: customer orders, client reservations, purchases, returns, and so forth.

No transactions are executed on the data warehouse; it is only used to obtain information. Figure 16.12 shows the relationship between the data warehouse and the operational databases in the enterprise. A process called Extract, Transform, and Load (ETL) takes data from the operational databases, cleans the data, and then loads it into the data warehouse. Notice, the direction is one way from operational databases to the data warehouse. New data and changes to the data warehouse come from the operational databases that are the source for the data warehouse. The ETL process is usually done on a schedule such as every night, once a week, or even once a month. The rationale is that to support the analysis being conducted, having real-time data are not required. For example, knowing the previous few years data are sufficient to generate a report detailing the demographic characteristics of customers grouped by credit scores.

The benefit of a data warehouse is it provides integrated data in a format optimized for analysis. The schemas used to design data warehouses are different than what is used for operational databases. A common schema is the star schema that allows a multi-dimensional organization of the data. The data are separated into fact tables or dimension tables. A fact table contains data about a business task that users would want to retrieve. Dimension tables contain data about qualifying characteristics of the fact data. For example, a data warehouse could have a sales fact table that contains all the sales data. The data warehouse can then have separate dimension tables to define different ways to analyze the sales data. Dimensions could be location, customer type, time, and product. A data warehouse integrates data, but only for OLAP, it has no integration benefits to the OLTP data that is used to run the business.

16.7 Application Integration Technologies

Enterprise Application Integration (EAI) is the unrestricted sharing of data and business processes among any connected applications and data sources in the enterprise. To do this, you should not have to make changes to the data structures or applications. Note, ERP has these properties within the ERP package but not necessarily with applications external to ERP (e.g., legacy systems).

An application programming interface (API) is a set of procedures that allow external applications to connect to an application and obtain data or services. The open database connectivity (ODBC) is an API that enables programmers to connect to databases. Using ODBC programmers can process SQL queries to obtain data from the database and then use it in their application. APIs are written for many applications. For example, a financial application might define an API to provide a service to obtain the latest currency exchange rates. When defining the API, the programmers specify the name of the commands to obtain the service, what input parameters are needed, how they are formatted, what output will be generated, and how it will be formatted.

Example API specification

The API specifies the format of the request command and its parameters. In this example, the request is `GetLatestRecord` and the input parameters are the three-letter currency abbreviations for the "from currency" and the "to currency."

```
GetLatestRate(FromCurrency, ToCurrency);
```

The application returns the exchange rate, the from currency, the to currency, and the time and date. If the user enters:

```
GetLatestRate(USD, EUR);
```

Then the API returns:

```
1.2754, USD, EUR, 10:05:00 EST, June 13, 2009
```

Because APIs are written by the programmers that make the application they can vary greatly on what data or services are made available. If an existing application has no API you can create them. This involves rewriting the code of the application. Essentially, you expose the business services that are probably accessible through the user interface. This strategy can be expensive because you have to add code to an existing application, and it may not be possible with commercial-off-the-shelf (COTS) applications. An alternative is to create a *wrapper*, which is software that encapsulates the application so that it can be accessed by other applications.

APIs provide access to business services, objects, or data. When available they greatly facilitate application integration. However, frequently, existing applications such as legacy systems and other packaged applications have no or insufficient APIs. One alternative is to change the application and write APIs for it or to "wrap" the application to provide access.

Middleware was mentioned briefly when discussing system architectures, but now we explore the technology in greater depth because it is the main means to achieve application integration. *Middleware* is software that lets systems talk to each other. It is connectivity software that consists of a set of enabling services that let multiple applications run on one or more machines to interact across a network. The services include: naming, event handling, transaction handling, and management services. Middleware hides the complexities of integrating applications so that application and system developers can work at a higher level of abstraction. There are many types of middleware; here we provide a brief description of each.

Remote Procedure Call

Remote procedure call (RPC) is a piece of code in the client application that invokes a procedure on the server application. It is not true middleware, but rather its predecessor. RPC essentially implements over a distributed network what is a common programming language concept of a procedure call – when you need to access a service offered by an application such as a database or operating system you call a procedure. RPC is a point-to-point solution that requires a direct, synchronous connection between applications. It has poor performance and does not scale well because of the amount of code required to program a PRC. Moreover, you must write a unique RPC for each application that needs to be integrated.

Object Request Broker

An object request broker (ORB) manages the sending and receiving of method invocations across a network so that applications can be distributed. Method invocation is when one software program requests a service of another software program. A standard for this approach is CORBA the Common Object Request Broker Architecture that was specified by the Object Management Group. Consequently, it is non-proprietary and has both platform and language independence (i.e., you can use C, C++, Java, etc.) It generalizes the remote procedure call (RPC) mechanism for independent programs, which may reside on different platforms. Each subsystem offers a set of services to other client subsystems. Each subsystem provides a stub to communicate with the ORB. The stub is written in the Interface Definition Language (IDL) to define interfaces in language neutral manner. Similar approaches to distributed software objects is the COM/DCOM model from Microsoft and Enterprise Java Beans.

Message-Based Application Integration

Message-oriented middleware (MOM) uses the notion of messages to communicate asynchronously between applications. The client sends a message that arrives to the queue of the server application. The server can process the message or continue with other processes until it is ready to process the message. Meanwhile, the client that sent the message does not need to wait for a response. Instead, the client can continue with other processes. Asynchronous communication via messages effectively decouples the client from the server. This makes development of the application a little easier because the applications are not blocked while waiting for a reply from the server. Figure 16.13 shows the sequence of events when Account A transfers funds to Account B.

16.7.1 Web Services

Web services provide a structured way to format data, a way to handle transactions, and a standard means to describe what the service does and make the services available to others via the Internet. Web services are based on the standard protocols of XML, SOAP, and WSDL, which allow them to interoperate across platforms and programming languages. Web services are discrete units of code; each handles a limited set of tasks. Unlike Web

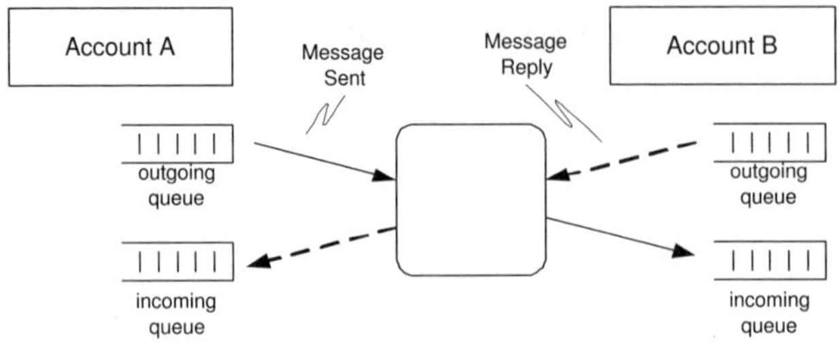

FIGURE 16.13
Message-oriented middleware.

sites, which are pictures of data designed to be viewed in a browser by a person, a Web service is designed to be accessed directly by another service or software application.

There are a suite of technologies to enable Web services:

- SOAP – Simple Object Access Protocol. A simple, XML-based protocol for exchanging structured data and type information on the World Wide Web. SOAP describes what is in the message, who should deal with it, and whether it is optional or mandatory. SOAP consists of four basic components:

    1. An envelope that defines a framework for describing message structure. An envelope is a wrapper containing a header, information to intermediaries (network nodes on the message path), and a body, the actual contents (depending on the application).
    2. A set of encoding rules for expressing instances of application-defined data types.
    3. A convention for representing remote procedure calls and responses.
    4. A set of rules for using SOAP with Hypertext Transfer Protocol (HTTP).

- WSDL – Web Service Description Language (WSDL) describes what functions are available from a specific Web service and what information must be passed to call them. WSDL is an XML-based language that allows formal descriptions of the interfaces of Web services: which interactions does the service provide? Which arguments and results are involved in the interactions? Which network addresses are used to locate the service? Which communication protocol should be used? Which data formats are the messages represented in?

- UDDI – Universal Description Discovery and Integration (UDDI) provides a directory enabling businesses to list services that they provide. It is an XML-based specification. The core component of UDDI is its business registration module: White pages denote the address, contact, and known identifiers. Yellow pages categorize services according to a standard taxonomy. Green pages denote technical information (e.g., WSDL).

All are based on XML and are standards defined by OMG. Microsoft's .Net is built around these technologies.

16.8 Summary

This chapter defined enterprise integration as the achievement of five types of enterprise integration: interconnectivity of the infrastructure, information sharing, interoperability of the applications, coordination of the processes, and alignment of the organization. Only by addressing all five integration types is full enterprise integration achieved. The chapter discussed the challenges facing a project team embarking on an enterprise integration project. These include both technical and management challenges. The chapter finished by reviewing the information technologies that are used to achieve enterprise integration.

Review Questions

1. Describe some of the benefits of achieving enterprise integration.
2. Think of an enterprise that you are familiar with (e.g., university, business you work for, social organization, ...). Select three technical challenges to integration and explain how they manifest themselves in that enterprise.
3. Think of an enterprise that you are familiar with (e.g., university, business you work for, social organization, ...). Select three management challenges to integration and explain how they manifest themselves in that enterprise.
4. Explain the reasons underlying the shift from centralized architectures to distributed architectures.
5. Contrast synchronous and asynchronous communication.
6. List and provide an example from your personal experience of each of the server types.
7. Explain one reason why architectures are designed in layers.
8. Describe a benefit promised by service-oriented architecture.
9. Define a protocol and provide an example of a protocol.
10. Contrast OLTP and OLAP.
11. Explain how client-server computing works.
12. Define API and explain how it works.
13. Explain why data integration is a popular approach to achieve enterprise integration.
14. Explain how message-oriented middleware works.
15. Contrast remote procedure call from message-oriented middleware.
16. Describe the different strategies of data integration and application integration.
17. Describe how Web services work.

Exercises

1. Use the Web to investigate the availability of Web services. Some useful sites might be Amazon.com, Google, or to search for common functions such as payment, credit card processing, or shopping cart. Describe the Web services that you find and whether you can find any implementations of them.
2. Investigate the type of system architecture in use at your university. Does the university have an ERP system? What are the main applications employed by the university? How are they arranged into an architecture?

Bibliography

[1] C. Batini, M. Lenzerini, and S.B. Navathe. A comparative analysis of methodologies for database schema integration. *ACM Computing Surveys*, 18(4):324–365, 1986.

[2] D. Comer. *Internetworking with TCP/IP: Principles, Protocols and Architecture.* Prentice Hall, Upper Saddle River, NJ, 1999.

[3] K.A. Crowston and T.W. Malone. The interdisciplinary study of coordination. *ACM Computing Surveys*, 26(1):87–119, 1994.

[4] M. Fowler and D. Rice. *Patterns of Enterprise Application Architecture.* Addison-Wesley Professional, New York, NY, 2003.

[5] R.E. Giachetti. A framework to review the information integration of the enterprise. *International Journal of Production Research*, 42(6):1147–1166, 2004.

[6] T. Gulledge. What is integration. *Industrial Management & Data Systems*, 106(1):5–20, 2006.

[7] IEEE. IEEE standard computer dictionary: A compilation of IEEE standard computer glossaries, 1990. Technical report, 1990.

[8] K. Kosanke, F. Vernadat, and M. Zelm. Cimosa: Enterprise engineering and integration. *Computers in Industry*, 40:83–97, 1999.

[9] A.P. Sheth and J.A Larson. Federated database systems for managing distributed, hetereogeneous, and autonomous databases. *ACM Computing Surveys*, 22(3):183–236, 1990.

[10] D. Smith, L. O'Brien, K. Kontogiannis, and M. Barbucci. The architect: Enterprise integration, http://www.interactive.sei.cmu.edu, 2002.

17

Enterprise Integration

"The way a team plays as a whole determines its success. You may have the greatest bunch of individual stars in the world, but if they don't play together, the club won't be worth a dime." – Babe Ruth (1895-1948), American baseball player.

The integration of the enterprise in terms of its data, processes, and organization is a priority throughout all phases of an enterprise engineering project. A prerequisite to enterprise integration is the existence of an enterprise architecture. The big picture provided by the enterprise architecture provides the guidance necessary to ensure that each project will work within not only the existing system infrastructure, but also in the future system infrastructure. The importance of an enterprise architecture to integration cannot be under-emphasized. The enterprise architecture shows how the enterprise is decomposed into subsystems and how all those subsystems work together. The architecture includes a design of the information technology infrastructure that connects all the subsystems. This vision guides all other integration efforts.

This chapter describes the points throughout the project when the team needs to consider integration issues and plans. How to perform the integration activities is divided into ensuring correspondence between the three views: process, information, and organization, and how to develop plans and execute each integration activity.

After completing this chapter, you should be able to:

- Describe when during the project, integration concerns should be examined.
- Develop activity-to-information, activity-to-organization, and information-to-organization matrices.
- Analyze the correspondence between the three views: process, information, and organization.
- Describe how to achieve data integration.
- Design how to achieve process integration.
- Explain how to conduct an interface analysis.
- Explain how to analyze and design the human-computer interface.
- Describe the tests for enterprise systems.

17.1 Introduction

Integration of the enterprise is done throughout the enterprise design process and should always be in the mind of the design team. However, there are identifiable points in the design life-cycle where consideration of integration is especially called for. These integration points are:

1. Strategy formulation – During strategy formulation the enterprise should ensure that all the enterprise strategies: marketing, operational, technology, etc. are aligned and the goals for each subsystem contribute to the enterprise strategy.
2. Enterprise architecture development – The enterprise architecture provides a high-level model of how all enterprise subsystems should work together. It establishes the enterprise integration needs.
3. Requirements analysis – The enterprise design team needs to explicitly identify, define, and analyze integration requirements for the enterprise project.
4. Integrate system design – The enterprise design is divided into the three views of Process, Information, and Organization Design. Section 6.7 discussed integrating the design as an explicit activity. This involves reviewing the integration requirements, the various design models, and ensuring there is no conflicts and that all integration requirements are satisfied.
5. View analysis and design – The three views of process, information, and organization are analyzed and designed separately and in parallel. The three views must all agree in terms of the enterprise design and they must all be consistent where the three views intersect. Throughout the design process, the sub-teams working on each view have liaisons that convey status of significant developments to the other sub-teams, they have periodic design reviews, and the project supervisors ensure coordination of the design efforts. At the end of the design phase the sub-teams have a formal design audit to verify and validate the integration of the three views.
6. Testing and quality assurance – The quality assurance team is responsible for all project products being delivered according to specification. The quality assurance team conducts the tests to verify compliance with the system requirements and design specifications.
7. Information technology infrastructure – The information technology infrastructure must be designed so that all the systems are interoperable and able to share needed data. The design teams conduct interface analysis to document the interoperability requirements between the systems. Additionally, the human-system interaction needs to be designed so that the systems are user-friendly and serve the business activities.

The following sections describe integration activities that occur to ensure the enterprise engineering project delivers an integrated system.

17.2 Correspondence between Views

The enterprise was decomposed into three views of process, information, and organization. We decompose the enterprise into these views to reduce the complexity with dealing with all aspects of the enterprise at once. All three views must provide a consistent definition of the enterprise. However, during analysis and design, it is possible that inconsistencies were introduced between the views. To analyze the congruence between the three views we use matrices that map one view into another view.

Process and Information

There are two versions of the *Process-Information Matrix* depending on whether the activities are modeled using a data flow diagram or IDEF0. When a data flow diagram (DFD) models the activities, then the matrix is a CRUD matrix, where CRUD stands for Create, Read, Update, and Delete that correspond to the four SQL commands that are done to relational databases. The data flows in the data flow diagram represent data structures, which are groupings of attributes that travel together. Each attribute in the data structure maps to an attribute in the entity-relationship model. Additionally, each data store in the data flow diagram maps to an entity or group of entities in the entity-relationship model. Figure 17.1 shows a partial data flow diagram and a partial entity-relationship model for the ordering process. The data structure for `Order Confirmation` is shown and how it maps to attributes in the data model. It is common that the attributes in a data structure do not all belong to the same entity. The reason is that through normalization and removing redundancy in the entity-relationship model the information is separated into many entities that are related to each other. However, much of this data must flow together as shown in the data flow diagram. In the same figure, there is a single data store for `Order`. During process modeling the modeling team did not have a complete entity-relationship model so it had created a single Order data store for the orders and the items in the order. So the `Order` data store maps to two entities: `Order` and `OrderItem`. At this later stage when both the data flow diagram and entity-relationship model are completed, the team might decide to revise the data flow diagram to show a second data store for `OrderItems`. This would help clarify the models when viewed together.

In the CRUD matrix every information object must have at least one "C" indicating it is created. Otherwise, this indicates a potential problem such as a missing process to create the information object. Each information object should also have at least one "R." Otherwise, this suggests the information is never used by any processes. This could occur if the information object is not required by the processes, if the processes inputs and outputs are not correctly defined, or if a process is missing. Doing this exercise ensures that the process model and the information model correspond to each other in representing the same system. The CRUD matrix shown in Figure 17.2 has some information objects that violate these rules, but the figure only shows a small portion of the entire matrix.

When the activities are modeled using IDEF0, then the Process-Information matrix (Figure 17.3) shows whether an information object is: `I` = an input to the activity, `C` = a control to the activity, or `O` = an output of the activity. The IDEF0 model does not show the CRUD attributes nor does it show data stores. To be complete, every information object must be an output of at least one activity. Otherwise, how is the information ever created? Every information object should also be an input or a control of at least one activity. Otherwise, the project team should question whether the information object is necessary since it is not used by any processes.

Organization to Information

The *Organization-Information Matrix* (Figure 17.4) shows which organization units: O = own the information object and U = users of the information object. The main objective here is to understand who the information owners are. An information owner is the organizational unit that has the authority to create the object, control the object, make changes to the object, or to grant that access to others. In large enterprises it is common that a department exercises control over a set of information objects that reside in databases they designed and created. Yet, other organizational units often need access to the information. For this

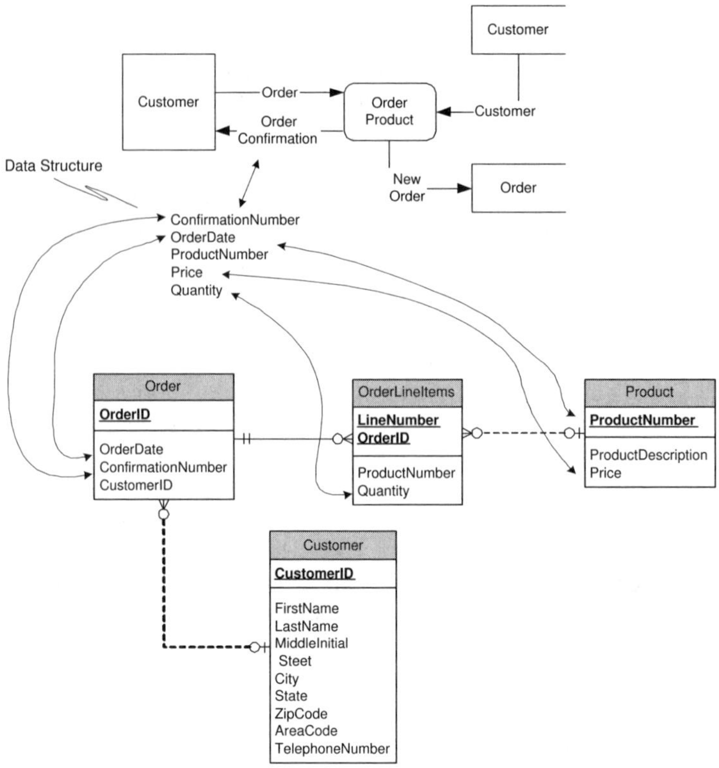

FIGURE 17.1
Correspondence between data flow diagram and entity-relationship diagram.

Activity (DFD) to Information Object	Search Cruise	Create Reservation	Modify Reservation	Cancel Reservation	Assign Ship to Itinerary	Plan Cruise Itinerary	Crew Scheduling	Capacity Planning	Order Supplies	Plan Entertainment
Reservation		C	R,U	D						
Ship	R	R	R		R	R	R	R	R	R
Cabin	R	R	R					R		
Departure Date	R	R	R		C,R,U	C,R,U	R		R	R
Guest		C,R,U	R,U	R						
Name		C,R,U	R,U	R						
Guest Invoice		C	R,U	R,D						
Credit Card		C,R,U	R,U	R,U						
Destination	R				R	C,R,U	R		R	R
Crew							C,R,U,D			
Crew Assignment							C,R,U,D			
Vendor									C,R,U,D	
Order									C,R,U,D	
Order Item									C,R,U,D	
Entertainer										C,R,U,D
Entertainer Assignment										C,R,U,D

FIGURE 17.2
Process-information matrix for data flow diagram.

Activity to Information Object	Search Cruise	Create Reservation	Modify Reservation	Cancel Reservation	Assign Ship to Itinerary	Plan Cruise Itinerary	Crew Scheduling	Capacity Planning	Order Supplies	Plan Entertainment
Reservation	C	O	I,O	I						
Ship	C	C	C		I	I	I	I	I	I
Cabin	C									
Departure Date	I	I	I,O		O					
Guest		O	I	I						
Name		O	I	I						
Guest Invoice		O	O	O						
Credit Card		I	I	I						
Destination					I	I	C		C	C
Crew							I,O			
Crew Assignment							O			
Vendor									I	I
Order									O	
Order Item									O	
Entertainer										I,O
Entertainer Assignment										O

FIGURE 17.3
Process-information matrix for IDEF0.

reason, data ownership was cited as a challenge in Chapter 16. As an information owner, the organizational unit is responsible for ensuring that:

- The information is correct and up-to-date.
- The information adheres to enterprise-wide policies on accuracy, privacy, security, and data storage.

Part of these responsibilities may be delegated to the IT department, but it is the business unit that maintains authority and control.

Process to Organization

The *Process-Organization Matrix* (Figure 17.5) shows which organizational units are: A = authority responsible for an activity, R = a resource for the activity, and C = concerned party about the execution or outcome of the activity. The process-organization matrix shows how the organizational design corresponds to the process design.

When an organizational unit is a resource for an activity, this means the employees of the organizational unit are part of the process's resource pool. The intervening design action is job design – to ensure the employees have the knowledge, skills, and tools to perform the activity work. When an organizational unit is also the authority responsible for an activity, this means it has decision control over the activity and its resources. The authority imposes requirements that the organizational unit can coordinate the activities to ensure integrated processes. Authority includes assigning resources to the activities, and directing the work of the activities. When the organizational unit is a concerned party, this indicates it receives the process output, provide process inputs, or interact in some way with the process. Figure 17.6

Organizational Unit to Information Object	Guest	Ship Captain	Ship Officers	Ship Steward	Ship Concierge	Housekeepers	Reservation Manager	Reservation Agent	Kitchen Staff	Purchasing Department	Entertainment Director	Entertainment Manager	Human Resources	Vendor	Ship Planning Department	Revenue Management Department	IT Department	Entertainer
Reservation	U				U		O,U	U							U	U	U	
Ship	U	U	U	U	U		U	U		U			U	U	U	U	U	U
Cabin	U				U	U	U	U								U	U	
Departure Date	U	U	U	U	U		U	U		U			U	U	O,U	U	U	
Guest	U				U		O,U	U								U	U	
Name	U				U		U	U								U	U	
Guest Invoice	U				U		U	U								U	U	
Credit Card	U				U		U	U									U	
Destination	U	U	U	U	U					U			U	U	O,U	U	U	U
Crew		U	U	U		U			U				O,U				U	
Crew Assignment		U	U	U		U			U				O,U				U	
Vendor			U	U	U					O,U				U			U	
Order			U	U	U					O,U				U			U	
Order Item			U	U	U					O,U				U			U	
Entertainer				U	U						O,U	U	U				U	U
Entertainer Assignment				U	U						O,U	U	U				U	U

FIGURE 17.4
Organization-information matrix.

Activity to Organizational Unit	Search Cruise	Create Reservation	Modify Reservation	Cancel Reservation	Assign Ship to Itinerary	Plan Cruise Itinerary	Crew Scheduling	Capacity Planning	Order Supplies	Plan Entertainment
Guest	R,C	R,C	R,C	R,C						
Ship Captain					R,C	C	C			
Ship Officers							C			
Ship Steward							C		R,C	
Ship Concierge							C			R,C
Housekeepers							C			
Reservation Manager	A	A			C					
Reservation Agent	R	R								
Kitchen Staff							C			
Purchasing Department					C				A,R	
Entertainment Director										R
Entertainment Manager										A,R
Human Resources					C		A,R			
Vendor									C	
Ship Planning Department		C			A,R			A,R		
Revenue Management Department	C	C	C	C	R,C			C		
IT Department	R	R	R	R					R	
Entertainer										C

FIGURE 17.5
Process-organization matrix.

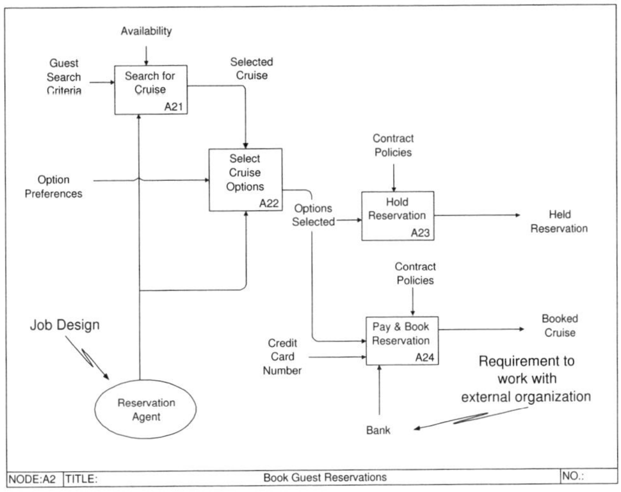

FIGURE 17.6
Using IDEF0 model to understand process-organization relationships.

shows one node of the IDEF0 diagram for the cruise reservation process. The mechanisms in IDEF0 are the resources. In this case the reservation agent is a resource for the activities and the bank is a resource and concerned party. The reservation agent job description must ensure its ability to do the tasks. The bank is an external organization, and integration is ensured by contracts and agreements between the cruise line and the bank.

17.3 Data Integration

Data integration is often the main type of application integration done in enterprises. Even when an enterprise chooses another integration strategy, it still needs to understand the various data sources. The main challenge in data integration is semantic heterogeneity or the need to reconcile many different data definitions for the same data entities and attributes. These semantic challenges were presented in Section 16.4.2.

Data integration is accomplished by taking the data structure and data definitions from the legacy systems, redesigning the structure and data definitions, and creating it in the new system. The conversion process is called extract, transform, and load (ETL). The data objects are extracted from the old data systems, cleansed, then transferred and loaded into the new database. Cleansing the data is necessary because many legacy systems had poor controls or were not designed properly and allowed bad data to be saved in the system. See the box entitled, "Data Cleansing," for an explanation of the problem.

To do the data conversion subproject, a data conversion team is assembled. A data conversion project is divided into two phases (see Figure 17.7): a planning phase with the objective of obtaining stakeholder buy-in to the project, and a data migration phase where the data sources are identified, cataloged, and transformed to the new integrated database. Obtaining buy-ins from the stakeholders is important because they control the various data sources and their participation is needed to identify the data, catalog the data, to decide on data definitions, and to decide on ownership of the data.

It is important that the conversion team work closely with functional and technical

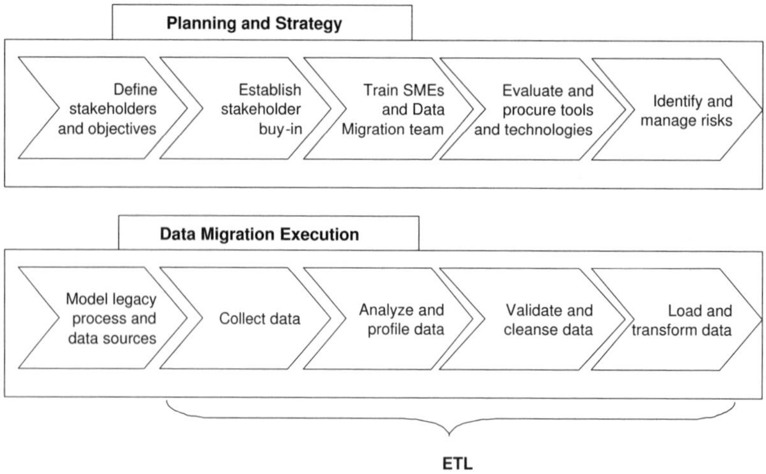

FIGURE 17.7
Data integration process.

experts from the old systems so that the data requirements are properly documented. A wide range of information, like the calculation for a field value or a field format, need to be documented so the conversion designs can be developed. For data sources not in control of the project team, they need to establish agreements about data quality, etc.

The conversion team creates a *Data Migration Register* for all the data objects that serves as a central point of information for data conversion. The Data Migration Register lists all data objects detailing:

1. Upload reference to tie to business process
2. Priority of the data object
3. Name
4. Owner responsible for data
5. Type of data (master or transaction)
6. Prerequisites for loading
7. Data volume
8. Upload procedure and reference
9. Clean up, mapping and conversion responsibilities and statuss (percentage completed)
10. QA status
11. Cut-over timings, dependencies and sequence for transferring the data to the production ERP
12. Additional comments such as changes required after go-live

Data Cleansing

It is surprising how "dirty" data can be. Cleansing the data is required to ensure it is correct, complete, consistent, and adheres to business rules. The figure below shows typical problems that can arise in databases that must be fixed before converting the data to a new system. The project team needs to document all the problems found and incorporate this knowledge into the design of the new system so as to avoid repeating the same data mistakes.

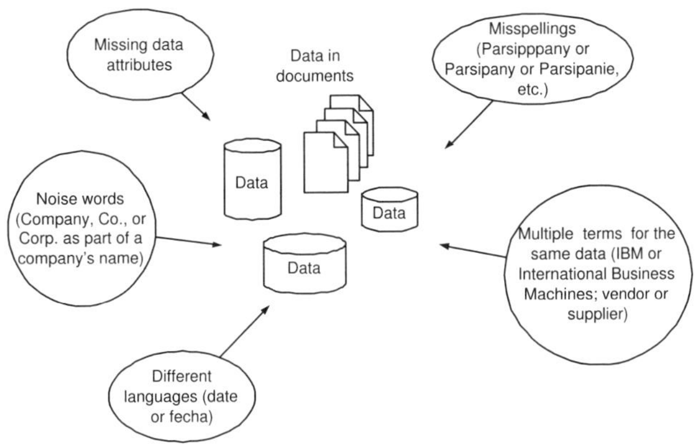

To conduct the actual data conversion, the project team identifies all the business rules that impact the data conversion. The schema differences between the new and old systems and their impact on the conversion are analyzed and agreements on data conversions are made. A data conversion method is developed for every data object. The data conversion and loading can be either manually with a team of clerks entering data into the system's front-end, or it can be automated with a set of computer tools (commercial and custom-built). The decision of which method depends on the lead time, volume of data, complexity of the data conversions, and available technologies. For automated conversions, a wide variety of languages exist for coding the programs. The code should be well documented and should return intelligent error messages so that debugging is made easier. For manual conversions, the development should include the definition of exact procedures to be followed by the users as well as a validation process for assuring the data are converted accurately. Manual data conversion is not unheard of – in a project to implement electronic medical records, a doctor's office hired a clerk who manually took the old patient records, entered some of the key data fields (patientID, first name, and last name) and then scanned the other medical records in the patient's folder. The data conversion took approximately six months to complete.

A best practice in data conversion is to provide for interim storage of the data between the legacy systems and the new system. The data are extracted from the legacy systems, cleansed, and loaded into an interim data repository. In the interim data repository the conversion can be validated and tested. Assuming the data are valid, they are transferred to the target system. Writing directly to database tables is highly complex because the conversion programs need to identify every database table and field that is affected by a transaction. Conversely, the use of APIs is always a slower course of action because the programs are dependent on the system processing speed.

Example Data Mapping

This is an extract of some of the data mappings a large, public, urban university made when it implemented the ERP package from PeopleSoft.

Legacy	PeopleSoft
Classification	Career/Level
Schools and Colleges	Academic Groups
Academic Depts.	Academic Organizations
Major	Plan
Tracks/Concentrations	Sub-Plan
Reference Number	Class Number
Course Prefix	Subject
Course Number	Catalog Number

Data Conversion at Cruise Line

A typical scenario invoking data integration is when one company merges or acquires another company. In order to achieve the efficiencies that mergers promise, the companies must consolidate and integrate their systems. Figure 17.8 shows partial data schemas for the Sunny Seas Cruise Line and Ocean Reef Cruise Line. The data integration issues are:

1. Sunny Seas Cruise Line identifies ships by their name, which is unique. The Ocean Reef Cruise Line has a `ShipID`. Negotiation among stakeholders can easily resolve this issue. A likely resolution is to assign `ShipID`s to all the ships.

2. Sunny Seas Cruise Line has the entity `Cabin` and Ocean Reef Cruise Line has the entity `StateRoom`. Again this can readily be resolved by agreeing on whether to call it a cabin or a stateroom. More problematic are the definitions of the primary keys for the cabins/staterooms. Sunny Seas uses both the `CabinNumber` and `DeckLevel`, whereas Ocean Reef uses just the `RoomNumber`, which has the deck level as part of the number. For example on Sunny Seas Cabin Number 55 may appear on many deck levels, whereas on Ocean Reef they would have state room number 155, 255, and 355 for state rooms 55 on deck 1, 2, and 3, respectively. Again standardizing on one approach would make sense. However, if they decide to go with just RoomNumbers, then this would entail replacing all the room numbers on the ships currently operated by Sunny Seas.

3. Sunny Seas has an entity `CabinClass` and assigned rates based on the cabin. Ocean Reef has an entity `Season` and a different policy of charging rates based on the season. This is more than a data integration issue. It involves business policies. The merged company needs to consider the two business policies and decide whether to standardize on one of the existing policies or create a new policy that may contain elements of the previous two policies. The decision is a prerequisite to the data integration issue, and may involve a significant redesign for the integrated information model.

This example highlights some of the data integration issues that are encountered when merging two separate information models. As the number of information models increases, the problem becomes more complex because many more competing definitions of data and how they are related are encountered.

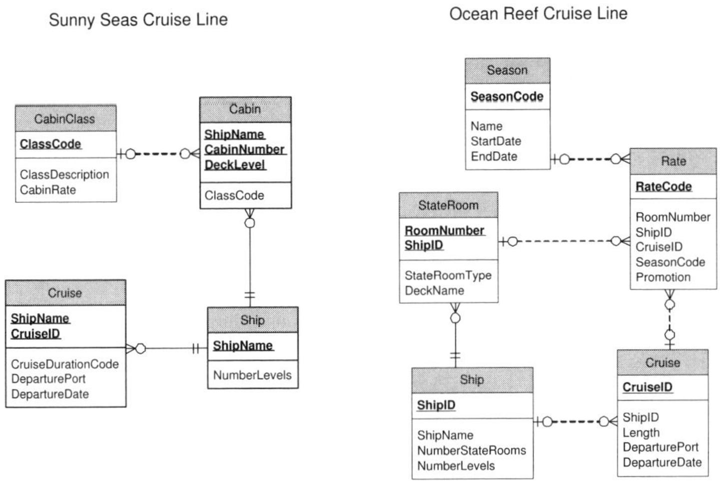

FIGURE 17.8
Integrating the information models of two cruise lines.

17.4 Process Integration

Process integration means the seamless flow of work items to satisfy customer needs. Good process integration involves the right materials at the right place, at the right time, and with the right quality. This suggests minimum work-in-process inventory, minimum waiting, and no process waste. Process integration thus described starts with good process design and also involves good information architecture, good information design, and good organization design.

Start by examining what is required to have the right materials at the right place and the right time. Materials refers to the inputs, controls, and resources of the process. If the inputs are physical materials, then this means they are available when needed in the correct quantity and quality. One way to accomplish this goal is to carry sufficient work-in-process inventory to ensure availability when needed. However, this countervails the goal of no process waste because inventory is considered a sign of waste. Many strategies have emerged to coordination and synchronize the flow of materials in a production facility such as just-in-time manufacturing and lean production. Detailed discussion of these techniques is beyond the scope of this book, the reader is referred to texts that discuss these strategies in detail [1, 4]. Many of these principles also apply when the process delivers a service. The process is designed to remove all waste such as wasted movements, wasted activities, and other non-value adding activities.

The minimization of waiting in a process is accomplished by the techniques discussed at length in Chapters 11 and 12. This includes:

- Decreasing variability in interarrival times and service times.
- Increasing the number of servers.

Additionally, the waiting time can be reduced if the arrivals and service capacity are synchronized. In most systems, the arrivals are not uniform throughout the day, week, or month. Restaurants clearly see more arrivals during lunchtime than a few hours earlier. Knowing the arrival patterns, the organization matches capacity by having more employees

TABLE 17.1
Interdependence and Coordination Analysis

Process	Interdependence Type	Coordination	Responsible
A22 - A24	Information sequential	Electronic transfer of data	Agent
	Controlled sequential	If failure, agent obtains payment information and places an agent hold in reservation for later batch processing	Agent

during the peak than the off-peak. It is also possible to match arrivals to the available service. Restaurants offer "early-bird specials" to customers who come to dinner before 6 pm. This is done to get customers to arrive during the off-peak hours when the restaurant has idle capacity. A flexible workforce can also help minimize waiting because it allows management to shift workers to the bottlenecks to relieve congestion. Finally, the pooling of resources that was analyzed in Chapter 12 can significantly reduce waiting time in a process.

To ensure synchronized, smooth flow of material, information, and customers the process interdependencies need to be identified, how they are to be coordinated, and who is responsible to ensure their coordination. Section 16.4.4 listed the interdependence types. To start the coordination analysis, the process models are examined to identify the interdependence types, which are listed in the coordination list. Then for each interdependence type the coordination mechanisms that are to manage the interdependence are described along with the responsible organizational unit. To illustrate the procedure, refer back to Figure 17.6 that shows a partial IDEF0 model of the reservation process. Node A22 has a controlled sequential and an information sequential dependence with node A24. How the information is controlled, and what happens if the control flow is interrupted needs to be coordinated. Table 17.1 demonstrates how to document these design decisions.

Many processes are information-intensive so the information flow needs to be carefully designed to support the process. The information needs to be provided to the process in a format that makes it easy to use. The above example highlights the importance of information flow between the reservation agent and the bank to have an efficient, integrated process.

17.5 Interface Analysis

The project team needs to identify, define, and document all interfaces. Interfaces exist between subsystems of the enterprise, and between the enterprise and external systems. The ability of the subsystems to interact and the enterprise to interact with their environment often determines overall enterprise performance. According to the Gartner Group, some 35-40% of all software project efforts are focused on developing and construction of system interfaces. In this section we examine system-to-system interfaces, in the next section we examine human-to-system interfaces.

An interface defines the boundary between a system and everything outside of the sys-

TABLE 17.2
Interface List

ID	Source	Destination	Data or Action	Frequency	Performance
I01	Bank	Payment function	Credit card authorization	Each guest reservation	Real-time
I02	Yield mgmt. system	Reservation system	Rates and promotions	Daily	None

tem. We will distinguish between *external interfaces*, between the enterprise and everything external to the enterprise, and *internal interfaces*, between subsystems or components internal to the enterprise. External interfaces are interfaces with: suppliers, partners, distributors, financial institutions, government agencies, and customers. External interfaces pose additional considerations because the enterprise exercises less control over external systems. If it is an output interface, there is the risk that the external recipient will change the interface or the connection to the interface will be disrupted. If it is an input interface the enterprise must design the interface to be robust to multiple external users, and be concerned with the security of the data. The establishment of external interfaces will require negotiation with the external agents on the type of interface, availability of the interface, standards governing the interface, and other business and technological policies. Internal interfaces have the same issues but the enterprise exercises control over both sides of the interface so resolutions are easier to come by. The exception might be internal interfaces that depend on vendor technology specifications that can change with changes in technology or markets.

The first step in interface analysis is to identify all the system interfaces. This can be a daunting task due both to the large number of interfaces and the fact that many interfaces may not be obvious. Interface identification is a continuous part of the project that subteams should be aware of and the project manager should monitor and track. This will help to avoid missing interfaces until system deployment. Interfaces can be identified by:

- Examining process inputs and outputs to see where information comes from and where it goes.
- Examining the data flow diagrams and other models identify interfaces.
- Examining the organizational to information matrix and where an organizational unit uses information owned by another unit there may be a need for an interface.

The process models, especially the DFD, provide a useful tool to identify system interfaces. The context diagram shows all the external interfaces. The data flows entering and exiting the system shown in Figure 17.9 indicate several interfaces. Likewise, lower-level DFD will show both external interfaces and internal interfaces. IDEF0 diagrams also indicate interfaces as the input, output, and control flows.

The interfaces are documented in the Interface List shown in Table 17.2. As a minimum, the Interface List documents the Source and Destination of each interface, describes the data that cross the interface, the frequency of use, and any performance requirements. The information helps the project manager prioritize work on the interfaces and make decisions about how much to invest in each interface. The project team might decide to create three separate Interface Lists, one for external interfaces, internal interfaces, and human-system interfaces. They are separated because the issues for each type are different and their monitoring and control can be handled separately from each other.

Each interface is assigned an interface owner, which can be an individual, organizational unit, or sub-team on the project. The interface owner is responsible for the interface control

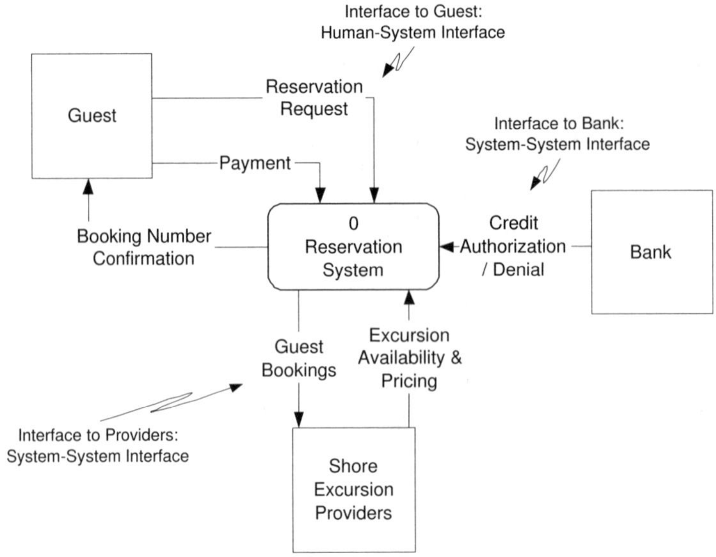

FIGURE 17.9
Context diagram to identify external interfaces.

document, and once it is baselined, the owner is responsible for reviewing any changes to the interface control document.

The project team creates an *Interface Control Document* for each interface to maintain configuration control over the interface. An Interface Control Document (ICD) describes the relationship between system components in terms of data items and messages passed, protocols observed and timing and sequencing of events. Interface control documents are a key element of systems engineering as they define and control the interface(s) of a system, and consequently places boundaries on the system's requirements. The ICD should include:

- Describe the interface(s) to a system or subsystem.
- Describe the inputs and outputs of a single system, the interface between two systems, the interface protocol from the lowest physical elements to the highest logical levels.
- Describe how to access the functions and services provided by a system via an interface.
- Describe whether the interface is uni-directional or bi-directional. Include whether the communication across the interface is synchronous or asynchronous.
- Include information about the size, format, and data.
- Describe the stability of the interface, whether it will change, perhaps due to technological changes, what control does the enterprise exercise over the interface, and whether it is subject to security risks.
- Provide a structured way to communicate information about subsystem interfaces between different design teams.

The Interface Control Document is base-lined and becomes a configuration item that is subject to configuration control. Additionally, the Interface Control Document is reviewed for potential project risks such as technical risk associated with an interface or business risk because, for example, the requirements for an external interface might change due to vendor issues.

The Interface Control Document states the interface requirements that the design team uses to specify the interface, and then the software engineers and programmers use to build the interface. The nature of the communication across the interface is used by the testing sub-team to determine test scripts to fully verify and validate the functioning of the interface.

Interface Access

The project team needs to determine what type of access is allowed for each interface to enforce security of the data and processes. Restricted access can be implemented by authorized user log-in accounts, data encryption, ID cards, or machine IDs. If establishing user accounts, policies are required for userID and passwords, types of accounts, level of access, and security measures related to authorized user accounts. Data encryption encodes the data while in transit so that if intercepted by unauthorized users, they cannot read it. Instead of users typing in accounts personal ID cards or similar devices can be used whereby the ID has a bar-code, magnetic strip, or other device that can be read to verify the user. One drawback of ID cards is they can be lost and somebody other than the authorized user can use them. It is also possible that the machine ID is used to grant levels of access. The enterprise can limit access from outside of the physical facilities using this approach.

17.6 Human System Interaction

Human system interaction is the study of the interface between humans and systems. In an enterprise much of this interaction is with computer systems, so it is called human-computer interaction (HCI). The design issue in HCI is how to allocate tasks between humans and the systems (or machines) in the enterprise. In general, the strengths and weaknesses of humans and machines are complementary. Understanding the relative strengths of humans versus machines can help in the correct allocation of tasks between the too. Humans exceed machine performance in the areas of:

- Value-based decisions.
- Prioritizing things.
- Resource allocations.
- Tasks requiring flexibility and creativity.
- Tasks that interact with other humans.
- Tasks that require adaptive behavior.

Machines on the other hand exceed human performance in the following areas:

- Processing large amounts of data.
- Repetitive tasks.
- Following complex algorithms that require no judgment.

- Maintaining system control.

Knowledge of when machines perform better than humans and vice versa is critical to job design and the human-computer interface. The human-computer interface is part of integrating the organization with both the process and the information. It also supports job design directly.

Extensive coverage of HCI principles to design the user interface is provided in references [2, 3]. Here we briefly mention the more important user interface design principles.

- The interface should be organized according to the normal way of reading left-to-right and top-to-bottom.
- The system should always confirm user input. Inform the user whether data were entered correctly or incorrectly. Inform the user whether an action request was completed or not completed.
- If a transaction takes time, then the system provides some indication that it is working. This can be the hourglass, a horizontal bar showing the percentage completed, or a clock saying how much time is left for processing.
- The system lets the user know what to do next, or what to do next should be apparent.
- The interface minimizes free-style input of users. Instead of a textbox to enter the state, it should be a drop-down box with the domain approved state abbreviations.
- All user input to the extent possible should be checked for correctness. Possible tests are:
  - Existence checks to determine whether data were entered into a required field.
  - Domain checks to determine whether the data entered conforms to the domain.
  - Data-type checks to ensure the data type is correct (e.g., number versus text).
  - Combination checks to determine whether data in separate input fields agrees (e.g., the departure date must be before the return date).
  - Self-checking digits such as used in credit cards whereby a formula exist to verify whether the number is correct.
  - Format checks to ensure the data are input in the correct format (e.g., date format as month-day-year versus day-month-year).
- The system interface is designed for multiple platforms such as MS Windows, Apple, Unix, and modern-day cell phones, PDAs, and other devices.
- The interface makes extensive use of default values to reduce user input needs.
- The interface avoids jargon and provides instructions and help for using the interface.
- The interface allows for users to go back and correct data entries.
- In multi-screen interfaces to implement a sequence of data entry and actions, the interface tells the user where in the process they are (e.g., a screen says at the top it is Step 5 of 8).

User interfaces are prototyped so that User-Acceptance Testing (UAT) can be performed on each interface. The UAT is designed so that all aspects of the interface are tested, including anticipated errors or other failure modes.

17.7 Testing

Enterprise integration is ensured by testing all the systems in isolation and as an integrated system prior to deployment. Testing falls within the quality assurance function of the project and the organization. The types of tests performed are:

- Unit Test – Test each individual component to ensure it is as defect-free as possible.
- Integration Test – Test functionally grouped subsystems to see they work together. It occurs between unit and system testing.
- User Acceptance Test – An independent test performed by the end user prior to accepting the delivered system; i.e., users sign off on test results.
- Interface Test – Test the interfaces in the end-to-end business process.
- System Test – Test the system as a whole.
- Security Test – Test users' security access provides proper authority for their roles in the business processes.
- Regression Test – Test that new changes do not adversely impact already tested components.

Unit Testing is the first level of testing for system development. It focuses on validating the system's smallest components meet specific functional test conditions. The purpose is to verify the unit is functioning correctly and no errors exist. Integration Testing (sometimes called business function testing) validates that the integration of the system's components that compose the business solution is complete and functioning completely. It is process-oriented testing of essential functional business processes and encompasses configuration, interfaces, critical reports, user exits, enhancements, and other development. User Acceptance Testing confirms that the requirements identified and accepted during the analysis phase have been realized. Business process owners will determine the criteria for acceptance prior to the start of testing. Process owners need to sign off on results at completion of tests. When the users conduct the test, they will need some training. Interface end-to-end testing is conducted with parties internal and external to the enterprise to verify the technical readiness of communications channels and that full production files can be successfully exchanged and processed. System testing validates the production readiness of the technical aspects of the system. This includes validation of the I/O devices, backup and restore testing, and testing of system administration functions. Security Testing validates that the security access created provides the proper authority for users to perform their business functions. Regression Testing verifies that changes made to the ERP have not adversely affected already-tested system functionality.

In designing the tests, every function, component, or object that is part of a configuration item under control is tested. This includes: subsystems, interfaces, databases, hardware, software, and processes. The testing procedure is developed by the quality assurance team. All tests should consider anomalous, failure states, and recovery states as well as normal operation. In these tests you use representative data and do the tests in the quality assurance environment. Each test is documented and becomes a configuration item. The project team and the enterprise sponsors sign off on the test documents and the test results. The end-result is a fully tested system that should function according to the specification and the requirements.

17.8 Summary

This chapter described the integration of the three views of the enterprise. Integration is a concern throughout the entire project life-cycle, but there are distinct milestones where formal analysis of enterprise integration is conducted. This chapter describes these points during the project life-cycle. One of the main tools in integrating the views are the matrices showing the correspondence between each view (Process-to-Information, Process-to-Organization, and Organization-to-Information).

This chapter described the data integration process to extract, transfer, and load data from legacy systems into the new system. Also, the chapter addressed the problem of merging two or more data models, and showed what problems to look for and how to address them. The process integration is mainly accomplished during process design. This chapter reviewed the main principles of process integration and how to accomplish them.

The interfaces pose a challenge for most large systems. The chapter explained how to identify interfaces; how to develop an interface list, and interface control document to manage the interfaces; and how to design the interfaces. The chapter also described the main concerns in human-computer interaction, which affects user interface design.

Lastly, the chapter briefly described the many tests that are performed to ensure the systems are functioning and integrated properly. Extensive testing enables the project team to deliver a high-quality product.

Review Questions

1. Identify phases in the project life-cycle where integration issues should be addressed.
2. Describe the different considerations needed for external interfaces compared to internal interfaces.
3. Explain a procedure to identify interfaces in a system.
4. Discuss when it is better to have a person perform a task compared to a computer.
5. Explain how security of the system's data and processes can be accomplished.
6. List the types of tests done to ensure functioning systems.
7. Explain why it is often necessary to cleanse data when taking it from legacy systems and inputting it to the new system.

Exercises

1. Construct a Process-Information matrix for the cruise line data flow diagram and information model provided in the earlier chapters.
2. Construct an Organization-Information matrix for the cruise line data flow diagram and information model provided in the earlier chapters.
3. Create an interface list for the cruise line data flow diagram.

4. Sketch the user interface for one of the data flows in the cruise line data flow diagram. Indicate what user interface design principles are implemented in the interface.
5. Identify a system interface for a process you are familiar with and create an interface control document for that interface.
6. Pick a process or system you are familiar with and apply the human-computer interaction advice to evaluate how well it is designed.
7. In Chapter 10 one of the exercises was to construct the context diagram for the Copper Canyon Tourist Office. For this context diagram create the Interface List.

Bibliography

[1] W.J. Hopp and M.L. Spearman. *Factory Physics.* Irwin McGraw-Hill, Boston, MA, 1996.

[2] J. Preece, Y. Rogers, H. Sharp, D. Benyon, S. Holland, and T. Carey. *Human-Computer Interaction: Concepts and Design.* Pearson Education Ltd, Harlow, England, 1994.

[3] B. Shneiderman and C. Plaisant. *Designing the User Interface: Strategies for Effective Human-Computer Interaction.* 4th ed., Addison Wesley, Reading, MA, 2004.

[4] J.P. Womack and D.T. Jones. *Lean Thinking.* Simon & Schuster, New York, NY, 2003.

Index